THE LIFE OF A REGIMENT

THE HISTORY *of the* GORDON HIGHLANDERS

VOLUME I

Original Uniform of Gordon Highlanders

Regimental Uniform, 1808-1810

Soldiers of the 1st Brigade, 2nd Division

50th, 60th, 71st and 92nd Regiments and Portuguese Infantry, 1812

Gordon Highlanders at Brussels, 1815

Soldiers of the French Army, 1815

Lancer — Cuirassier — Carabineer — Hussar

Light Infantry — Infantry of Line — Old Guard

Soldiers of the British Army, 1815

Trooper	Gunner	Private	Trooper	Private	Sergt. Battn. Coy.
1st Royal Dragoons	Royal Horse Artillery	1st Foot	Royal Horse Guards	95th Rifles	Coldstream Guards

THE LIFE *of* A REGIMENT

THE
HISTORY OF THE GORDON HIGHLANDERS
FROM ITS FORMATION IN 1794 TO 1816

BY

LT.-COLONEL C. GREENHILL GARDYNE

PREFACE TO FIRST EDITION

IN the year 1896 I was asked by the Gordon Highlanders' Association in Aberdeen to give a lecture on the origin and history of their regiment. It was afterwards suggested by the late General J. C. Hay, Lieut.-Colonel the Hon. J. Scott Napier, commanding the 2nd Battalion, and others, that I should write such a history of the regiment as would tend to keep up the memory of individual officers and soldiers, and of the traditions which, in the days of long service, were told by veterans to recruits on the line of march or round the barrack-room fire, but which are apt to be forgotten when generations of soldiers succeed each other more quickly, and when their time and thoughts are taken up by the more varied intellectual and physical occupations of modern military life.

It happened that in my youth I heard many stories of the French War ; I took delight in the tales of old officers and soldiers (my first salmon was caught under the auspices of a Waterloo sergeant), and when I joined the 92nd fifty years ago, my instructors had been themselves drilled by Peninsular and Waterloo men, so that the traditions of those times were still well known in the ranks ; and though I afterwards served much longer in another regiment, the impression left by these tales never altogether faded from my memory.

The late Mrs Cameron Campbell of Inverawe ; Sir Charles Seton, Bart. ; Lieut.-Colonel Stewart of Achnacone ; and Mr Innes have kindly lent me letters written during the various campaigns. I have received great assistance from Mrs MacDonell of Keppoch, who knew many of the old officers personally, the Earl of March, Sir F. C. MacKenzie, the late Colonel Ewen MacPherson, C.B., of Cluny ; Rev. Canon H. MacColl, the Rev. T. Sinton, Dores ; Mr F. J. Grant of the Lyon Office ; Mr MacPherson, banker, Kingussie ; and others. The original Order Books, which had been lost, were found by General Alastair MacIan MacDonald and entrusted to me. I had some years ago read a journal kept by Quartermaster-sergeant MacCombie, and I have in my possession that kept by Sergeant Duncan Robertson, a most intelligent Athole man. Also the " Military Memoir " of a 92nd officer, printed at Edinburgh in 1823. The name is not given, but I have reason to believe he was Lieutenant James Hope, nephew to Lord Hopetoun, Colonel of the 92nd, who was promoted ensign from volunteer private in 1809. From these and similar sources I have taken the life and ideas of the regimental family at various times.

I have been at pains to have the illustrations of uniform and costume correct, and they are taken as far as possible from contemporary sketches in the British Museum, by British and French artists, the drawings being done by Messrs R. Hope, H. Payne, and R. Simkin. I am indebted to the courtesy of Messrs Blackwood & Sons for the permission to use the plans of Vittoria, Orthes, Quatre-Bras, and Waterloo from Alison's "History of Europe," plans being also taken from Napier's "History of the Peninsular War," and others.

In matters of general history, it has been my object merely to give the reason for the various expeditions in which the regiment took part, and in describing the operations, to confine myself as far as possible to the part taken by it. Many of the details as to nationality, dress, messing, and recruiting are of little interest to the general public, but they often show how the interior economy and discipline of a Highland regiment were carried on, and the tone of good feeling which prevailed among officers and men. The book is intended principally for the present and succeeding generations of the Gordon Highlanders, though as a by-way of Highland and military history, it may have some attraction for those interested in such subjects. It is my intention to carry on the story of the 92nd to 1881; then to give an account of the 75th Regiment from its formation till it became the First Battalion of the Gordon Highlanders, and afterwards to continue the history of both battalions to the present time.

Those who read will not require to be told that I have no claim to be a practised writer, but I believe I have been accurate as to facts and details. If these tend to increase the respect for the Highland soldiers of former days, and to stimulate their successors to imitate their gentleness in peace and their manliness in war, I shall not altogether have lost my time.

C. G. GARDYNE.

GLENFORSA, ISLE OF MULL,
November 1900.

PREFACE TO SECOND EDITION

FOR some time the Officers, past and present, of the Gordon Highlanders have had under consideration the question of bringing up to date the History of their Regiment, which had been written up to the year 1898 in the original work on the subject: *The Life of a Regiment.*

In 1926 they appointed a committee of their number, under the presidency of General Sir Aylmer Haldane, G.C.M.G., K.C.B., etc., to formulate a scheme for a complete History; and the original book having been out of print for many years, it seemed imperative that the first step should be to publish the new edition of it which is here presented.

The work is reproduced without more variation than the excision of a few errors, the incorporation in the text of the numerous notes, additions, and amendments collected by the late author in the years since 1903, and of a little material from other sources, new matter of over six lines being indicated by indentation of type on left side; but it has been found possible to add to the original illustrations portraits of several illustrious officers and others connected with one or other battalion, while the print and paper have been improved.

The two volumes now published should be considered as the first portion of the complete History of the regular and other battalions of the Gordon Highlanders up to the reconstruction of the Army after the Great War.

A. D. G. GARDYNE.

GLENFORSA, ISLE OF MULL,
September 1928.

CONTENTS

CHAPTER I

CHAPTER II. 1794

CHAPTER III. 1794-1798

CHAPTER IV. 1798

CHAPTER VII. 1801

CHAPTER VIII. 1802

CHAPTER IX. 1807

CHAPTER X. 1808

CHAPTER XI. 1809

CHAPTER XII. 1809

CHAPTER XVI. 1812

CHAPTER XVII. 1812

CHAPTER XXIII. 1815

CHAPTER XXIV. 1815

CHAPTER XXV. 1815

APPENDICES

LIST OF ILLUSTRATIONS

The Spanish and Portuguese uniforms and costumes are taken from "Sketches in Portugal and Spain," 1808-9, *by the Rev.* —— *Bradford, Chaplain to the Forces, and from Booth's "Military Costumes of Europe."*

The French uniforms are from Booth's "Military Costumes of Europe," 1812, *and "Collection des Uniformes des Armées Françaises," by H. Vernet et Eug. Lamie, and others.*

The British uniforms are from "Booth," Hamilton Smith's "Costumes of the British Army," 1815, *and other contemporary drawings.*

MAPS AND PLANS

CHAPTER I

BEFORE entering on the history of a Highland regiment, it may not be out of place to consider the state of the Highlands at the period when it was raised.

The mountainous and thinly peopled part of Scotland known as the "Highlands" included the Western Isles from Lewis in the north to Bute and Arran in the south, and the mainland north and west of the "Highland line." This was drawn from Bute, by the Firth of Clyde, to Leven in Dumbartonshire; by Ardoch and Drymen in Stirlingshire, to near Doune, Crieff, and Dunkeld in Perthshire; Airlie, Prosen, and Glenesk in Angus; Glenmuick and Edinglassie in Aberdeen, Ballindalloch and Craigellachie in Banff; and, taking in part of Moray and nearly the whole of Nairn, included along with these parts of counties the whole of Argyll, Inverness, Ross and Cromarty, Sutherland and Caithness, with, perhaps, the exception of the district immediately around Wick.[1] This was the line of demarcation between Highlands and Lowlands adopted by the Government in considering the treatment of the forfeited estates and their proprietors after the failure of the rising in 1745, and was chiefly determined, apparently, by the use of the Gaelic language and the Highland dress.

Till 1745—"Bliadhna Thearlaich,"[2] as it is called in the Highlands—the chiefs and gentry exercised almost absolute power on their estates. Their quarrels were frequently settled as "in the good old times of yore, when Buckler was defendant and plaintiff was Claymore"; and their people unhesitatingly backed them without the slightest inquiry into the merits of the case. Political opinions were expressed in blows rather than by votes, and the tenants seldom differed, or, at any rate, seldom expressed a different opinion from the laird. Such instances did occasionally occur, but practically there was no law but the laird's above the "Highland line."

The population was kept down by fighting, by smallpox and other diseases—more fatal when doctors were few and dwellings insanitary; by periodical famines and generally hard conditions of life, which prevented any but the strongest children from coming to maturity.

After the "Forty-five," the hereditary jurisdiction was taken from the chiefs, the estates of those who had followed Prince Charlie being forfeited to the Crown. The Highland dress was abolished by Act of Parliament,[3] on purpose to break the clan spirit, the only

[1] Stewart's "Highlanders of Scotland."

[2] The "Year of Charles."

[3] This Act was passed against the advice of Lord President Forbes, who stated that the Lowland clothing was not suited to the habits of a pastoral people in a country without roads or bridges, where they constantly had to ford rivers, and sleep out in the hills, and that in his opinion they could not carry on their business if they were not allowed to use their own dress.

exception made by the Act being in favour of "Officers and soldiers in His Majesty's army," Highland regiments being always allowed to wear it, as they have done ever since.

For a civilian who persisted in wearing the kilt the punishment was six months' imprisonment for the first offence, and for the second, transportation to the American plantations, which was a life of slavery. It is hardly credible in the present day that the Government should care what clothes a man wore; but in 1747 troops were detached in many parts of the north, who patrolled certain districts, and made prisoner every man they met wearing the forbidden garb. In the "Life of General Wolfe," the hero of Quebec, and elsewhere, there are some curious reports by the officers of detachments. For instance, an officer at Fort-William writes (I quote from memory), "Sir, I have the honour to report that the patrol under my command this day took a man wearing the philabeg. He pled that it was not made of clan colours, but of plain stuff. As this was a moot point, we took it from off him, the sergeant cut it in pieces with his sword, and we let him go free." So the poor gentleman had to walk home in his shirt! Another reports from Morven (Argyllshire) that the people of the district have given up the philabeg, but have not adopted the breeches as ordered, but a sort of loose dress sewn in the middle, and leaving the knees bare. The officer asks if this dress is objectionable, or if it may be allowed.

It would probably give nearly as much trouble to make the people of Morven return to the kilt to-day as it took to make their grandfathers leave it off!

This edict was not enforced very strictly for more than ten years, though it remained in force for thirty-five years; and one officer complains of the Sheriffs of Perthshire for not convicting; and another at Braemar, mentioning a man who had been taken wearing the Highland dress, reports that "one Shaw, a half Laird," and his servant Allan Coutts, had encouraged the mob in Irish to rescue him, adding, "We want magistrates that will and dare exert themselves." Captain Beckwith, stationed at the head of Loch Arkaig, reports—"On the 24th of last month one of my men brought me a man to all appearance in a philabeg, but on close examination I found it to be a woman's petticoat (which answers every end of that part of the Highland dress). I sent him to the Sheriff-Substitute, who dismissed him." Thus it seems probable that in some parts comparatively little notice was taken of it.[1] Still, there was always

[1] In 1757 a man named M'Alpine or Drummond M'Gregor was acquitted in Perthshire, on his proving that his kilt had been stitched in the middle. Some men, wishing to keep within the law, which had not specified on what part of the body the breeches were to be worn, carried a pair over their shoulders on their sticks!—Stewart's "Highlanders of Scotland."

the possibility of being punished, and when, about 1782, the "Dis-kilting Act," as it was called, was repealed by the influence of the Duke of Montrose, there were great rejoicings; the event was celebrated by the famous poet Duncan Ban M'Intyre and others, and the dress was again worn, but not so universally as it had been before, as the people gradually learnt to be artisans, and to adopt habits of trade and industry, for which it is not so well adapted as for war, the chase, or pastoral pursuits. Still, at the latter end of the 18th century, when the Gordon Highlanders were raised, it was in most districts the ordinary wear. Of recent years it has been worn chiefly by soldiers and sportsmen. It is the recognised military uniform of the north.

At this period (1794) the land continued, as formerly, to be let out in large tracts by the lairds to gentlemen "tacksmen,"[1] often their kinsmen; and the smaller tenants paid rent in money, kind, or labour, not to the laird, but to the tacksmen, who were, in fact, middlemen, and had great power over the tenants and cottars on their farms. Cattle were the principal stock of the country, sheep-farming being introduced only about this time, and not becoming general until much later. Drovers collected great herds, and employed a number of active gillies to drive them to the southern markets. There was little other employment, and little money; even well on in the 19th century ploughmen's wages were £2 10s. for the half-year. Few could speak English.

The Caledonian Canal was not made till after Waterloo. There was no field for emigration except Canada, and to get there was difficult and costly. There were no county or burgh police, now a favourite employment with young Highlanders; the Post Office was in its infancy,[2] and the railways, which now employ an army of men, did not exist. The kelp industry and the herring fishery occupied part of the island population; illicit distillation (for whisky had begun to replace ale as the favourite drink of the Highlanders), oak-barking in spring, or shearing the corn in the neighbouring lowlands in autumn, gave a questionable or occasional employment to some on the mainland, but there was practically no outlet except the army for enterprising young men.

In 1793, the Rev. J. L. Buchanan published an account of the state of the Western Isles, where he had spent nine years as missionary minister from the Church of Scotland, commissioned by the S.P.C.K. In the introduction he apologises to his readers for any grammatical errors, as he had seldom spoken English during that

[1] Tacksmen, from having a "tack" or lease of the lands, in distinction to the small tenants, who had no leases. The forfeited estates were restored in 1784.

[2] At this period only three postmen were employed in Liverpool, at 7s. a week each, and four in Glasgow. The salary of the postmaster at Arbroath was only £20 a year, and of a clerk in the Glasgow Post Office £30.—Hyde's "One Hundred Years by Post."

period. He describes the huts of the small tenants as "remarkably naked, open, destitute of furniture"; they sleep in a blanket in any corner; cows, goats, and poultry have the common benefit of the fire. The windows are but holes made through the thatch immediately above the side walls. In gentlemen's kitchens, which are separate from the main house, men and women all sleep together. Men's money wages are from 10s. to 40s. a year, and out of it to pay for damage by carelessness. Two meals daily is usual for small tenants and *scallags*.[1] Salt is very scarce, and their diet of fish and potatoes, or sometimes, if in good circumstances, broth with bread, potatoes, and mutton, is often eaten without salt; barley or bere meal the only bread, and not always that. The land is worked with the "cas chrom" and "cas direach."[2] The *scallag* is sometimes formally tied up and flogged. Mr Buchanan gives credit to the minister of Tygheary, in North Uist, who has also a large farm, for "never having been known to kick, beat, or scourge, or in any way lift his hand against his *scallags* in the whole course of his life." The *scallag* builds his own house with sods and wood. If he is sent to another part of the farm, he carries his cabers with him and forms a new hut. He works four or five days for his master, and on the sixth cultivates a patch of land for himself; he is allowed brogues of horse or cow hide or sealskin to wear in carrying seaware over rocks, etc., but often goes barefoot, with perhaps "mogans," *i.e.* hose-legs and bare feet. He is also given tartan hose, a coarse coat, and a blanket or two. Mr Buchanan says the large landowners are generally more considerate to their people than the smaller lairds or tacksmen, though many of these are very kindly. Among the proprietors he mentions as kind *masters*—the word applied by small tenants to the laird—are the Duke of Argyll and MacKenzie of Torridon. He notices the change from servitude to freedom in Lewis, brought about by MacKenzie, "the present noble-minded proprietor." He alludes to the example set by Lord MacDonald of Sleat, which he hopes will soon be followed by others, of taking the small tenants out of the control of the tacksmen, and giving them holdings direct from himself at a fixed rent in money or kind, which makes them, he remarks, much more comfortable and independent. This seems to be the beginning of "crofter holdings" in the west; but the word "crofter" is not mentioned, and appears to be unknown till introduced from the south at a later period. He also praises MacDonald of Boisdale as an honourable gentleman and a great agricultural improver, who distributes justice and preserves

[1] Farm-servants.

[2] These are primitive Highland instruments, the one a peculiar spade or dibble, the other a primitive foot plough; a straight stick with a short pin or shoulder some twelve inches from the point on which the foot presses so as to drive the share into the ground.

peace and order among his people, like a prudent and kind master of a family, and is loved and esteemed accordingly.[1]

In describing the people generally he says, " They have a fine vein of poetry and music, vocal and instrumental ; had the language been more generally understood, the Gaelic music would have been introduced on every stage on which taste and elegance prevailed." They are also spirited dancers, using the violin for dancing in the house, the pipe for weddings, funerals, etc., and in great houses the piper plays before the door at meal-times. He mentions that the Highland Society of London gives prizes for pipe-music. The men were handy at making implements of husbandry, etc. They wear, he says, the short coat, the philabeg and short hose, bonnets sewn with black ribbon around the brim and a slit behind, with the same ribbon in a knot. Their coats are commonly tartan, striped with black, red, and other colours, after a pattern made by themselves or other ingenious contriver ; waistcoats commonly of the same, but the " feilabegs "[2] are often of fine Stirling plaid, if their money can afford them. When going herring fishing they dress something like sailors. They tan their brogues with the root of the tormentilla, which they dig from the hillocks by the sea ; the poorer sort often go barefoot even in winter. The people, he continues, retain a certain dignity of manner, constantly addressing each other as " duineuasal " and " bheanuasal " (gentleman and gentlewoman), raising their bonnets on meeting.

He complains of the scarcity of church services, mentioning one parish church where there was only occasional worship of any sort, and where the Communion had not been dispensed for several years. A hundred years ago there was only one Established Church in a parish perhaps thirty miles long, so that attendance could not be very regular. The Roman Catholic Church was then, as it still is, strong in certain districts of Inverness, Banff, and Aberdeen, as was the Episcopal Church in some parts of the Highlands where it has now few adherents.[3] But though imbued with a strong religious sentiment, I doubt if the Highlanders of that time could be called " kirk-greedy."

Mr Buchanan's remarks refer to the West Highlands and Islands, but the above description applies more or less to the whole Highlands, where the life was one of hardship, tempered by intelligent and even intellectual amusements—music, poetry, dancing,

[1] A son of Boisdale raised many men in Uist for the Gordon Highlanders, of which he was major.

[2] Kilts.

[3] After the " Forty-five," the military were employed not only in putting down the Highland dress, but in arresting Episcopal and Roman Catholic clergy if found holding divine service. This was not on account of their religious tenets, but because they were adherents of the house of Stuart.

and, in the long winter evenings, story-telling by the dim light of the peat fire, tales of warlike deeds handed down from generation to generation. I have heard them told in the present day, in probably the self-same words as were used in the "Fifteen," the "Forty-five," round the bivouac fires in the Peninsula, or in the herring boats off Barra Head.

The Highlanders were distinguished from their Lowland neighbours by a natural courtesy of manner, which is remarked upon by a French traveller who visited the Highlands in 1786. He describes the people as "poor but honest"; he had never seen "such civility without the shadow of servility, such plain frankness without the least rudeness, such a poverty and such contentment." He also mentions their pride in being an old and unconquered race.[1] They were not, however, without the love of gain common to mountaineers, and were perhaps deficient in that strong sense of fairplay which is an attribute of the Saxon.

In summer the young men and women went to the "aris" in the higher grazings to make cheese and butter for winter use, practising feats of strength and agility in the evening, or playing the fiddle, the pipe, or the trump; rather despising, perhaps, than respecting the "dignity of labour,"[2] but obedient to parents and to those whose authority they recognised and respected. A people, in short, whose feelings, traditions, and present circumstances rendered them ready to join the army as a profession at once honourable, profitable, and suited to their inclinations.

Encouraged, no doubt, by the gallant service rendered by the first regiment of Highlanders, the 42nd, Mr Pitt, when Prime Minister, about 1757, had recommended the King to attach the Highlanders to his Government by employing them in his service,[3] and in his celebrated speech in Parliament nine years later, he says—"I sought for merit wherever it was to be found; it is my boast that I was the first Minister who looked for it and found it in the mountains of the North. I called it forth, and drew into your service a hardy and intrepid race of men, who, when left by your jealousy, became a prey to the artifice of your enemies, and had gone nigh to have overturned the State in the war before the last. These men in the last war were brought to combat on your side; they served with fidelity, as they fought with valour, and conquered for you in every part of the world."

[1] *Leisure Hour*, August 1899.

[2] A woman whom I met in Sutherland in 1859, who was then 105 years old, and still active and in full possession of her senses, told me that when she was a young woman the only trades thought worthy of a man were soldiering and droving. Another at Acharacle, Ardnamurchan, who was grown up before Waterloo, said as there was little work, numbers went to the army; that the kilt was commonly worn, and young men turned out very smart on Sunday in tartan coats and often red waistcoats.

[3] Stewart's "Highlanders of Scotland."

JEAN, DUCHESS OF GORDON

After the painting by Sir Joshua Reynolds in the possession of The Duke of Richmond and Gordon at Goodwood

Having thus briefly touched upon the state of the Highlands in the latter part of the 18th century, we will now consider the connection of the ducal house of Gordon with these districts. Soon after the reign of Robert Bruce, the Gordons obtained, in addition to their property in the Lowlands, the great possessions in Badenoch and Strathspey which had formerly belonged to the Cummings. They also secured, by purchase or royal grants, estates and superiorities in Lochaber formerly possessed by the once powerful Lords of the Isles, so that the Gordon lands and lordships extended from the shores of Loch Eil on the west coast of Scotland, to Speymouth on the east. Consequently, the political power of the family was enormous, and its influence among the neighbouring Highland gentry very great.[1] In the case of the fourth Duke, who raised the Gordon Highlanders, his personal popularity among all classes, added to his territorial influence, ensured the success of the efforts by which three regiments were raised by him between 1759 and 1793.

The first of these was the 89th Highland Regiment, raised by the family influence of the young Duke, who was a captain in it, while his two brothers, Lord William and Lord George, were respectively lieutenant and ensign. The regiment was commanded by his step-father, Major Morris.

Upwards of 900 men were asembled at Gordon Castle in December 1759, and marched to Aberdeen. They went from there to Portsmouth and embarked for the East Indies, December 1760. The Duke wished to accompany them as his brothers did, but King George II objected to his doing so, saying that a Scottish Duke had more important duties at home than the command of a company in India.

A detachment of this regiment, under Major Hector Munro, took an active part in suppressing the mutiny at Patna. The 89th distinguished itself at the battle of Buxar in 1764, where the enemy lost 6000 killed and wounded and 130 pieces of cannon. Soon after, the regiment was ordered home, and it was reduced in 1765. The men of this corps were remarkable for their fidelity and good conduct.

At this time, though the Militia had long existed in England, there was none in Scotland ; but what were called Fencible Regiments were raised at various times, some for service in Scotland only, others for the defence of the three kingdoms while the regular army was engaged abroad, and in 1799 some were even raised for service in Europe and America.[2] They were disciplined and armed in the

[1] The Duke of Gordon was called by the Highlanders "Coileach an taobh tuath," *i.e.*, "The Cock of the North." He was also hereditary Constable of the Castle of Inverness.

[2] *Military Journal*, November 1800, Vol. II. p. 659.

same manner as troops of the line ; Fencible officers ranked junior to those of the line, but took precedence of the Militia.[1]

In 1778 the Duke of Gordon raised a Fencible Regiment in the counties of Inverness, Moray, Banff, and Aberdeen—a very efficient corps, which was reduced in 1783.[2]

The country being again in danger in 1793, the Duke raised another regiment of Gordon Fencibles, and his commission as colonel was dated March 3rd. The uniform was the full Highland garb. Upwards of 300 men were raised on the Gordon estates in Lochaber, Badenoch, and Strathspey, an equal number on other estates in these neighbourhoods, and 150 men from the Lowland parts of Aberdeenshire, Banff, and Elgin.[3]

The following account is taken from the personal reminiscences of an officer of the Royal Marines, who acted as quartermaster to this regiment, and who set out to Aberdeen to join, after making arrangements for accoutrements, etc., for the regiment in London. " I found Lieutenant-Colonel Woodford, formerly of the Guards, an active clever officer and a great disciplinarian, in command. Recruits daily arriving, the clans of Cameron, M'Pherson, M'Intosh, and Frazer had joined their standards to the Gordons. In a month we were ready for inspection, 600 strong, and formed a fine body of young men. We went to Edinburgh, where the forming of flank companies excited no little jealousy among several of the Highland officers, especially one young chief, who had no conception, when he brought fourscore of his clan as volunteers, that they were to be disunited, and said in the mess-room, ' If the commanding officer dared to draft any of his men to other companies, he would order his piper to sound his gathering, and march them back to Lochaber ' ; that his men were gentlemen, and he would not have them associate with ' Botich nam brikis.' " [4] It had to be explained that his men were now soldiers and must go to the company they suited, and that a court-martial might prove a disagreeable commencement to his own military career.

In the spring of 1795, on account of the alarm of invasion, it was thought necessary to order the regiment to England, along with seven other regiments of Fencibles raised for the defence of Scotland. This was done without consulting the colonels, though the express condition in their letters of service was that they were only to be ordered out of Scotland " in case of actual invasion of the

[1] Militia was introduced in Scotland by Act of Parliament in 1797, and when put in operation later was opposed by riotous proceedings in the Highlands, under the erroneous impression that the ballot was used to enable the Crown to remove the people from Scotland.—" Military Forces of the Crown " (C. M. Clode).

[2] Stewart's " Highlanders of Scotland."

[3] *Ibid.* This regiment was also called the " North Fencibles."

[4] " Churls with breeches."

island." The men therefore objected to go, and were to be seen in knots talking in Gaelic with an air of mystery. The lieutenant-colonel marched them by detachments to the chapel and, ascending the pulpit, lectured them, but without any effect in allaying the suspicion that they were being imposed upon. He did not understand the Highland character. At last an express was sent to Gordon Castle. The Duke arrived in forty-eight hours ; the regiment was paraded ; he explained in a few well-chosen words

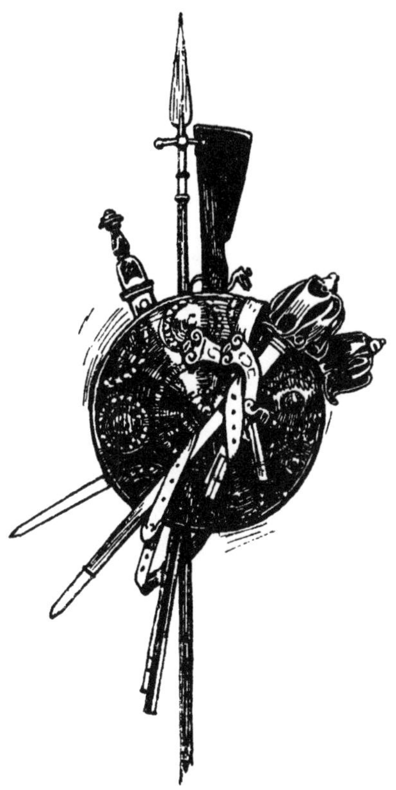

the nature of the service required, that they were called for the defence of their country, and he trusted that any soldier who was such a dastard as to refuse such service would step out of the ranks and he should have his discharge. Though raised for the defence of Scotland, England was now in danger, and none but cowards would refuse the call. They would find him at their head. The men waved their bonnets, crying that they were ready to follow his Grace to the world's end, and they embarked at Leith in high spirits.

The King, never having seen a Highland regiment, ordered them to London (in 1796), where he reviewed them, and expressed himself much gratified at their appearance. The Duchess of

Gordon and her daughters were present among the royal group, wearing Highland bonnets and Gordon tartan plaids. The royal family particularly noticed the sergeant-major, Dugald Campbell,[1] who is described as "a most superb specimen of the human race."

This regiment returned to Scotland in 1796, and later was reduced.

The three corps mentioned above may be considered as the forbears of the distinguished national regiment whose conduct in war has done much to preserve the blessings of peace at home, whose conduct in peace has upheld the character of their country wherever duty has called them, and whose career of upwards of a hundred years it will be my endeavour to describe.

[1] Sergeant-major Dugald Campbell was promoted ensign, and was afterwards appointed adjutant of the 92nd (Gordon Highlanders), in which he became a celebrated character.

CHAPTER II

THE French Revolution was at its height when, in February 1793, the Republic declared war against Great Britain and Holland. It became necessary to increase the British army, and again the patriotism of the Duke of Gordon induced him to come to the aid of the Government, by offering to raise a fourth regiment; this time, as in the case of the 89th Highlanders, for general service.

He received authority to do so on the 10th of February 1794, and the command was given to his son, the Marquis of Huntly, who had served as captain in the 42nd Royal Highlanders, and was then a captain and lieutenant-colonel in the 3rd Guards.[1] The Duke himself, and his son, took a personal interest in the recruiting, and the celebrated Duchess Jean, still a beautiful woman, lent to it all the prestige of her high position, and all the grace and charm of manner for which she was famed alike in Court and cottage.

She rode to the country fairs in Highland bonnet[2] and regimental jacket (it was not unusual, in those days of military enthusiasm, for ladies to wear the uniform of their husbands' or brothers' regiments).[3] It is told how she gave a kiss to the men she enlisted —a fee more valued than the coin by which it was accompanied, as in the case of a smart young farmer at Huntly market, who took the shilling and the kiss, and then paid "smart,"[4] saying, "A kiss from your Grace is well worth a pound note."

Sometimes she is said to have placed a guinea between her lips.[5] There was in a Highland village a young blacksmith, remarkable for his strength and good looks. Recruiters for the Guards and Line had in vain tried to enlist him, but he could not resist her Grace! He took the kiss and the guinea; but to show it was not the gold that tempted him, he tossed the guinea among the crowd.[6]

Commissions were given to gentlemen in the north for raising a certain number of men, the field officers and generally the captains having previous army service. But so great had been the drain on the Highland population for some years, that it was no longer easy to recruit a battalion quickly from them alone. In the present

[1] Now Scots Guards.

[2] The bonnet she wore, though denuded of its ornaments and feathers, is now in possession of the 2nd Battalion Gordon Highlanders (late 92nd).

[3] Kay's "Portraits."

[4] If a recruit repented his bargain before being sworn in, he paid £1, called "smart money."

[5] The Duchess's recruits were proud of being enlisted by one who was the greatest lady in their world, and it is natural they should have boasted of the fact; when, in after years, one of them was wounded, a Highland comrade would cry, "Och cha n'eil ach pog eile o'n Bhan Diuc!" or, as an Aberdonian would facetiously put it, "Mind, lad, ye got a kiss o' the Duchess o' Gordon for that!"

[6] Told by General Sir John Gordon, K.C.B. See also Appendix XIV.

instance it was rendered the more difficult, because, as we have seen, a Fencible Regiment had been raised by the Duke of Gordon in the preceding year, and the 79th, or Cameron Highlanders, had just been raised by Mr Allan Cameron, son of the tacksman of Erracht, largely in the same districts where the Duke of Gordon's influence lay. This regiment has had a similar glorious career to the Gordon Highlanders, and a lively feeling of friendship has always existed between the two corps. Lochiel, Chief of the Cameron Clan, however, had not approved the venture of his kinsman of Erracht, but used all his influence in favour of the Gordon regiment,[1] the Duke being to a certain extent his feudal superior, his Inverness-shire lands being held from the latter, and also his personal friend. In this Lochiel was greatly assisted by his uncle, Mr Cameron, tacksman of Fassiefern, whose son John received a captain's commission in the Gordons, and afterwards became its most distinguished commander. He was closely connected with such Highland chiefs and gentlemen as Cluny MacPherson, MacNeil of Barra, MacDonald of Glencoe, Campbell of Barcaldine, and others, who helped him in raising his quota of men.[2]

It is not surprising, therefore, to find that, such strong influence being exerted in its favour, the Gordon regiment obtained more recruits in the Highlands than the 79th had done in 1793,[3] but still some of the officers had to complete their quotas from other parts of the country.

When a new regiment was raised, the custom was to embody it, not in the country districts where the men were enlisted, but in some garrison town where they could conveniently be inspected by a general officer, and passed as fit for service. In this instance, Aberdeen was the rendezvous.

One can imagine the difficulty of bringing lads from the distant islands of Barra, Uist, and Skye in open boats to the mainland, to join those from the "rough bounds" of Arisaig, Ardnamurchan, and Lochiel; the long marches, cheered by song or pipe, through Highlands and Lowlands, ere their tired feet trod the "plainstanes" of the Granite City. How they would gaze with astonished admiration at the buildings and the shops of a great town!

[1] Letter from Donald Cameron, Esq. of Lochiel.

[2] "Your recruits leave this to-morrow if the wind proves fair. They will answer well, as they are all smart young boys. In case of their being reduced, I beg you to be at pains to prevent their enlisting in any of the old regiments, but to pack them off home as soon as possible, as it would be a great satisfaction to their parents, in case of peace. It was by way of a great favour to us that they allowed them to go." [*N.B.*—There were twelve recruits in this batch.]—Letter from Mrs. MacNeil of Barra to Captain John Cameron, dated at the Island of Barra, June 1794.

[3] The "79th Register," compiled by John Ford, lieutenant and adjutant 79th Regiment, from documents in possession of the regiment, in 1817, gives the number of Highlanders as 278, out of the 600 passed when they were embodied at Stirling, February 1794.

Here the embryo soldiers probably had their first experience of the tented field, as there was at this time a considerable encampment near Aberdeen.[1]

The regiment was embodied on the 24th of June 1794,[2] and was inspected the following day by Lieutenant-General Sir Hector Munro, the same officer who served as major in the first regiment raised by the influence of the Duke of Gordon at the battle of Buxar, thirty years before. At this inspection, Sergeant D. Nicol[3] mentions in his journal that each man passed the General separately, running 50 paces, and that only one man was rejected for age."

This first parade must have presented an appearance more picturesque than military, for, except some of the officers and sergeants, they had no uniform. There were the lads from the Highlands in their tartan coats and kilts, with "cuarans," *i.e.* brogues of home-dressed skin, on their feet; the Lowland loons in grey breeches, ribbed stockings, and low-heeled shoes; and all with the round blue bonnet, then universally worn in the north of Scotland, while here and there an Irishman's "caubeen" would give variety to the headgear. There was something of the same feeling that was exhibited when the Gordon Fencible Regiment was embodied in the preceding year. The men who had come from Highland districts with a captain they knew and confided in, objected to be removed from his immediate command, and these officers had to explain the necessity of the flank companies being selected from the whole, promising still to look after their interests.[4]

When their Centenary was celebrated by the Gordon Highlanders in Glasgow in 1894, a newspaper much read in the Highlands commented upon the small number of Gordons among its members, and inferred that when raised the regiment was to a large extent composed of men of that name. This is an entire mistake. It was called the "Gordon Highlanders," not because it was a regiment of Gordons, but because it was a regiment of Highlanders raised by the Duke of Gordon. Among the non-commissioned officers and soldiers there were, in 1794, just twenty Gordons.

[1] There were at this time hardly any barracks in Scotland, except those in the Highlands, built 1718 to 1728, viz., at Kilahimen (Fort-Augustus), for 300 men, afterwards increased for a larger number; at Ruthven, in Badenoch, for 250 men; at Bernera, in Ross, for 150 men; at Inversnaid, for 100 men, and two or three smaller ones. Rooms to be made 18 feet by 17 feet, five beds for ten men in each.—Clode's "Military Forces of the Crown." The Aberdeen (Castlehill) Barracks were being built in 1794.

[2] Cannon's "Historical Record."

[3] In 1905, Mr MacLean, Architect, Edinburgh, sent the author a most interesting journal kept by his grandfather, D. Nicol, who enlisted in the Gordon Highlanders early in 1794 and served as private and segeant in both battalions until taken prisoner at Talavera.

[4] "Memoir of Colonel John Cameron," 4to, 1858. Privately printed.

MacDonald was the name most largely represented in the ranks, Cameron the next[1]; there were several Andersons and other similar cognomens, and the names of all the Highland Clans are to be found on the first muster-roll.

I have been unable to find any parade state or muster-roll during the few days the regiment remained in Aberdeen; but I counted the recruits of 1794 from the Description Roll, and I found that, out of 749 whose places of birth or enlistment are given, 241 were from Inverness-shire, 92 from Aberdeenshire, 58 from Banffshire, 33 from Argyll, with considerable contingents from Caithness, Sutherland, Ross, Moray, Nairn, Perth, and Stirling shires. Kincardine, though now part of the new "territorial" district, produced only four men, while nearly every county in Scotland had one or two recruits to its credit. There were 15 Englishmen, 1 Welshman, 45 from Ireland, 26 from Edinburgh, and 54 from Glasgow[2]; but many of those from the last two places appear from their names to have been Highlanders in search of employment, and a few of those from Ireland, bearing such names as M'Millan, Grant, Balfour, may have really been from the Scottish Fencibles quartered there. There were besides the above-mentioned recruits, 23 whose places of birth or enlistment are not given. Their names are Cash, Cameron (2), Connacher, Connan, Christie, Campbell, Gordon (2), Grant (2), Gunn, Kenedy, Kincaid, M'Nicol, M'Kay, M'Arthur, Mackie, Mark, Munro, Murray—making altogether 772, which, if not absolutely correct, is as nearly so as possible. Inverness-shire, therefore, can boast of having supplied nearly a third of the whole.

Indeed, the number of soldiers furnished by this county during 1793-4 seems perfectly wonderful; for not only did it supply nearly the whole of the "Gordon Fencibles," a large number to the Cameron Highlanders, and nearly one-third of the Gordon Highlanders, but also the majority of the "Fraser Fencibles," and a considerable part of the "Inverness Fencibles," and of the 97th or "Strathspey" Regiment, and, besides all these, gave recruits to the 42nd and other regiments. At the present day, though the county possesses one excellent and thoroughly Highland battalion of militia, it gives but very few recruits to the army,[3] the reason for this falling

[1] In the muster-roll, December 1797, three years later, there appear six Evan Camerons, six Alexander Camerons, six John Camerons, besides two or three each of many other Christian names among the n.c. officers and soldiers.

[2] The municipality of Glasgow had for many years encouraged the immigration of young Highlanders. Gaelic churches, both Protestant and Roman Catholic, had been established for their benefit, and the influx of Highlanders to the city had been ever increasing.

[3] In 1893 the number of natives of Inverness-shire enlisted in the county for the regular army was 22; in 1894, 11; in 1895, 8.—*Return from Regimental District.* (No doubt more natives of the county were enlisted in Glasgow and other places.)

RAISING OF THE GORDON HIGHLANDERS

From a picture in possession of F. J. Usher, Esq. of Dunglass, by Skeoch Cumming

PART OF
EUROPE
SHOWING THE BOUNDARIES OF
FRANCE AND ADJOINING COUNTRIES
before the
REVOLUTION IN 1789
A. K. JOHNSTON, F.R.G.S.
Scale of English Miles
GERMAN OCEAN
ENGLISH CHANNEL
BAY OF BISCAY
CORSICA
SARDINIA
Balearic Isles
ALGIERS
TUNIS
CONSTANTINOPLE
Malta
W. & A. K. Johnston Limited, Edinburgh

RECRUITS ON THE WAY TO JOIN

(*From description in Buchanan's "Hebrides Isles," 1793, and drawings of the time*)

off being not a diminished population,[1] but is to be looked for in the changed circumstances and ideas of the people.

General David Stewart, in his "Highlanders of Scotland" (which was written about 1820, while many of the original officers and soldiers were living, who was himself in garrison at Gibraltar along with the Gordons in 1796, and who had, therefore, good information), puts the proportion of *Highlanders* among the original recruits at three-fourths of the whole. It is difficult from the Description Roll to decide exactly, because, though the place of birth is generally given, it seems to have been often taken for granted that the recruit was born where he was enlisted, especially in such towns as Glasgow or Edinburgh. For instance, a man named Colin M'Kenzie is entered as a native of Minorca, Spain! But I afterwards discovered him to be a transfer from a regiment quartered there. A proportion of those enlisted in the south were, from their names, evidently Highlanders; while, as there was at that time no attraction for strangers in the Highlands, those enlisted there were no doubt natives. Still, judging from the roll, I would put the proportion of Highlanders at rather less than three-fourths.

The first recruits of the Gordon Highlanders were not tall men on the average. They were aged from eighteen to thirty-five years; but I find several younger, and one man of forty.

They were generally of respectable families[2] and excellent character. General David Stewart describes them as "Moral, well-principled, and brave; they have never failed in any duty entrusted to them"—a character which, with rare exceptions, has always been kept up by their successors. Many of them were the sons of those who had been out with Prince Charlie, and were filled with the traditions of that chivalrous period. Most of them were unaccustomed to what would now be called ordinary comfort, much less to luxury, and were accordingly well fitted to bear the hardships of a European campaign.

The bounty[3] appears to have been three guineas to ordinary recruits, and five guineas to men who had served in Fencible Regiments.

The following is a list of the original officers who were, with

[1] The population of Inverness-shire was, in 1801, 72,672; in 1891, 88,362.—*Table of Census for Scotland.*

[2] From contemporary letters and oral evidence.

[3] It was not uncommon in the Highlands for officers enlisting men from their own or their fathers' estates or farms to arrange that the parents of the recruit should enjoy some little advantage, such as grazing for an extra cow, etc. A correspondence between the Marquis of Huntly and Major Alexander Stewart, 95th (Rifles), January 1804 (formerly of Gordon Highlanders), shows that an ensign, for a lieutenancy, raised eighteen men, and received five guineas for each who was approved, and the officer paid for any deficiency £20 per man. See also Appendix XIV.

perhaps one exception, Scotsmen, the majority of them being Highlanders :—

LIEUTENANT-COLONEL COMMANDANT

George, Marquis of Huntly, had served in the 42nd Royal Highlanders and 3rd Guards in Flanders, and at the siege of Valenciennes ; afterwards general ; and colonel successively of the 92nd, 42nd, 1st Royal Scots, and 3rd Guards ; died, 5th Duke of Gordon, 1836.

MAJORS

Charles Erskine of Cardross, had served in the 25th, 16th, and 77th Regiments in the war against Tippoo Sahib in India, and against the French at Martinique ; killed in Egypt, 1801.

Donald MacDonald of Boisdale, had served in the 76th Regiment (or MacDonald's Highlanders) ; died 1795.

CAPTAINS

Alexander Napier of Blackstone, from lieutenant 7th Fusiliers ; killed at Corunna, 1809.

John Cameron, son of Fassiefern, had been ensign in the 26th Regiment, and lieutenant in Captain Campbell of Ardchattan's Independent Highland Company ; killed at Quatre-Bras, 1815.

Honourable John Ramsay, son of the Earl of Dalhousie, from lieutenant 57th Regiment ; died 1842, lieutenant-general, and colonel of the 79th Highlanders.

Andrew Paton, from lieutenant 10th Regiment ; retired.

William MacIntosh of Aberarder ; killed in Holland, 1799.

Alexander Gordon, son of Lord Rockville,[1] killed at Talavera, 1809, when lieutenant-colonel 83rd Regiment.

Simon MacDonald [2] of Morar, had served in the American War in MacDonald's Highlanders (then the 76th Regiment) ; retired as major, 92nd.

CAPTAIN-LIEUTENANT [3]

John Gordon, came from 81st Regiment ; retired as major.

LIEUTENANTS

Peter Grant, came from a Fusilier regiment ; died at Keith, major half-pay, 1817.

Archibald MacDonell, Inch, Lochaber, came from 79th Highlanders ; died in Lochaber, lieutenant-colonel of Veterans, 1813.

Alexander Stewart of Achnacone, afterwards in 67th Regiment and 95th Rifles ; died in Appin, 1854, lieutenant-colonel.

[1] Lord Rockville, a Lord of Session, was a brother of the Earl of Aberdeen.

[2] On one occasion, when a charge was ordered, a man of his company rushed in front of this officer. Lieutenant MacDonald having asked what brought him there, he answered, "You know that when I engaged to be a soldier I promised to be faithful to the King and to you ; and while I stand here neither bullet nor bayonet shall touch you except through my body." This chivalrous soldier was born on his officer's estate, and was descended of the same family.—Stewart's "Highlanders of Scotland."

[3] Captain-Lieutenant seems to have been senior lieutenant, and commanded the colonel's company.

John MacLean of Dochgarroch, promoted from 1st Royals, lieutenant-colonel 27th Regiment, 1808; died, lieutenant-general, K.C.B., 1825.
Peter Gordon; died 1806.
Thomas Forbes, Newe; killed at Toulouse when lieutenant-colonel 45th Regiment, 1814.
Ewan MacPherson, Ovie, Badenoch, came from 78th Highlanders; died 1823, lieutenant-colonel retired, Governor of Sheerness.
George Gordon (son of 4th Duke of Gordon), came from 6th Dragoons, retired colonel; died at Glentromie, Badenoch.

ENSIGNS

Charles Dowle; killed in Egypt, 1801.
George Davidson; killed at Quatre-Bras, when captain 42nd, 1815.
Archibald MacDonald, Garvabeg, Badenoch, retired lieutenant; died in Canada.
Alexander Fraser; killed in Holland, 1799.
William Todd, son of the factor at Fochabers, retired.
James Mitchell, Auchindaul, Lochaber, lieutenant-colonel 92nd Gordon Highlanders, C.B., retired 1819; died in Lochaber, 1847.

CHAPLAIN

William Gordon.

ADJUTANT [1]

James Henderson, came from an Independent Company; died 1796.

QUARTERMASTER

Peter Wilkie; died 1806.

SURGEON

William Findlay; died in Egypt, 1801.

N.B.—Several of the officers were not actually gazetted till later.

The men having received a few necessaries at Aberdeen, where they lived well, boarding out for two shillings or half a crown per week, the regiment marched on June 27th for Fort-George, where it embarked on July 9th, and landed at Southampton on the 16th of August, after a voyage of five weeks! Considerably longer than would now be occupied by a voyage to India. Sergeant Nicol in his journal mentions that Lord Kintore and the Duke of Gordon entertained the regiment on the march to Fort-George. It crossed the Spey on a raft at Fochabers and was reviewed at Nairn by Sir Robert Sinclair, Governor of the Fort, whose wife, Lady Madelina, Lord Huntly's sister, was present in regimental bonnet, jacket, and dirk. She gave each man a present of money. She and her sisters had helped in recruiting. The regiment was eventually

[1] Adjutant was a separate rank at this time. There were no regimental paymasters prior to 1798. The first in Gordon Highlanders was Archibald Campbell, appointed January 1800. He was of the Lochnell family, born at Killiechronan, Mull; afterwards an army agent.

encamped on Netley Common, and brigaded with the 78th, 90th, and 97th Regiments.[1] Here the men were instructed in the first principles of discipline and drill, and received their arms, clothing, and appointments, which were as follows :—

They wore the full Highland dress or " breacan an-fhéilidh "—that is, plaid and kilt in one, called in Regimental Orders " the belted plaid." [2] The officers had twelve yards, or rather six yards double-width, of Gordon tartan, but of rather a smaller *sett* than was afterwards used. The rank and file had a smaller quantity. The officers' purse was badger-skin, having a silver rim round the top, and six silver-mounted white tassels.[3] The purses of the rank and file were of grey goat-skin, with six white tassels.[4] The hose of all ranks were cut out of the strong red and white tartan cloth known in the Highlands as " cathdath " or " battle colour," which was worn by all Highland corps, and by the better class of civilians when in Highland dress. The rosettes and garters were scarlet ; the sash was crimson, and was worn over the left shoulder by both officers and sergeants. The officers wore a gilt gorget. All ranks had long hair, tied behind with a black ribbon, and powdered on Sundays and special occasions, and on guard. Neither moustache nor whisker was worn.

The head-dress consisted of the round bonnet then commonly worn in Scotland, but cocked and ornamented with ostrich feathers, and having a diced border of red, white, and green, said to represent the " fess chequé " in the arms of the Stuart kings. It had a hackle fastened over the left ear by a black cockade, with regimental button on it. This hackle was white for the Grenadier Company, green

[1] There were in Netley Camp the 7th Dragoon Guards, the 5th, 9th, 13th, 17th, and 24th Dragoons, and the 78th, 90th, 97th, 98th, and 100th Regiments of foot. The Barrack Department was instituted in 1792. At that date all the barrack accommodation in Great Britain and the Channel Islands (in forty-three fortresses and garrisons) would only house 20,847 men of artillery and infantry. The troops were constantly under canvas, even in winter, causing loud complaints, or they were billeted. When in billets, either on the march or in stationary quarters in England, the cavalry soldier paid 6d. a day to the innkeeper out of his pay of 8d. ; the infantry soldier paid 4d. out of his pay of 6d., for which the innkeeper had to supply them with food and beer. In camp, instead of beer, they had bread at reduced price, and an allowance in money was also given in camp and quarters for necessaries.—*Military Journal*, Vol. I., p. 477.

[2] Few people in the present day know how this ancient garb was worn. It was put on in the following manner :—The belt was laid down and the plaid over it, the centre of the plaid being over the belt ; it was then neatly pleated across the belt, but leaving a part at each end unpleated ; the belt was then fastened round the waist, so that the lower half of the plaid formed the "feile" or kilt, of which the unpleated part became the apron, and the upper half, falling over the belt, formed the "breacan" or plaid, which was fastened on the left shoulder, or it could be thrown round the shoulders as a cloak. By loosening the belt, the whole became a blanket or plaid. ("Plaide" is the Gaelic word for blanket.)

[3] Cannon's " Record."

[4] Described to me in 1852 as "speckled" by Pensioner Corporal John Innes, Tom a Mhulin, Glenlivat, one of the recruits of 1794. Cloth hose were worn in the regiment till 1849-50, when the material, but not the colour, was altered.

for the Light Company, and for the battalion companies the lower half was red and the upper white.

The jacket was scarlet for officers and sergeants, and red for the rank and file, with lapels turned back with yellow, showing the waistcoat, and laced two and two ; lace silver with blue thread in the centre ; silver or plated buttons, with the number of the regiment in the centre. Epaulettes—two for all ranks of officers,[1] of silver bullion, having two stripes of yellow silk in the centre of the strap, with a gold-embroidered thistle and a binding of blue round the edge. The n.c. officers and men had white tufts for battalion companies and wings for flank companies. The officers' waistcoat was scarlet, laced with silver, the n.c. officers' and men's white, these being the origin of the mess-waistcoat and white jacket worn now by officers and men respectively.[2]

The officers were armed with the Highland claymore, worn at the back, in a buff belt, fastened by an oval breast-plate of silver, having a crown and thistle, surrounded by the words "Gordon Highlanders" ; they had also a silver-mounted dirk. The sergeants with claymore and pike. The rank and file carried flint-lock muskets, the barrels brightly polished, and bayonets. All musicians were armed with the claymore. The knapsacks were of goat-skin.

Perhaps some extracts from Garrison and Regimental Orders may show better than words of mine the condition and development of the corps.

The first Regimental Orders I find are dated August 17th, 1794. They are about a captain and subaltern of the day being appointed in future, and as to their duties, and that each officer is to supply himself with a book, in which he is to insert in his own handwriting all Regimental Orders, and so not plead the neglect of the orderly sergeant for being ignorant of them. These books to be inspected by the commanding officer every Sunday.

The men to dine at one o'clock, till the regiment gets into a regular way of messing. An officer of a company to attend to see the dinners are well cooked.

Lord Huntly is sorry that the present undisciplined state of the regiment makes it necessary for him to order the officers to attend strictly to everything concerning their companies, but hopes that by their own attention he will soon be able to relieve them of a great part of their trouble.

1 Except in Highland regiments, only field officers wore two epaulettes.

2 I have no evidence of the dress of the drummers and band in 1794, but they probably wore then, as they certainly did a few years later, yellow jackets turned up with red for drummers, and white with yellow facings for the band. Pipers wore then, and till about 1860, the same uniform as the rank and file.

Minute directions as to telling off and sizing a company,[1] "as some officers may be unacquainted with the method of sizing them."

Officers commanding companies to let the quartermaster know how much hose tartan they require to complete every man to three pairs.

At this time they were numbered the Hundredth Regiment. The title "Gordon Highlanders" was continued along with the number in all Regimental Orders and Returns, but in Horse Guards or General Orders and in the Army List it was not used—simply 100th Regiment. The first use of the number was in G.O., August 20th, 1794—"The 100th Regiment gives the guards and orderlies to-morrow." No doubt a certain number of n.c. officers and men who had served before were capable of this duty.

All the officers were drilled every day by the adjutant, and they are recommended to get some person to instruct them in marching and giving the word of command.

A sickness among the men is attributed to want of salt, and "companies that have already cleared with the paymaster are to supply themselves with that article," while officers of companies in a less solvent position are "to advance 6d. each to their men, which they will stop when their settlement takes place."

Orders as to squad drills and sergeants' duties explained.

The cost of the men's rations was—meat (¾ lb.), 1/9 a week; bread, 8½d.; meal or vegetables, 4d.; and 8½d. for washing, hair-powder, etc.

G.O., August 23rd, 1794.—"Officers commanding corps will see that divine worship is not omitted to-morrow. Infantry to wear powder on Sundays. Recruits unclothed not excepted."

R.O. of same date, as to officers being properly dressed by the 31st, and their "hair cut agreeable to Lord Huntly's order." Officers commanding companies to supply immediately men that want them with regimental kilt and hose. "Those who have at present breeches or trousers to be first provided"—evidently most of the men had their own kilts.

At this time the first court-martial sat, a Sergeant Munro being reduced to the rank and pay of a private sentinel.

On the 24th they were supplied with "nabsacks" (*sic*) and leather stocks. "No velvet stocks nor any that are tied in front are to be allowed."

They had "small horses" placed by commanders of companies for the men to dry their belts on, "and they are not to hang belts on the horses for the arms."

There seems to have been great difficulty in enforcing cleanliness

[1] There were ten companies at this period. The colonel, lieutenant-colonel, and senior major of a regiment each had a company.

in the camp. By the Orders of the 25th August, it appears there were married men enlisted, and the ladies seem to have required instruction as much as their lords.

At this time, while Lord Huntly was in London, an evil-disposed person made the men believe that he had sold the regiment to the East India Company. The Highlanders assembled apart from the officers, and some proposed marching to Scotland, as the 42nd had tried to do fifty years before under somewhat similar circumstances. On Huntly's return, however, they were at once reassured and returned to their duty. In Orders of September 3rd he refers to this " misunderstanding " and offers a reward of twenty guineas to any man, soldier or civilian, who can discover the giver of the bad advice which occasioned it.

Trooper, Private of Light Infantry Regiment, and Officer of Infantry, 1794

CHAPTER III

WE now come to the second period in the life of the youthful regiment, when service in the Mediterranean was gradually to prepare it for the prominent part it afterwards played in the stirring events of the time.

On the 5th September they embarked at Southampton for Gibraltar, where they arrived on the 26th, and disembarked on the morning of the 27th, landing at the "Ragged Staff."[1] The effective strength was 3 field-officers, 7 captains, 9 lieutenants, 6 ensigns, 4 staff, 29 sergeants, 21 drummers, 727 rank and file. Here their education in drill and interior economy was continued. White trousers were served out to the men,[2] who seem previously to have had only the Highland dress. Only the flank companies had as yet got feathers for their bonnets. Mention is made of officers and men parading in their "belted plaids." The orderly officer wore white breeches and boots.

Having in the last chapter given the names of the officers of 1794, with any particulars which are known regarding them, it may be interesting to the regiment to have a list of the sergeants of that year also. I have accordingly copied their names from "The Muster Roll of H.M. 100th Regiment of Foot (or Gordon Highlanders), 183 days, from June 25th to December 24th, 1794, included," and have added all the particulars I could gather from other regimental papers.

They seem nearly all to have been drawn from remote districts, and it speaks well for the education given at the period in the parish schools of Scotland, that no less than five of their number were afterwards promoted to commissions in the line or embodied militia.

The list, which gives a fair idea of the districts from which the regiment was recruited, is as follows :—

RANK.	NAME.	TOWN OR PARISH.	COUNTY.	REMARKS.
Sergeant-major	Thomas Thomson	Annan	Dumfries	Came from 61st Regiment, promoted adjutant Nov. 9th, 1796
Qmr.-sergeant	Lachlan MacPherson	Laggan	Inverness	
Sergeant	Dugald Cameron	Kilmallie	Inverness	Promoted ensign, East Middlesex Militia, May 1801
"	Alexr. Cameron	Kilmallie	Inverness	
"	John Davie	Terynissle	Aberdeen	
"	Alexr. Dobie	Bellie	Elgin	Killed at Egmont-op-Zee, 1799 (as private)

[1] Garrison Orders, Gibraltar, September 26th, 1794.

[2] Major Henry (Knight of Windsor), who enlisted in 79th when raised, told me that fatigue trousers were first served out when they went abroad. A great joke was to get the Highland lads who had never worn them to put them on backside foremost for the officer's first inspection.

Rank.	Name.	Town or Parish.	County.	Remarks.
Sergeant	Peter Fergusson	Kilmallie	Inverness	Promoted ensign, Renfrew Militia
"	George Gordon	Cabrach	Aberdeen	
"	William Hay	Rhynie	Aberdeen	
"	Andrew Hollingsworth	Dublin	Dublin	Came from 46th Regiment, musician
"	John Kennedy	Kingussie	Inverness	
"	Angus MacBain	Moy	Inverness	Came from 42nd Royal Highlanders
"	Alexr. MacGregor	Laggan	Inverness	
"	Alexr. MacIntyre	Ardentalen	Argyle	
"	Angus MacIntosh	Petty	Inverness	
"	John MacIntosh	Abertarf	Inverness	Promoted in West India Regiment, 1805
"	Lachlan MacIntosh	Laggan	Inverness	Drowned in Gibraltar Bay, Oct. 18th, 1800
"	Kenneth MacIntosh	Durness	Caithness	Came from 46th Regiment, killed at Egmont-op-Zee, 1799
"	Duncan MacKinley	Lochgoil	Argyle	
"	James MacPherson	Laggan	Inverness	
"	Donald MacPherson	Kingussie	Inverness	
"	John MacKenzie	Boleskine	Inverness	
"	James Martin	Kinethmont	Aberdeen	
"	John Mitchell	Boro'ness	Linlithgow	
"	James Ross	Rosemeven	Ross	
"	John Reynolds	Cabrach	Banff	
"	Donald Stewart	Blair in Athole	Perth	Promoted adjutant, 29th May 1806
"	Alexr. Stewart	Kilmonivaig	Inverness	
"	William Sinclair	Reay	Caithness	Drowned at the landing in Holland, 1799 (as private)
"	Alexr. Smith	Dunlichity	Inverness	
"	George Still	Aberdeen	Aberdeen	
"	James Wilson	Bellie	Banff	

Extracts from Regimental Orders, 100th Regiment (or Gordon Highlanders), 29th October 1794, Gibraltar. Officers to set their watches by the adjutant's, that they may be exact to time on parade.

N.c. officers or soldiers never to go into the town without a pass.

The companies always to be inspected by their respective officers, and they will be pleased to proceed with them as follows :—When the company is fallen in and dressed, it will receive the word of command—" Fix bayonets ; shoulder arms ; open pans ; slope arms." The officer will then go through the company and inspect the locks. When that is done, he will order, " Carry arms ; shut pans ; order arms ; to the right face ; draw ramrods." Each man will then draw his ramrod, and as the officer comes opposite to him he will ram it down, and after he has returned it, he will face to his front. The officer will then inspect every part of the men's dress, and Lord Huntly particularly orders that the men's faces and hands are all washed and their hair combed, as there can be no excuse for dirtiness. After each man has been looked at, the company will stand easy.

November 1st.—A fatigue party of twenty men to be employed every morning in cleaning barracks. Lord Huntly was ashamed of the state they were in.

Roll to be called at 8 p.m.

Officers in future, when they mount guard, will be pleased to wear gloves ; a pattern will be seen at the adjutant's quarters.

November 2nd.—In order to accommodate officers, Lord Huntly has no objection to their making up frock jackets, which they may wear when off duty ; a pattern will be shown. They may likewise wear round Hatts (*sic*), and a small Durk (*sic*), such as Lord Huntly has got. Bathing parades and strict orders as to sanitary matters.

November 6th.—Lord Huntly, understanding that most of the officers wish to mount guard in the belted plaid, he shall give orders for their mounting in that dress.

The Grenadiers [1] and Light Infantry will mount guard in bonnets, and the battalion companies in leather caps. The Grenadiers will have a white feather, and Captain Cameron will endeavour to procure green ones for the Light Infantry.

November 9th.—Each man to pay 1d. a week to the sergeant-major for pipe-clay ; also to pay for hair-powder, soap, and oil for light.

December 29th.—Every officer attending the general parade to be dressed in the belted plaid.

The quartermaster will divide all the (hose) tartan equally among the companies. The men will be charged for these hose, and the two pairs they are annually entitled to from the colonel will be served out to them when the convoy arrives.

In December 1794, the regiment received its colours on Windmill Hill. After their consecration by the garrison chaplain, the Marquis of Huntly made an impressive address, calling the attention of the officers and men to the duties which their King and country expected from them, and to the honours which he trusted they would acquire under these banners.

The first, or King's colour, was the great Union ; the second, or regimental colour, was of yellow silk. In the centre of both, the number "100," surmounted by a crown and the words "Gordon Highlanders," the whole within a wreath of thistles and roses.[2]

[1] Grenadiers originated in the French army about 1670. They carried twelve or fifteen cast-iron shells in a leather bag. These were lit at close quarters by a slow match and hurled into the ranks of the enemy. Grenadiers were selected on account of their height, as able to throw further. The shell to which they owed their name became obsolete about 1760, but they continued to be *corps d'élite*, stood on the right of the line, and headed all storming parties. There were horse as well as foot Grenadiers. (Grosse's "Military Antiquities.")

[2] Cannon's "Historical Record."

It appears the men were often employed on various working parties, for which they were paid, but which must have interfered with their military education.

R.O., Gibraltar, January 29th, 1795.—There will be no parade this afternoon, as the men are employed heaving down the *La Lutine* frigate.[1]

The n.c. officers are positively ordered to discontinue the practice of selling liquors in their own houses.

Gibraltar, February 8th, 1795.—No officer is to appear with a white stock or neckcloth, and their jackets are to be hooked through the frill at the top and across the breast, not only on parade, but at all other times.

February 15th.—No officer will appear on court-martial duty except in regimental waistcoat, breeches, and boots. The first officer who disobeys this order shall be desired to wear the above dress on all occasions.

In March, Lord Huntly rebukes n.c. officers for allowing sentries to relieve one another.

Bounties to recruits had risen by the competition of officers raising men, and this practice is checked by War Office Circular, February 21st, 1795 :—

My Lord,—The high bounties offered to the new levies being judged extremely prejudicial to the recruiting for the navy, and His Majesty having in consequence thereof signified his commands that the bounty given to recruits enlisted for general service shall not exceed fifteen guineas, and that the bounty to recruits for the Fencibles shall not exceed ten guineas,[2]

. . . .

I have the honour to be,
My Lord,
Your Lordship's most obedient Servant,
(Signed) M. Lewis.

To Lieut.-Colonel the Marquis of Huntly,
100th Regiment of Foot.

April 3rd.—Lord Huntly hopes the companies will vie in trying which shall be in the best order, and it is recommended to the men to punish one another when any berth in the barrack-room is found dirtier than it ought to be. He repeats the order that no man is to appear in his new jacket or belted plaid but on duty. He expects the officers at parade to be clean and dressed like officers ; he will not punish the whole for the slovenliness of one, and anyone who comes improperly dressed may expect to be ordered to wear upon all occasions the uniform directed for regimental duties. No officer

[1] *La Lutine* was afterwards wrecked, and all on board were lost, together with £200,000 of treasure which she was carrying.

[2] See also Appendix XIV.

is to be seen with strings in his shoes, except with the belted plaid.[1]

April 12th.—Officers to be particular that the men are well dressed, caps or bonnets well put on and jackets hooked through the shirt at the top across the breast, that uniformity may be established as much and as soon as possible.

May 4th.—Sergeant Fergusson of the Light Company, on account of his not having made up his books agreeably to order, is removed to the 6th Company, and Sergeant Stewart appointed in his room ; and as Lord Huntly considers it an honour belonging to a flank company, he will remove a n.c. officer who at any time behaves improperly.

May 8th.—The men to have their shirt frills well ironed and hair well powdered for the review.

The Highland dress appears not to have been considered a correct costume for dancing, as the Orders of May 9th desire " the officers to appear in belted plaids at the Commander-in-Chief's dinner [2] on Monday, but Lord Huntly begs them to appear at the ball on Monday evening in white breeches."

June 4th.—At the request of the captains of companies, the men were allowed a bottle of black strap [3] each and fresh provisions, it being the King's birthday.

R.O., June 9th.—Lord Huntly appoints John MacPherson of the Grenadier Company and Alex. Clark of the Light Company to do duty as volunteers.[4]

At this time the regiment was held in readiness to proceed to the Island of Corsica.

The regiment, consisting of ten companies, embarked on the 11th June, and in Garrison Orders of the 10th the Commander-in-Chief expresses " his approbation of their good conduct and regularity of discipline since he has had the honour of commanding them." They landed at Bastia on the 11th July.[5]

Corsica, the birthplace of Napoleon Bonaparte,[6] had separated from France, and in 1794 tendered the sovereignty of the island to King George III., and united itself to Great Britain. A body of

[1] The meaning of this is not clear, but in Duncan Ban's well-known Gaelic song of that time in praise of the Highland dress, he mentions thongs as the proper fastening of the brogues, and alludes to shoe buckles as an innovation.

[2] The dinner-hour was in the afternoon.

[3] A sort of port, stronger, no doubt, than their daily ration of wine.

[4] Volunteers in this sense appear to have been men of superior education, who enlisted and served in the ranks as private soldiers, but without pay and at their own expense, in the hope of getting commissions. They were included in the rank and file. The two men in question seem, from their attestations, to have been till this time ordinary privates, and were enlisted at Laggan and Kingussie, but volunteers generally enlisted as such.

[5] One transport had " very nigh been taken by the French fleet, who passed us in the night."—*Letter from Volunteer C. Cameron, 100th Regiment.*

[6] Napoleon was born at Ajaccio, in Corsica, August 15th, 1769.

Corsican troops was added to our army, their officers ranking junior to those of the regular British army. The island was governed by a Viceroy (Sir Gilbert Elliot), who, on the 27th July, inspected the regiment in belted plaids. Judging from the Regimental Orders, that form of the Highland dress was seldom, if ever, again worn by the rank and file, though it is mentioned later for the officers. Afterwards the "feile-bheag," or little kilt, was used, that is, the lower half of the belted plaid, permanently pleated and stitched; but so essential to the full Highland dress was the "breacan"[1] or upper half considered, that the remembrance of it was kept up by the ornamental plaid still worn by all ranks in full dress; though it is not made, as it formerly was, and as it might still be, to form a light cloak for the shoulders when required.

Highland soldiers seem to have borne a character for honesty in Corsica, according to Carr's "Caledonian Sketches," published in 1807, where it is stated that "While Corsica was in our possession, the Governor's butler, on great fêtes and public occasions, used to request that some of the men without breeches might have charge of the plate during the entertainments." Grapes, however, seem to have been a great temptation to the young soldiers, and they were cautioned against trespassing in the vineyards, under pain of court-martial.

Lord Huntly keeps his eye on all ranks. "Officers will be pleased to put their swords in their belts, and not walk about with them in their hands."

Ensign MacDonald, who seems to have been slack in his orderly duty, "is to be officer of the day till further orders."

Any woman refusing to wash for the men is to be turned out of barracks, and her rations stopped; 6d. a week to be paid for each man's washing.

Some of the young Highlanders seem to have thought being on sentry over French prisoners was much the same sort of duty as herding cattle from the corn at home, and Lord Huntly cautions them against such unsoldierlike conduct as sitting at their posts.

Officers when not on duty or parade may wear any kind of black hat they please with a feather in it. No soldier to appear in the streets without side arms. Volunteers in future to attend all parades.

Fault is found with want of cleanliness in hospital and barrack-rooms, and officers are desired to be very particular, on account of the men's health.

Grape-stealing and other slight irregularities having ceased,

[1] I met a pensioner, in 1857, who had worn the "breacan an fheile," or belted plaid, in the 42nd. He said it looked well on a tall man, but rather like a bundle of clothes on a little man.

Lord Huntly, in Regimental Orders of August 23rd, returns his warmest thanks to the officers for their great attention to their companies, and is certain that, as long as they exert themselves, he must have great satisfaction and credit from the regiment. Lord Huntly is happy to see that for some days past the men have returned to their former good behaviour, and hopes they will continue behaving in the same manner, as they will then be a credit to their country, and may depend on having every indulgence and kindness shown them that is in his power.

The regiment was now in a state to perform all duties, and detachments were sent to various places. Major Erskine is Commandant of St Fiorenza, where prisoners, both French and Turks, were confined.

A War Office circular, October 2nd, 1795, permits n.c. officers and soldiers to send letters home for 1d.,[1] signed outside by the commanding officer.

September 9th.—In consequence of the good conduct and regular behaviour of John Cameron of the 1st Company, ever since he joined the 100th Regiment, he is again appointed sergeant, and Lord Huntly hopes that in future sergeants will be more careful of the language they make use of to one another, and that they will not trouble their officers with every trifling dispute they may have together.

At this time bonnets to complete the regiment arrived, and Lord Huntly orders officers commanding companies immediately to provide feathers, that they may be made up. " Lord Huntly will make a present of one feather to each man."

R.O., Bastia, December 18th, 1795.—As Lord Huntly expects to leave this in a day or two, he begs leave to return to the officers his warmest acknowledgments for their uniform support and attention to the regiment ; and also to express to the n.c. officers and soldiers his most grateful thanks for their regular and good conduct since he had the honour of commanding them, and to assure the officers and soldiers that their interest will always be his study, and that no exertions on his part will be wanting to prevent the Gordon Highlanders from being drafted.[2]

That he kept his word is proved by the fact that no instance can be found of men from the Gordon Highlanders being drafted,

[1] The charge for civilians' letters from the Mediterranean was—for a single sheet, 2s. 1d. ; and for 1 oz., 8s. 4d.—" One Hundred Years by Post," by J. W. Hyde.

[2] At this period the Government were in the habit of drafting men from one regiment to another, and to this the Highlanders strongly objected. It was, therefore, a great advantage to a corps to have at its head a man of sufficient influence at headquarters to prevent this. It appears that there had been a risk of the Gordon Highlanders being drafted, as, in a letter dated November 28th, 1795, Captain J. Gordon writes :—" It is exceedingly lucky that we have escaped being drafted ; the Marquis would have been in a perfect fury."

though several regiments in garrison with them at Gibraltar shortly after, including the 42nd Royal Highlanders and 90th Regiment, were so treated.[1]

R.O., Bastia, December 22nd, 1795.—Major Erskine takes the earliest opportunity of communicating to the officers his anxious wish that the discipline of the 100th Regiment should be carried on as pleasantly as possible during the short time he is to have the honour of commanding it, and he begs to assure them that nothing in his power shall be wanting that can contribute to their comfort and happiness. He takes the same opportunity of informing the n.c. officers and soldiers that he will be happy to show them every indulgence consistent with propriety, to make up in some degree for the loss they will sustain by the absence of Lord Huntly,[2] to whom they should ever consider themselves indebted for his exertions in their behalf.

Mention is now made of the "musick"[3] practising regularly; and whereas hitherto each company had its own tailor and shoemaker, each charging his own price, there was now a tailors' shop and shoemakers' shop, the former presided over by a master tailor, while the shoemakers were looked after by Sergeant MacIntosh, and their charges were fixed "as in the 42nd and other Highland regiments."

R.O., February 1st, 1796.—As to-morrow is Lord Huntly's birthday, all the stopt wine will be issued to the men to drink his Lordship's health.[4]

It appears from the Regimental Return, February 1st, that there were over 100 men sick.

Hairdressing must have been a serious business in those days. Each company had a barber, who found soap and hair-powder, and was paid 3d. a week by each man. The hair was well greased, combed out, powdered, and tied behind. The soldiers sat on a bench one behind the other, and tied each other's tails, taking it in turn to be powdered. So the junior had to be up betimes, the senior being left to the last. "Hair to be just long enough at the sides to friz a little, and behind an inch below the rosette." Both officers and men were clean shaved.

R.O., May 21st.—The companies to parade every afternoon at four, in any dress they please, that their hair may be inspected by an officer. Lord Huntly's company is excepted from this parade, on account of having paid more attention to their hair than others.

[1] Garrison Orders, 19th May 1797.

[2] The ship on which Lord Huntly was going home was taken by a French privateer; but he was released on paying a ransom.

[3] Band.

[4] A common minor punishment was to stop a man's daily allowance of liquor, which, according to the station, consisted of wine, beer, or rum.

On the 3rd May 1796, Lieut.-Colonel the Marquis of Huntly was promoted colonel of the regiment, and Major Erskine, lieut.-colonel commandant. On the 15th, the greater part of the regiment marched to Corte to suppress a rebellion which had broken out there. They had much severe marching in a difficult mountainous country, and were thanked in General Orders for their exertions and good conduct during the above service.

At this time there was an opinion prevalent with the military authorities that the Highland dress was unhealthy for soldiers in a hot climate, and Lieut.-Colonel Erskine appears to have shared this prejudice. While at Bastia, the men's belted plaids were made into tartan trousers, to be worn with buckled shoes on guard, and into tartan jackets for fatigue. They had also grey trousers, or white in the hot weather. It appears that the change of dress did not conduce to the health of the men, but, according to General David Stewart, had the contrary effect. It was certainly disliked by the men, who, after a time, made known to the colonel their wish to return to the kilt, and it was resumed.

A draft of recruits joined at Bastia, under Captain M'Donald. In July Lieut.-Colonel Erskine obtained leave of absence.

R.O., Bastia, July 28th, 1796.—Lieut.-Colonel Erskine having made a most favourable report to Lord Huntly of the great attention of the officers, and of the general good conduct of the n.c. officers and soldiers during his absence, he takes this opportunity of returning them his best thanks, and trusts the reports which he shall from time to time receive from Major Napier will be equally good.

Major Napier accordingly took command, and in Orders of July 29th he "hopes that whenever Lord Huntly or Colonel Erskine meet the regiment again, they may find it in the same good order as when they left it. The regiment is so much obliged to Lord Huntly for his great exertions for them, that Major Napier hopes they will take a particular pride in behaving in such a manner as they are certain will always give him great pleasure."

On the 14th August, a detachment of 1 field officer, 2 captains, 4 subalterns, 7 sergeants, and 208 rank and file, was held in readiness to embark on a secret expedition. In consequence of the unanimous wish of the officers and men, Major Napier requested the Commander-in-Chief to permit the whole regiment to take part in this service, to which he received the following reply:—"The Commander-in-Chief is sensible of the zeal and laudable motives which have induced the officers and men of the 100th Regiment to offer their services on the present occasion, and he desires Major Napier to express his best thanks to them, as well as his assurance that he will be ready at all times to testify his satisfaction at their general good conduct and appearance, although circumstances will not, at this time, allow him to avail himself of their services to the extent they offer them."

The expedition proved to be against Porto Ferraio in the Isle of Elba. It was completely successful, and without loss to the troops engaged, the Island of Elba remaining in possession of Great Britain till given up at the Treaty of Amiens, 1802.

R.O., *August 21st.*—The major having found the barrack-rooms, without exception, in the most infamous condition, he directs the orderly officer to be particularly attentive to their being perfectly clean and swept, and not allow old hose, shoes, etc., to be under the beds. Orderly n.c. officers of dirty rooms to attend drill till further orders.

R.O., *August 25th.*—As the regiment is to embark immediately for Gibraltar, officers commanding companies will be pleased to provide anything they think will be wanted for the men during the voyage.

The success of the revolutionary party in France, and the brilliant career of their countryman, Napoleon Bonaparte, had encouraged the French partisans among the Corsicans, and the inhabitants began to repent their union with Britain. The British Government, therefore, seeing that the advantage of possessing the island was small, while the cost of keeping it would be great, decided to sever the connection, and gradually to withdraw the troops.

Garrison Orders, *Bastia*, *September 5th*, 1796.—The 100th Regiment being to depart for Gibraltar, Lieut.-General de Burg cannot suffer them to leave Corsica without testifying his approbation and satisfaction of their general good conduct and soldier-like behaviour since he has had the honour to command them. At the same time, he desires they will accept his best wishes for their success and welfare on all occasions. Lieut.-General de Burg pledges himself to those men of the 100th Regiment who are detained as artificers, that he has no intention of placing them in any other corps whatever, and that he will forward them to their regiment the moment they can be spared from the public works they are now employed in.[1]

On September 6th they embarked—1 field officer, 5 captains, 7 lieutenants, 5 ensigns, 3 staff, 37 sergeants, 21 drummers, and 706 rank and file—on the transports *Granby*, *Esther*, *Elizabeth*, and *Borus*, protected by seven men-of-war under Admiral Mann. In those days of slow post and no telegraph, the Admiral was unaware

[1] This refers to a number of tradesmen—smiths, carpenters, masons, etc.—who were employed, and who had evidently been afraid they might be drafted to another corps if once separated from their own.

that war had been declared by Spain against Great Britain,[1] till he was fired upon by the Spanish fleet, which was greatly superior in force. Our Admiral made all sail for Gibraltar, but the *Granby* was taken, having on board 2 staff officers, 3 sergeants, and 48 rank and file of the regiment,[2] who were made prisoners of war. Thus the first loss sustained by the Gordon Highlanders was in a naval engagement.

The following account of this affair is given by Captain Vivian of the 28th Regiment, then at Gibraltar, in a letter dated October 22nd, 1796 :—

> The Spanish fleet had passed by this a fortnight since. At that time war was not declared, and they pretended to say it was not about to be ; but in a few days they met Admiral Mann and immediately gave chase to him, but he happened to be nearer the Rock than they were, and owing to their prime sailers being afraid to engage without the whole fleet, he got off. An 80-gun ship, an uncommon sailer, was ahead of her fleet and coming up fast with a transport, when Admiral Mann made a signal to the *Hector* to put about and take her in tow. Though the *Hector* was a league astern of the fleet, as soon as the 80-gun ship saw her put about, she did the same, and ran for the Spanish fleet. They had fifteen sail to the British seven. The only thing they took was an hospital ship belonging to the 100th Regiment.[3]

After this adventurous voyage, the regiment landed at Gibraltar on the 4th October.

R.O., October 5th.—Each company to have a set of types to mark their clothing.

October 7th.—Officers will be pleased to complete all their men's bonnets with black ribbon of the same kind they formerly had, that they may be enabled to wear their bonnets properly, and not down on their heads like a night-cap.

As it is uncertain if the men are to be charged for their provisions on board ship, officers of companies are to charge them 3d. for each day, which will be repaid if the provisions are not charged for.

October 20th.—The men's hair not to be powdered, but well greased, and tied the same as if they were powdered. The scarcity of flour, which was often used by the men as hair-powder, seems to account for this order.

[1] In 1795, Prussia had concluded a peace with the French Republic, and in consequence of the United Provinces of Holland having leagued with France, Great Britain had taken possession of the Cape of Good Hope and Ceylon (then Dutch colonies). The former allies of Great Britain had now been converted into enemies. War was declared by Holland, which had been constituted the Batavian Republic, against Great Britain in May, and Spain followed the example in October.—Cannon's "Historical Record."

[2] In the Orders at Bastia, August 31st, for the distribution of companies to ships, is—*Granby*—The Marquis' and Captain M'Donald's, 154 men. This may, however, have been changed, the above being the number given in Cannon's "Historical Record" as being taken prisoners.

[3] "Lord Vivian : A Memoir." By the Hon. Claude Vivian.

The regiment was under canvas. A marquee was allowed for each field-officer and captain, to 3 lieutenants, and to 4 ensigns or surgeon's mate. One tent to a n.c. officer and 9 men, or 3 married couples. Vivian writes—"The troops are encamped on the south front for fear of the town being set on fire by the Spaniards." Many men were employed on the works and in other ways, receiving extra pay, which must have made the duty heavy on others.

R.O., October 18th.—All constant workmen and servants not with officers of the regiment to pay two reals a week pistreen money.[1]

R.O., October 27th.—In consequence of the hard work the men have had in cleaning their encamping ground, and also of their good behaviour since they landed, the major desires that the quarter-master will be pleased to buy wine, to be paid for out of the pistreen fund. Each man of the fatigue party to get a pint.

The pipers are to attend all fatigue parties. This is the first mention of pipers in orders. At this time, though the pipers were a most important part of a Highland regiment, and were held in the highest esteem by officers and men, they were not recognised by the authorities, and had no separate rank as the drummers had. They were purely a regimental, not an army, institution.[2]

A dispute as to precedence having arisen between the piper and the drummer of a company of a Highland regiment, and the captain deciding in favour of the latter, the piper expostulated, saying—"Oh, sir, shall a little rascal that beats a sheepskin take the right hand of me that am a *musician?*"[3]

November 29th.—To-morrow being St Andrew's Day, officers paying companies will be pleased to advance one day's pay immediately after parade, which will be at two o'clock.

In the Garrison Orders, Gibraltar, the promotions are given, and among them, December 9th, 1797:—"100th Regiment—Sergeant-major Thomas Thomson to be adjutant, *vice* Henderson deceased."

The same Gazette shows that the position of army chaplain might be purchased in those days:—"42nd Foot—M. Calder, clerk, to be chaplain, by purchase, *vice* Watson, who retired 23rd November."

Sergeant John M'Intosh was made acting sergeant-major, his appointment being confirmed in March, "to receive pay from the date of Adjutant Thomson's promotion, and also the additional pay which Adjutant Thomson received from the regimental fund."

R.O., Gibraltar, May 26th, 1797.—Major Napier, wishing to

[1] What was pistreen money? Probably from "pistrina," a bake-house. It was doubtless used for the benefit of the men, in bread or otherwise, and seems to have been peculiar to Gibraltar. "Pistareen" is mentioned as the name of a coin in "Mexico of To-Day," by C. F. Lummis.

[2] Pipers received rank and pay the same as drummers only about 1853.

[3] Carr's "Caledonian Sketches."

establish in the regiment a badge of honour for those men who have never been found guilty by a court-martial of any crime or irregularity since the regiment was embodied, requests officers commanding companies will send in a list to his quarters of such as come under that description.[1]

From the Garrison and Regimental Orders at this time, it appears that the soldiers of the 42nd and 100th Regiments were served out with round hats with feathers to be worn on guard. The officers to supply themselves with the same, and a foraging cap for each man. The hats [2] to be worn on guard, but bonnets at evening parade. The men had pads and false tails (clubs of hair), for which they paid 3 reals 12 quartos each, but false shirt frills are forbidden.

In April a draft of recruits arrived.

By Regimental Orders of March 15th, 1797, a Sergeants' Mess was established for the first time, "and the major hopes they will always conduct it in so proper a manner as never to require his interference."

On the 6th April, Lieut.-Colonel Erskine rejoined, and gives credit to Major Napier for the "improved appearance of the corps," but regrets that his first orders should be for a court-martial on some men for "rioting in the streets and very unsoldierlike conduct." There were great temptations to excess at Gibraltar, where wine was cheap, and free indulgence in it affected the conduct of many men of the garrison. In the journal of a private of the Gordon Highlanders, he laments that many of his comrades had taken to gambling, and I find Regimental Orders on the subject.[3] There were many general and district courts-martial in Garrison Orders, but few, however, on men of the 100th.

A soldier of the 28th is sentenced to receive 1000 lashes for desertion. These brutal punishments in the army were only a reflection of those in civil life, at a time when men were hung for stealing a sheep, and when an unfortunate debtor was confined in a crowded and loathsome jail till his debt was paid, which often meant till his life's end. Colonel Erskine and his officers seem to have had a great aversion to the infliction of corporal punishment, and though the law obliged regimental courts-martial to order it, the culprit's captain often begged him off, and the commanding officer would appear to have been glad of a good reason for remitting that part of the sentence.

[1] Good-conduct badges or pay did not then exist.

[2] General D. Stewart says these hats were found by the 42nd to be very useless and inconvenient.

[3] A military writer of the day regrets that more attention is not paid to encouraging athletic exercises and games, such as would give a healthy excitement and amusement. The men had gardens, however, and the Orders show that the King sent a present of seeds for the soldiers' gardens.

R.O., *May* 1797.—Lieut-.Colonel Erskine has always pleasure in attending to the requests of officers in favour of any man of general good character who may be unfortunately confined under sentence of court-martial, but begs they will not apply in favour of one whose crime is such as makes it incompatible with his duty as commanding officer to forgive. He has told the regiment a thousand times, and he once more repeats it, that he will forgive no man whatever who is guilty of being drunk on guard, sleeping on his post, of being abusive to n.c. officers, or of stealing.

R.O., *August 26th.*—At the regimental court-martial held this day, was tried Alex. Kenedy of the Light Company for refusing to go sentry when ordered by the corporal, for which crime he was ordered to receive 300 lashes. On account of his character, the commanding officer pardons him on this occasion; but he desires the regiment to remember that nothing but Kenedy's remarkable good character would induce him to forgive his disobedience of orders, which no plea whatever can justify, and any soldier who receives an order that may be improper, will always get redress by complaining through the regular channel.

Again.—Robert Robertson and Daniel Ross of Major MacDonald's company are sentenced to 250 and 200 lashes respectively; but at the request of Major MacDonald, the corporal punishment is remitted by the commanding officer, but both men are sent to the Prevost.

Recruits lately joined appear never to have received belted plaids, and are credited with one year's allowance of tartan, deducting what was allowed them in their kilts. Sergeants' tartan cost 1s. 9d. and privates' 1s. 6d. a yard.

R.O., *June 29th.*—Dirty barrack-rooms to be cleaned by the pioneers, and the n.c. officers and men off duty belonging to the dirty room are to pay—sergeants 2s. 6d., drummers 1s., privates 6d. ; the money to be equally divided between pioneers and regimental hospital.

June 25th, 1797.—The King, considering the low rate of pay of subalterns of infantry, which is inadequate to their necessary expenses, orders an increase of 1s. a day to all lieutenants, ensigns, adjutants, and quartermasters. And their pay is no longer to be subject to deductions for poundage, hospital, and agency as heretofore.

The pay of the infantry soldier, which had been 6d.,[1] as the old song says—

> O happy the soldier who lives on his pay,
> And spends half a crown out of sixpence a day—

[1] Before 1794 the infantry soldier's pay was 6d., but he had an allowance of 2½d. daily for necessaries and bread money. In 1795 these were increased to 4d., but the soldier was

was this year increased to 1s., and the liberal allowance of small beer was continued ; there was a stoppage, however, on home service of 6$\frac{5}{7}$d. per day for food, and on foreign service of 6d. a day when rationed and 3½d. a day when not rationed by the public.

R.O., Gazette, July 5th, 1797.—100th Regiment. Volunteer Gordon MacHardy to be ensign without purchase, *vice* M'Donald, promoted in the 51st Foot.[1]

August 30th.—Lieut.-Colonel Erskine hopes that the melancholy accident that happened last night to Robert Chisholm of Captain Gore's company, and the consequences that are likely to follow to William Henderson, who was the cause of his death, will prevent soldiers of the regiment making free use of their hands ; for even if Henderson should escape with his life, for which he must be tried, it must be dreadful for any honest man that, by unjustifiable violence, he has been the cause of another man's death.

Henderson was found not guilty of wilful murder, and returned to his duty.

At this time Captain J. Cameron, Lieutenants Arch. M'Donnell and Jas. Henderson, Ensigns W. M'Pherson and John M'Pherson, and five sergeants were employed to recruit in Scotland.

R.O., September 10th.—The major, in going round barracks, is requested to give any directions he may think necessary for the health and comfort of the men.

October 22nd.—The recruits this day drawn into companies must at first be treated at drill and upon every occasion with the greatest lenity.

November 16th.—Lieut.-Colonel Erskine, having heard that the men wish to wear the Highland dress, cannot agree to it at present, being afraid of their health, and because they have lately paid for blue pantaloons, but he promises that as soon as possible their request shall be granted.

December 16th.—That any man who may unfortunately have got into a scrape, or who has merited to be returned as a bad character, may have an opportunity of recovering himself, Lieut.-Colonel Erskine will call for a return on the 4th June next, when he and every officer will be glad to hear of any reformation, and every man will be returned as of good character who has behaved well in the intermediate period.

liable to deduction for his mess of 3s. a week, or 5d. a day. In 1797 these allowances were increased by 2d., the whole being consolidated into 1s. a day, the deduction for mess being increased to a sum not exceeding 4s. a week.

[1] Gordon MacHardy was the son of parents in good circumstances in the neighbourhood of Tomintoul, Banffshire. He was educated in France for the Roman Catholic priesthood, but, hearing that his brothers had enlisted in the Gordon Highlanders, he gave up the idea of taking holy orders, and, following their example, joined as a volunteer. His superior education and conduct gained a commission.—*Told by his grand-nephew, James MacHardy, proprietor of the Gordon Arms Hotel, Tomintoul.*

At this period the men were ordered to be inoculated,[1] and officers are to explain to those averse to it the nature of the operation and the danger of smallpox.

R.O., February 22nd, 1798.—The men to be supplied with proper puffs for powdering their hair.

R.O., Gibraltar, March 4th, 1798.—As the regiment is to embark on Friday, Lieut.-Colonel Erskine trusts the men will conduct themselves with the same regularity they have hitherto done

GIBRALTAR
(From an old drawing.)

on such occasions. Upon no occasion is the discipline of a regiment so conspicuous as in embarking and disembarking. Lieut.-Colonel Erskine is convinced of the regard the soldiers of the 100th Regiment have for their own character and the credit of their regiment, and he is confident they will conduct themselves as before. He is sorry to say that there are a few men who have no regard for their own character or that of the regiment, and if such men indulge in intemperance, they will be seriously punished on board ship.

[1] Inoculation had been recently introduced into England from Turkey, by the influence of Lady Mary Wortley Montagu, and was, till the discovery of vaccination, the best known means of mitigating the scourge of smallpox.

On the 16th March 1798, the regiment embarked on board of the transports *Alliance*, *Queen*, *Singleton*, *Minerva*, *Mary*, and *Jane*. Tempestuous weather drove them for shelter into the Tagus, and it was only after a boisterous passage of two months that the regiment disembarked at Portsmouth on the 15th May.

Thus ended their first tour of Mediterranean service. The men were no longer " Johnny Raws " ; the reports of General Officers were all to the effect that the Gordon Highlanders were orderly and regular in quarters, and exemplary in the performance of duty. The Regimental Orders show that this was accomplished by continual attention on the part of Lord Huntly and Lieut.-Colonel Erskine, who commanded them, and whose motto seems to have been, " Suaviter in modo, fortiter in re."

CHAPTER IV

AT Hilsea Barracks, Lieut.-Colonel Erskine granted the request of the regiment to return to the Highland uniform, which has never since been departed from, and on May 19th the quartermaster is desired to issue to the companies a sufficient quantity of tartan for kilts and hose. The kilt of those days had at least the advantage of lightness and economy, but was a scant covering compared with the voluminous garment which is now worn. The quantity ordered " as sufficient for a kilt " was 3¾ yards for the Grenadier Company, and 3½ yards for the other companies ; ¾ yard for a pair of hose.

Meanwhile it would appear that the long voyage had given some officers a taste for naval costume.

R.O.—Lieut.-Colonel Erskine begs to remind those officers who have bad memories that nankeen pantaloons and half-boots is the uniform in which they were ordered to appear on regimental parade. He hopes he will never see a repetition of what he saw last night—an officer coming to inspect his company dressed like a sailor.

In July 1797, Lord Malmesbury had been sent for the second time to Paris to negotiate a peace, but the demands of the French Directory rendered the attempt abortive. On the 17th October a treaty of peace was signed between France and Austria, and Great Britain was thus left to continue the contest single-handed against France and her allies.

In 1796 the French had already turned their eyes to Ireland as the part of the British Isles where they could most seriously harass the British Government. Accordingly a fleet was fitted out at Brest, having on board about 14,000 troops, which sailed December 17th, and, evading the British cruisers, neared Bantry Bay on the 22nd, the plan being to land there and march to Cork, forty-five miles distant, before the Irish Government had notice of the expedition. Heavy gales from the east, however, scattered the fleet, and prevented a landing. Meanwhile news of their arrival having been received in Dublin, steps were taken to oppose them. A council of war on board the French fleet decided that, under these circumstances, and as the inhabitants of that part of Ireland showed no disposition to revolt, they should sail towards the Shannon ; but a fearful storm arose on the 28th, the scattered vessels steered for France, and the danger of invasion was averted for the time.[1] But French emissaries travelled through the country, who fanned into flame the discontent which smouldered among a large part of the

[1] " Autobiography of Wolfe Tone." This attempt on Ireland was connected with a scheme for the invasion of England. A body of about 1200 Frenchmen were to have ascended the Avon and burned Bristol, but having landed in Pembrokeshire, they surrendered to a force of Fencibles and Militia under the Earl of Cawdor. The two frigates that brought them were captured.—Hume's " History of England " (Continuation of). Murray, 1865.

population, and culminated in the Irish Rebellion of "the year ninety-eight, when our troubles were great."

In consequence of these unhappy circumstances, the Gordon Highlanders were hurriedly ordered to Ireland, and, embarking in ships of war at Southsea Beach on May 31st, they arrived at Dublin on the 15th June, and were quartered in barracks. Here they formed part of a large garrison of regular and fencible corps. Among the latter were such now forgotten regiments as the Angus and the Reay Fencibles, the Dumfries Dragoons, the Ancient British Dragoons, etc. The troops were held ready to march at a moment's notice, and all was carried on as in an enemy's country. Soldiers were ordered not to speak to any man in plain clothes unless buying in a shop, and only to be out of barracks before 10 a.m. Officers are desired to pay great attention to the men being well fed, as they have so much duty.

A.G. Office, Dublin, June 16th, 1798.—His Majesty has been pleased to appoint Colonel the Marquis of Huntly, 100th Regiment, to serve as brigadier-general on the staff of the army of this kingdom. It is, therefore, General Lake's orders that he be obeyed as such.

The Gordon Highlanders formed part of the brigade commanded by their colonel.

R.O., June 20th.—On account of there being so few officers with the regiment, Lieut.-Colonel Erskine directs that Volunteers M'Pherson, MacDonald, Clark, and Cameron shall in future fall in as supernumerary officers, and are to be obeyed as such.

No officer to appear in the streets without his cross-belt and sword, except field-officers, who, for convenience in riding, may wear a waist-belt.

On June 21st, Lieut.-General Lord Lake defeated the rebels at Vinegar Hill, but they were encouraged to continue their resistance by expected aid from France.

On July 2nd, the regiment marched at a moment's notice, being conveyed on carriages, and encamped at Gorey on the 7th. The carriages raced on the journey so that some were upset. One man's firelock was bent and the country people chaffed him, asking what sort of gun he had. "It 's made to shoot rebels round the corner when they run away" would be the answer.

The General Orders are stringent that the peaceful part of the inhabitants shall be well treated; that horses captured from the rebels, having generally been taken by them from the peaceful inhabitants, are not to belong to the captors, but are to be returned to the rightful owners; and the Regimental Orders desire that the people be treated with kindness, and all milk, etc., paid for; but the men are discouraged from making acquaintances.

R.O.—No man to be allowed to marry a woman of the country. Lieut.-Colonel Erskine informs the soldiers that no marriage without his consent can be lawful, and no woman will be allowed to embark with the regiment but those that came with it.

The dreadful cruelties of the rebels had given rise to reprisals on the part of the troops, and especially of the yeomanry of the country, and the peasantry were subjected to all the horrors of civil war. The exceptional conduct of the Gordon Highlanders at this time, as witnessed by the following letter, has been a matter of honest pride to the regiment ever since.

R.O., August 19th, 1798.—Lieut.-Colonel Erskine is extremely happy that the following letter from the Dean of Ferns to the Marquis of Huntly, expressive of the good conduct of the regiment during their stay at Gorey, should be inserted in the Regimental Orderly Book :—

My Lord,—I have the honour of enclosing to you that part of the proceedings in the last Vestry held in Gorey wherein your Lordship and your regiment are mentioned. This mark of our respect and gratitude should have been sooner expressed and conveyed to you, had not our calamitous situation delayed the calling of a Vestry, which we conceived the most regular mode of expressing our sentiments collectively. It may be pleasing to your Lordship to hear that, in the attendance of my parish, I have heard all the poor loud in the praise of the honesty and humanity of the privates of your regiment. They not only did not rob them of the wretched pittance that was left by the rebels, but refused such trifling presents (of provisions, etc.) as were offered them,[1] saying their King paid them nobly,[2] and enabled them to supply every want at their own expense.

I have the honour to be,
With great respect,
Your Lordship's obedient Servant,
(Signed) Peter Brown,
Dean of Ferns.

"We, the loyal inhabitants of the parish and vicinity of Gorey, in Vestry assembled, beg leave thus publicly to acknowledge the goodness and humanity evinced by the Marquis of Huntly during his short stay amongst us. We are proud to add that during that short stay rapine ceased to be a system, and the confidence of the people in the honour of government began to revive. We should be wanting in gratitude if we omitted our testimony that the humanity of the colonel was emulated by the soldiers, and we request the 100th Regiment

[1] The journal of a private soldier of the regiment mentions the poverty of the people, and that his comrades, though they often had the opportunity of taking by force what they liked, never took even a drink of butter-milk without paying for it.

[2] The pay and allowances of a soldier no doubt were "noble," compared to those of Post Office employés in the Highlands. In 1799 the runner from Inverness to Loch Carron, a distance of over fifty miles, was paid 5s., and from Inverness to Dunvegan in Skye, a much greater distance, 7s. 6d. for the week's work.—J. W. Hyde ("One Hundred Years by Post").

to accept our thanks for the moderation and honour which marked the conduct of every individual officer and private who composed it.

(Signed) PETER BROWN, Rector.
J. JERMAN, Churchwarden."

Further testimony to the high character of the Gordon Highlanders is given in the "History of the Rebellion." After describing the devastation and plundering sustained by the inhabitants, without distinction of loyalist or "croppy," the author continues :—"On the arrival of the Marquis of Huntly, however, with his regiment of Scottish Highlanders, the scene was totally altered ; its behaviour was such as, if it were universal among soldiers, would render a military government amiable. To the astonishment of the (until then miserably harassed) peasantry, not the smallest trifle would any of these Highlanders accept, without payment of at least the full value."[1]

July 29th.—The regiment marched and encamped in the Glen of Emall, where fifty men were discharged. On August 10th, the lieut.-colonel and 300 men marched to Torbay and encamped ; on August 24th, the regiment marched to Blessington. Sergeant D. Nicol, in his journal, relates that at Emall, Huntly invited the inhabitants to visit the camp and on Sunday afternoon it was like a fair. The priest of Davidstown was great on these occasions, dancing jigs and taking his glass. When a girl refused to dance with the Marquis, his reverence called her a d——d foolish jade for refusing to dance with a Lord. Apropos of the above is the following extract from Sir John Moore's diary :—"The good conduct of the 100th Regiment (Highlanders) to the people and the affable manners of Lord Huntly did much towards reconciling them and bringing them back to their habitations."

R.O., Blessington, August 24th, 1798.—The long roll will beat to-morrow morning at three o'clock, when the tents will be immediately struck and the carts loaded. At half-past three the pipes will go along the line, when the battalion will fall in.

The French Directory having undertaken to support the Rebellion with men and arms, General Humbert with about 1200 troops, all experienced soldiers, landed at Killala, bringing arms and uniforms for the rebels, whom they expected to join them. They behaved with great moderation and civility to the loyal inhabitants, but were disappointed in the number and quality of the Irish volunteers. Comparatively few joined their standard, but with them they marched to Castlebar, where they defeated the British troops opposed to them.

[1] "History of the Rebellion in Ireland in the year 1798." 2nd Edition.

The Gordon Highlanders were kept constantly on the move, every precaution being taken on the march to preserve discipline and to keep the men in condition. On one occasion the regiment marched ninety-six Irish miles in three successive days, with arms, ammunition, and knapsacks.[1] The officers of companies were never to be all out of camp together, soldiers without a pass not after 10 a.m. The Orders are dated at Ballinamoin, Ballinahinn, Carrick-on-Shannon, Crophill, etc., moving every day till September 16th, when they camped at Moat. The men's shoes being worn out, Huntly presented each man with two pairs.

Meanwhile, the united French and rebel forces, after routing the British troops under General Hutchinson at Castlebar, had been defeated by Lieut.-General Lord Lake on the 8th September at Ballinamuck, where the French troops surrendered as prisoners of war. The regiment on 5th September passed the troops defeated by the French at Castlebar. These consisted of the 6th Dragoon Guards, Lord Jocelyn's Light Horse (Fox Hunters), and Fraser's Fencibles, who had behaved well but lost a colour. They all appeared rather downcast. A week later the *Hoche*, ship of the line, with eight frigates and 3000 troops sailed from Brest, but was intercepted off Lough Swilly, and all but two frigates were taken by the British. A third French expedition succeeded in reaching Ireland, but, learning the fate of its predecessors, returned without landing its troops, and the famous Irish Rebellion was practically at an end.

The regiment was not engaged against the French at Castlebar or Ballinamuck, and their different affairs in pursuit of the rebels are too various to be given in detail ; but their conduct and discipline were so remarkable that Major-General John Moore, in Orders of September 25th, 1798, after rebuking the irregularities committed by the troops generally, added :—" The major-general must, in justice to the 100th Regiment, state that hitherto he has had no complaint of any one of them, nor has he ever met them in town after the retreat beating." [2]

The thanks of both Houses of Parliament were given to the troops " for their meritorious exertions on the present important crisis."

A.G. Office, Dublin, September 21st, 1798.—Ensign Dugald

[1] Stewart's " Highlanders of Scotland."—Forty English miles a day seems almost beyond the power of even a body of picked men ; at the same time, Sir J. Sinclair, in his " Observations on the Military System of Great Britain " (1793), says that in Highland regiments the men are so peculiarly active and fit for encountering the hardships of war, and overcoming the difficulties of marching in a mountainous country, that they ought, perhaps, to be formed into light infantry battalions. He also mentions that they are inclined to deprive themselves in order to send money to their parents.

[2] This was the beginning of the mutual respect between General Moore and the Gordon Highlanders, which lasted till he fell at Corunna.

Campbell,[1] of the North Fencibles, is appointed to do duty as adjutant of the 100th Regiment till His Majesty's pleasure is known, and to be obeyed as such.

In the Orders of October 11th, the coats were ordered to be buttoned down to the waist; till this period they had been open, showing the waistcoat. The men's white waistcoats seem after this to have been used with sleeves as fatigue jackets, though still worn under the coat in cold weather.

While the regiment was at Moat, the then 91st, 92nd, 93rd, 94th, 95th, 96th, 97th, and 99th Regiments were disbanded, the 98th or Argyllshire Highlanders became the 91st, and the Gordon Highlanders became the 92nd, "under which number they have often distinguished themselves, and on twenty-six occasions on which they met the enemy, from 1799 to 1815, the latter invariably gave way before them. This fact has, in a very particular manner, attracted the notice of the brave and experienced enemy to whom this country was so long opposed."[2]

In honour of the occasion, Huntly gave each man a gill of whisky to drink success to the new number, and promised to do all he could to prevent them being drafted.

A.G. Office, Dublin, October 16th, 1798.—It is His Majesty's pleasure that the 100th Regiment shall in future be numbered the 92nd, and that it be placed on the same establishment with regard to field-officers as other regiments of the line.

(Signed) G. Hewitt, Adjutant-General.

Three field-officers had companies, as appears from—

R.O., October 21st.—Corporal Stronach of the colonel's to be sergeant in the same company. Corporal Cameron of the major's to be sergeant in Captain Cameron's company. Corporal M'Kinnon of the lieut.-colonel's to be attached to the major's.

On October 30th, the 92nd marched to Athlone, and were quartered partly in barracks and partly in billets, the pipes being ordered to play round the town half an hour before parade. General Moore was in command of the garrison.

October 24th.—Volunteer Charles Cameron, 92nd Foot, is appointed adjutant of the 4th Light Infantry Battalion.

At Athlone there was what Lieut.-Colonel Erskine calls "an

[1] This is the handsome sergeant-major mentioned in the account of the North Fencibles raised by the Duke of Gordon in 1793. Pryse Gordon, in his "Memoirs and Reminiscences," mentions that he was visiting at Gordon Castle when Dugald, deserting the forge to wield the claymore, came from Lochaber to enlist, and says that in his Highland dress he was a perfect model for a sculptor. His father was a dirk and pistol maker at Fort-William, and also rented a farm of the Duke of Gordon. He soon rose to corporal, sergeant, sergeant-major, and ensign, and became a famous character in the 92nd, with which he served in all the campaigns till after Waterloo, retiring on full pay as brevet-major. He was celebrated in the regiment for his worship both of Bacchus and Mars.

[2] Stewart's "Highlanders of Scotland."

amazing sick list," which he believes proceeds from drinking, particularly whisky, and cautions the men against excess.

Gazette of October 11th.—Brevet Lieut.-Colonel James Robertson, from the 72nd Foot, to be lieut.-colonel in the 92nd Foot.[1]

The officers apparently still mount guard in the belted plaid, and one is ordered to fall in every day " till he knows how to march off a guard like an officer."

R.O.—The men's clothing being much worn by exposure in the late severe duty, officers are to get cloth to patch it, till the new coats and kilts are taken into wear.

Gazette, November 22nd.—Adjutant Donald MacDonald, from half-pay of the late 100th Regiment, to be adjutant 92nd Foot, *vice* Thomson, who exchanges.

December 10th.—Ensign Dugald Campbell to be adjutant in 92nd Regiment, *vice* MacDonald, whose appointment does not take place.

The Perthshire Highland Regiment (Fencibles) being disbanded at Limerick in February, but allowed to enlist in any other regiment with a bounty as for recruits, Captain MacLean [2] and a party were sent to enlist such as should volunteer for the 92nd, and returned with 64 men.

R.O., January 14th, 1799.—Lieut.-Colonel Erskine particularly desires that officers and n.c. officers will pay the greatest attention to pointing out to those men of the Perthshire Fencibles who have just joined the duties that will be required of them, and that, until such time as they are perfectly acquainted with their characters, they will treat them with that kindness and lenity which every recruit and every man not acquainted with the system of the regiment is entitled to. They are all to receive enough of plaid tartan for a kilt, and of hose tartan for two pairs of hose. They, or the recruits lately joined from Scotland,[3] are not to be charged with the kilts and hose served out to them. They are also to be supplied with false tails, and have their hair cut in the regimental form.

Men were also received from the Reay Fencibles.

A number of men were at this time discharged unfit for service, being sent to Chatham ; the commanding officer desires officers to see they have good coats to go away in.

[1] This officer must have been on the Staff, as Lieut.-Colonel Erskine continues to command the regiment.

[2] Captain MacLean had been in temporary command of the Grenadiers during Captain Ramsay's absence, and the commanding officer, in a later order, assures Captain MacLean that, on account of the great attention he had paid to that company, and the actions he had used in getting so many men from the Perthshire Fencibles, he shall be appointed to the Grenadiers, should they ever happen to be vacant while he commands. There were several Englishmen among these Perthshire Volunteers.

[3] There is no mention of a depôt all this time. Recruits seem to have remained with the parties, which were under officers, till sent direct to the regiment.

Gazette.—George Cumming,[1] of the Grenadier Company, was, on February 1st, 1799, promoted ensign in the North Fencible Highlanders.

The dislike entertained by Colonel Erskine and his officers to the infliction of corporal punishment is again evident. At Athlone several soldiers are sentenced to it by regimental court-martial, but it generally happened that, on "the strong application of his captain," or "on account of many recommendations in his favour," the corporal punishment was remitted. The lieut.-colonel, at the same time, took these opportunities to request officers not to induce him to refuse their requests, by applying in favour of any of their men who have been guilty of a serious breach of military discipline, and he generally ended with good advice to the men. One of the kind colonel's reasons for not flogging a man seems rather original:—

R.O., May 19th, 1799.—Corporal Buchanan of the Light Company is reduced by court-martial for allowing sentries to relieve each other. For particular reasons which appeared, the court did not sentence him to corporal punishment, which the commission of such an unmilitary crime would certainly have merited. The lieut.-colonel is sorry that a corporal, of whom he entertained so good an opinion, should have so far forgot himself, and hopes it will be a warning to him in future. At the same court-martial was tried Donald M'Kinnon, for allowing himself to be relieved without the corporal, and he was sentenced to one hundred lashes. The punishment awarded is so trifling [2] for an offence attended with so bad consequences, that the commanding officer remits it entirely.

He then goes on to explain what might be the consequences of such conduct. Lieut.-Colonel Erskine seems to have been quite as attentive to the health and comfort of his men as he was to their discipline and appearance.

Dinner was then the only regular meal ; the men just took a snack as they chose in the morning and evening ; and it was only in the spring of 1799 that all the companies are "to be provided with knives and forks as the Grenadiers have."

Sergeant D. Nicol in his journal mentions that Lieut.-Colonel Erskine refused his men permission to marry, unless the girl produced a good character and a fortune of at least £20.

At this time Volunteer Charles Cameron,[3] who had been

[1] It does not appear that this man was a sergeant. He belonged to Kirkmichael, Banffshire.

[2] One hundred lashes would hardly be considered a *trifling* punishment in modern days !

[3] Charles Cameron became a most distinguished officer in the 3rd Regiment. He belonged to Kilmallie, Inverness-shire. (Note on Description Roll signed by Colonel J. MacDonald.) Alexander Clark belonged to Kingussie, Inverness ; John MacPherson to Laggan, Inverness. There is also a man named Ronald M'Donald, from South Uist, Inverness, opposite whose name in the Description Roll is "to ensign," but no date or regiment given. He enlisted May 1794. These men seem originally to have enlisted as ordinary recruits.

primarily made adjutant of a light battalion, was appointed ensign in the 92nd, *vice* Dowd. Volunteer Alexander Clark to be ensign 92nd, *vice* William MacPherson. Volunteer John MacPherson to be ensign 92nd, *vice* John MacPherson.

The nationality of officers of the regiment is given in a General Return of the name, country, age, and service of the officers of His Majesty's 92nd Regiment or Gordon Highlanders, reviewed at Athlone, April 14th, 1799 :—

English.	Scottish.	Irish.	
4	31	2	
0	1	0	Chaplain.
0	1	0	Adjutant.
0	1	0	Quartermaster.
1	1	0	Surgeons.
5	35	2	

And of n.c. officers and soldiers at the same date :—

Years of Age.	Men.	Country.	Men.	Years Service.	Men.
55	2	English	47	30	1
50	18	Scots	754	25	4
45	34	Irish	37	20	7
40	48	Foreigners [1]	3	15	20
35	69			10	29
30	104			8	2
25	204			7	4
20	264			6	6
18	98			5	588
				4	61
				3	19
				2	41
				1	59
Totals,	841		841		841

R.O., March 7th, 1799.—Lieut.-Colonel Erskine is extremely sorry that the charges to be made against the men on the 24th inst. shall be so heavy, but as he knows every good soldier prides himself on the neatness of his head-dress, and particularly every Highlander on the smartness of his bonnet, he is confident they would wish that their money should be disposed in that way, rather than in many

[1] The "foreigners" seem to be Colin M'Kenzie, Alexander M'Arthur, and John M'Pherson, entered as from the parish of Minorca, Spain, labourers, but had served in the 73rd, 44th, and 66th Regiments, and been transferred from them.

others, in which neither their own pride nor the credit of their regiment is concerned. He is the more convinced of this from the expense of the bonnets not being nearly so great as in any other Highland regiment that he has heard of, which will show that he has attended as much as possible to their cheapness. The following is the expense of the bonnets, with different articles that are charged:—

	s.	d.
To 1 large feather,	3	0
,, 6 small do. at 1s. 4d.,	8	0
,, 1 hackle,	1	5
,, 1 bear-skin, dyeing,	0	3½
,, Canvas, wire, thread, and work,	0	8
	13	4½

Officers to be careful that deductions are made for any feathers a man has supplied himself with. Forage caps, 2s. 6d. each; rosettes, 2d.; all in British currency.[1]

R.O., *March 21st*.—The captain of the day reported the barrack-room of the Light Infantry Company to be especially dirty. The orderly-sergeant of that company was tried by court-martial and sentenced to be suspended for two months. To be read in the presence of all the n.c. officers.

R.O.—Lieut.-Colonel Erskine will go round barracks to see the different messes sit down to dinner. All officers to be with their companies.

R.O., *March*.—Extra shoes to be kept for the men, in case of the regiment being suddenly ordered to take the field.

R.O.—Knapsacks not to be put on too low, that the men may more easily get at their powder in priming and loading.

R.O., *April 1st*.—Number of flints to be specified to complete the battalion to two flints a man.

On April 1st the regiment wanted to complete, 172 rank and file.

R.O., *April 7th*.—The regiment will parade to-morrow to fire ball, twelve rounds a man.[2]

On the 6th May, His Majesty signified his pleasure that the establishment of the 92nd Regiment should be forthwith augmented to 1000 rank and file.

R.O.—Men are to be subsisted at the rate of 5s. 5d. Irish per week; no man to be subsisted at less than 5s. Irish.

Those officers who wish to have the new regulation Highland sword [3] will be pleased to send in their names to the Orderly Room.

In May 1798, General Bonaparte had been sent by the French Directory on an expedition to Egypt. On the way, Malta, then

[1] At this period, and until 1825, Irish coinage was depreciated to the extent of 8⅓ per cent.

[2] This is the first notice of firing ball, though blank cartridge is previously mentioned.

[3] Probably brass-mounted, instead of steel.

governed by the Grand Master and Knights of the Order of St John, was surprised and seized, and thus one of the strongest fortresses in Europe, so essential to his communications from Egypt to France, was placed in his hands. He took Alexandria after slight resistance, and the victory of the Pyramids opened to him the gates of Cairo. Egypt belonged to Turkey, and the Sublime Porte, incensed by the unprovoked invasion, declared war against France, and formed an alliance with Russia. The fleet which had conveyed the French troops to Egypt was destroyed by Admiral Nelson at Aboukir on the 1st August, so that a large portion of the army of France was cut off from Europe, and confined to the territory it had conquered on the Nile.

In November 1798, the Island of Minorca surrendered to the British arms.[1]

In March 1799, the French declared war against Austria, and the Russians, under Suwarrow, rapidly recovered all the conquests made by Bonaparte in Italy, except Genoa. The Irish Rebellion having been put down, and it being unlikely that the French would again risk an invasion of Ireland, the British Government felt secure in that quarter, and was enabled to continue the contest elsewhere. In December 1798, Great Britain secured the co-operation of Russia for the prosecution of a Continental expedition.

In the unfortunate campaign in Holland of 1794, the Prince of Orange and the Duke of York had been compelled to retire before the overwhelming armies of France, and, though the Dutch at first tried to defend themselves, a large portion of the nation was willing to fraternise with the French. The Stadtholder and a great number of the better class fled to England, and Holland submitted to the dominion of France almost without resistance.

Of all the military undertakings calculated to cripple the Republic of France, and to turn the fortune of war in favour of Great Britain, none appeared more attractive to the Cabinet of St James's than the restoration of the Stadtholder. This measure was not only directed against every remaining succour and resource that France could draw from Holland, but would force her to employ a considerable body of troops intended to augment the army of the Rhine, would make an important diversion by opening the campaign in the Low Countries with a force attached to the Orange interest; and, if successful, a political separation between France and Holland must take place, and Britain once more secure the guarantee of her influence on the Continent.[2]

A joint expedition against Holland was accordingly prepared, and it was agreed that Great Britain should furnish 13,000 troops

[1] Minorca had been in possession of the British at various periods, but was at this time part of the Spanish dominions.

[2] The analysis of military events relative to the late expedition to Holland.—"Ann. Journal," Vol. II.

and a fleet, and Russia 17,000 men. The Gordon Highlanders were selected to take part in this important service; and the young regiment, which had gained such golden opinions for its honourable conduct from the population among whom it had been sent to keep order, was for the first time to have the opportunity of showing that it could be equally trusted to uphold the honour of its country before a foreign foe.

G.O., Athlone, 12th June 1799.—The 92nd Regiment to march agreeable to route to Cork. 1st Division to-morrow morning at whatever hour Lieut.-Colonel Erskine will appoint.

The regiment marched in two divisions, and encamped at Monkstown on the 24th and 25th of June. It was relieved at Athlone by the Glengarry Fencibles.

The following address and answer were published in Orders at Monkstown:—

"At a meeting of the inhabitants of Athlone and vicinity, on the 15th June 1799, Thomas Mitchell, Esq., in the chair, the following address to Lieut.-Colonel Erskine, commanding His Majesty's 92nd Regiment in this garrison, was unanimously agreed to.

Sir,—We heard with concern that His Majesty's 92nd Regiment, which you have commanded in this garrison, has been ordered to march for the purpose of joining those troops intended for a foreign expedition, but however we may regret your departure, we are not surprised that a regiment so eminently conspicuous for its steadiness and discipline should be selected for an arduous enterprise. We have during your continuance amongst us experienced a polite attention from the officers of your regiment; and the uninterrupted peace and tranquillity which have prevailed in this town and neighbourhood evince the attention of the soldiers under your command.

Permit us, therefore, to return you our thanks, and to request that you will convey the same to the officers, n.c. officers, and soldiers of your regiment.

By order of the meeting,
(Signed) Thomas Mitchell.

Sir,—I have the honour of receiving the address of the inhabitants of Athlone and its vicinity. I shall not fail to communicate to the officers, n.c. officers, and soldiers of the 92nd Regiment the approbation which their conduct has met with, and which must be most flattering and pleasing to them, as it is to me, who has the honour to command them.

I have the honour to be,
with respect,
Sir, your most obedient humble Servant,
(Signed) Charles Erskine,
Lieut.-Colonel, 92nd Regt."

Thos. Mitchell,
Vice-Sovereign,
Athlone.

The above address, with Lieut.-Colonel Erskine's answer, was, by resolution of the Corporation, published three times in the Athlone *Herald* and Dublin *Evening Post*.

R.O., Camp at Monkstown, July 15th, 1799.—The regiment

will hold itself in readiness to embark to-morrow on the following transports (five ships). It embarked at Cove of Cork, landed at Dover on July 30th, and encamped at Barham Downs, where the troops for the expedition to Holland were assembling under the orders of Lieut.-General Sir Ralph Abercromby. The regiment was placed in the 4th Brigade, consisting of the 2nd Battalion 1st Royals, and the 25th, 49th, 79th, and 92nd Regiments, under Major-General Moore. In describing the composition of the Army and of his own Brigade encamped at Barham Downs, General Moore, in his diary, says:—"The Guards are certainly a fine body of men. The regiments of the line are in general but poor, and few of them are formed or disciplined. The 92nd (Highlanders) are an exception—they are excellent."

R.O., Camp, Barham Downs, 30th July 1799.—The regiment to parade at ten to-morrow with all necessaries, ammunition, blankets, and camp kit, all to be inspected in the most minute manner.

Officers to wear their bonnets on duty and on all parades.

The regiment was now put on the English establishment.

On August 1st there were present and fit for duty 717 men out of a total of 764, and at this time they received four men from the 9th Foot, two from the 63rd, and fifteen from the 62nd.

When the volunteers from English regiments found that the 92nd wore Highland dress, some repented of their bargain, and Huntly allowed those who wished to do so to go to the Royals. Sixteen remained with the 92nd.

PIPER, 1798

CHAPTER V

THE regiment embarked at Ramsgate on the 8th August, with an effective strength of 1 colonel, 1 lieut.-colonel, 2 majors, 6 captains, 12 lieutenants, 7 ensigns, 5 staff, 40 sergeants, 21 drummers, and 730 rank and file.

Women were not allowed to go with the transports, and General Moore mentions in Orders that a " sett of gentlemen " who were present at the embarkation of the 4th Brigade had offered to pay all the soldiers' wives' passages to London, and also to give 2s. 6d. to each.

Colonel the Marquis of Huntly joined his regiment on this expedition, taking command in the field ; but he afterwards, in Regimental Orders, " desires that all reports of every description may be made to Lieut.-Colonel Erskine, in like manner as if he had the whole command of the regiment. The lieut.-colonel will fill up all vacancies of non-commissioned officers, and in every respect consider himself as commanding, except with respect to signing returns, which must be done by Lord Huntly himself."

The army effected a landing below Patten on the Helder, on the 27th August, in the face of a considerable body of French and Dutch troops. A melancholy accident occurred during the landing, one of the boats being upset, and Sergeant Evan Cameron and fourteen rank and file of the Gordon Highlanders being drowned.

The enemy having abandoned his fortifications, and evacuated the town of Helder, it was occupied by our troops. A numerous train of heavy and field artillery was found in this important post, and the troops were warmly thanked by Sir Ralph Abercromby. The 92nd, though among the first ashore, were not actively engaged with the enemy, and suffered no loss, except from the misfortune mentioned above. On the 30th, the Dutch fleet in the Texel surrendered, and hoisted the colours of the Prince of Orange.

The thanks of both Houses of Parliament were afterwards unanimously accorded to Sir R. Abercromby and the officers, non-commissioned officers, and soldiers for the action of September 27th, " thereby securing the command of the principal port and arsenal of the Dutch Republic, and affording H.M. fleet the means of rescuing from the French the naval force in the Texel."

On the 2nd September, the army took up a position on the Zuyder-Zee, and the troops were placed in cantonments, the 92nd being in advance of the right, at Oude Sluys. Here, as is customary in the British army, strict orders were given for the protection of the lives and property of the inhabitants.

The service on which the army was to be employed did not permit camp equipage to be carried, but only such small articles as are usually carried by soldiers, and one small portmanteau for each officer. The Commander-in-Chief " rests assured that every temporary sacrifice will be borne with the cheerfulness which tends to secure the success of the British army," and adds that when the army marches without camp equipage, more than the usual allowance of spirits will be made to the men. There seems to have been no regular commissariat, as commanding officers of regiments are recommended to purchase bullocks to be driven along with them for the use of their respective corps.

During the time the army remained here, the usual attention to arms and appearance on parade was required of the regiment. The false tails must have been very troublesome appendages—many had lost them, but their hair was to be neatly combed back and tied. At evening parade, after a day's march, the battalion companies were to put on their clubs, and the flank companies their plaits, but on the line of march they need not wear them. At the same time, more warlike duties and precautions were not neglected ; pickets were advanced in front of the army at night, and neither officer nor soldier was allowed to lie down to sleep, but to sit with their arms in their hands.

At daylight on the morning of the 10th September, the French and Dutch forces under General Brune attacked the pickets, and the action soon became general ; they were repulsed by the British, and in Orders from headquarters, Shagan Bourg, of that evening, Sir Ralph Abercromby thanks his troops " for their noble and steady conduct," and Major-General Moore, who was himself wounded, told the 92nd that their conduct made him proud of being their countryman.

On this occasion the Gordon Highlanders lost their first comrade " killed in action," Private Malcolm Ferguson.[1] Captain the Hon. John Ramsay and six or seven rank and file were wounded.

The Russians joined the expedition about September 13th, and on September 14th, Field-Marshal His Royal Highness the Duke of York landed and took command of the army, General Sir Ralph Abercromby remaining in command of the division, in which was Major-General Moore's Brigade, of which the Gordon Highlanders formed a part.

In General Orders of the 16th, a letter was published addressed to Sir Ralph, by express command of His Majesty, and signed " Henry Dundas," expressing the anxiety felt by His Majesty and by the public respecting the expedition, " the sense His Majesty entertains of the steady and enterprising bravery of the army under your command, in the arduous and ever memorable action of the 27th ult." " High as the British army stood before this, it will be impossible that the landing at the Helder Point, preceded and

[1] Ferguson was a native of North Uist.

attended by so many untoward difficulties, and the battle by which it was immediately followed, should not attract the admiration of Europe, and raise their character still higher in every part of the world, as it has already done in the eyes of their sovereign and their countrymen." The King goes on to lament the loss of valuable officers and soldiers which attended the glory of that day, and expresses his anxiety for the wounded. He compliments Sir Ralph on the "cool judgment, military order, and superior abilities" he had displayed, and praises the "perfect harmony and unanimity which prevailed between the land and sea forces."

The Duke of York, being at the head of 35,000 men, and being aware that extensive reinforcements were advancing to support the Republicans, resolved to move forward and attack the enemy.

The force was divided into four columns, the nature of the ground precluding the employment of large masses. The first, 8000 Russians and a British brigade, was to advance by the Sand-dyke against Brune's left, resting on the sea. The second, 7000 men, of whom 5000 were British, was charged with the attack of the French centre. The third, under Sir J. Pulteney, was intended rather to make a diversion than a real attack, unless in case of unlooked-for success; and the fourth, under Sir Ralph Abercromby, was destined to turn the enemy's right.

On the morning of the 19th September the troops stood to their arms at four o'clock, and the army advanced to the attack, which was at first successful at all points; but the Russians, after having captured Bergen in the most gallant style, failed to hold it; tempted by plunder they broke their ranks, and they being driven back, the places acquired by the other columns had to be abandoned, and the army withdrew to its former position at Zype.

In this action 60 officers and 3000 men and 16 pieces of cannon were taken from the enemy, which sufficiently proves that no advantage had been gained by them.[1]

In General Orders of the 20th September, H.R.H. the Duke of York compliments the troops on "the distinguished and spirited exertions which added new lustre to the British arms. The column under Sir Ralph Abercromby, after a most fatiguing march, possessed the city of Hoorn; but, from the doubtful situation of affairs on the right, could not in prudence advance further. H.R.H. will not fail to represent to His Majesty the sense he entertains of the services of the generals and different corps engaged on this occasion."

On the 22nd was published to the army a letter of thanks from the City of London. "To Lieut.-General Sir Ralph Abercromby, K.B., and the generals, officers, and soldiers under his command, for their gallant and essential services in effecting a landing on the

[1] "Alison's History," Vol. IV. p. 151.

coast of Holland, driving the enemy from their strongly fortified situation, thereby rendering it practicable for the squadron of His Majesty's ships to compel the surrender of the Dutch fleet in the Texel, and that he be presented by the Court (of Common Council), as a token of their sense of the important service thereby rendered to this country, with the freedom of the city, and with a sword of the value of 100 guineas."

Moore's Brigade occupied farmhouses and barns, and he and the regimental officers spared no pains to ensure the comfort of their men so far as the nature of the service allowed.

By General Order of the 27th, all pieces of ordnance, colours, and tumbrils taken from the enemy are to be delivered to the British artillery, and the paymaster of the artillery is to pay to the regiment that took them the following rewards:—

For each common howitzer, £20; for each tumbril, £10; for each colour standard, £10. All horses taken are to be paid for if fit for the public service, but no small-arms are to be paid for except those of deserters, who are to be paid for their arms.

From the 20th September the opposing armies remained within their entrenchments, strengthening their lines of defence; the enemy having meanwhile increased his numbers by the arrival of French reinforcements.

G.O., October 1st.—Regiments are immediately to cook one day's provisions, and commanding officers will take measures to be able to move at short notice.

R.O., October 1st.—The regiment will fall in to-morrow so as to close upon the 25th Regiment, and enable the brigade to move forward at three o'clock. The same orders and directions as when the regiment last marched.

At this date the 92nd had present and fit for duty 30 sergeants, 21 drummers, and 662 rank and file.

On the 2nd October the Duke of York made another attempt on the French and Dutch position between Bergen and Egmont-op-Zee. The combined attacks were made in four columns. "That under Sir Ralph Abercromby, composed of 13 battalions, 8½ squadrons of cavalry, a half-troop Royal Horse Artillery, and some field-pieces, formed on the dyke and beach which connect the sand-hills of Patten to those of Camperdown. By 6.30 a.m. the infantry stood in column of companies, the artillery on the right, and the cavalry on the right of all. At seven o'clock General Coote's, General Hutchinson's, Colonel MacDonald's, and General Moore's Brigades moved. The first turned at Camperdown immediately to the left, and proceeded on the road to Schoreldam. The second moved on the ridge of the sandhills which commanded that road; the third entered and marched in the centre of the sandhills, inclining to the left. General Moore's Brigade formed the more immediate advance guard of the column by penetrating into the sandhills directly, and keeping continually his right flank on the hills which rose from the beach.

"The French, though in possession of Camperdown Hill, an elevation of about 300 feet, did not make any resistance to this first movement, but fired a signal gun and retired skirmishing, when the advanced brigades got possession of the entrance to the sandhills, the main column proceeded forward, occasionally reducing its front when the beach became very narrow on account of the tide. The right flank of the cavalry was continually in the water. The column proceeded in this way for six or seven miles, the troops much harassed and fatigued in consequence of the nature of the soil, which yielded at every step up to their ankles.

"The French had now lined some high sandhills with a body of riflemen, who began to keep up a very smart fire upon the British; shortly they were considerably reinforced, and they galled our troops from almost every eminence and outlet of the multitude of sandhills. In spite of all, our troops advanced with that ardour and perseverance which so eminently distinguish the British soldier. Though perfectly unacquainted with the system of sharp-shooting (and it is impossible not to lament the want of that species of warfare in our army), though galled on all sides by offensive weapons that did their mischief partly unseen and always at a distance; though momentarily deprived of the encouraging presence of their officers by wounds they received, and though they were themselves neither equipped for light service, nor had the advantage of a light body for that purpose; notwithstanding this combination of unfavourable circumstances, our brave countrymen persevered and fought their way for four miles. We should be unjust to omit on this occasion the honourable testimony which has been given from every quarter of the personal courage and good example of their leaders.

"It was a country most favourable to the French system of making war, and gave them a decided advantage over an invading enemy. If we except their Grenadiers, the troops employed on this service were under the size of our rear rank battalion men.

"General Moore's Brigade having suffered immensely both in men and officers, regiments from the main column were continually thrown into the hills, which measure became the more indispensable, as the force of the enemy hourly increased upon us. Only the 92nd Regiment and some of the Guards remained with the cavalry and artillery on the beach. The French, taking advantage of every strong post which commanded the beach, kept up a galling fire; they brought two guns from Egmont-op-Zee, which were advantageously planted, and cannonaded the column on the beach and the cavalry. A strong body of French also appeared on the heights above Bergen. The British on the sandhills were exhausted by fatigue and want of water, and weakened by loss of men. In vain did they attempt to storm the enemy's position, they were beaten back, but their innate intrepidity seemed to rise in proportion as the resistance they met became more formidable and destructive; they repeated the attack with unabated fury, and though their ranks were thinned, no symptom of fear or disorder appeared amongst them. The remainder of the column was ordered to charge, the whole instantly pushed forward against the post, which formed towards the beach an amphitheatre of hills, the tops of which were defended by the enemy. Our troops rush through a most tremendous fire of musketry, gain possession of the heights and drive the enemy to a considerable distance. By this time Colonel MacDonald's Brigade had arrived, and became instantly engaged on the left. About half-past four o'clock the enemy gave way and retired on all sides, but as Bergen was not taken and the British were considerably advanced, Sir Ralph determined to take up his position for the night, and not push forward to Egmont-op-Zee till the day following.[1]

"Our guns had been advanced to check the French artillery, and two troops of the 15th Light Dragoons guarded them, being concealed by sandhills, when 500 French cavalry, thinking the guns unprotected, charged them, and were actually engaged with the gunners when the two troops of the 15th dashed into the French, and drove them off. The French, ashamed of being repulsed by such a handful of men, rallied and advanced again, when a third troop of the 15th, which had been ordered to advance, came up, charged, and drove them off half a mile. The whole British cavalry had now reached the scene of action, but it was too late to attempt more. When the last charge was made it was near six o'clock. The cavalry passed the night on the beach in line with the infantry on the sandhills, where neither horses nor men had any water."

The Duke of York, in his dispatch of October 4th, states—"The points where this well-fought battle were principally contested were from the sea-shore in front of Egmont, extending along the sandy desert or hills to the heights above Bergen, and it was sustained by the British columns under those highly distinguished officers, General Sir Ralph Abercromby and Lieut.-General Dundas, whose exertions, as well as the gallantry of the brave troops they led, cannot have been surpassed by any former instance of British valour."

The following description of the part more immediately played by the Gordon Highlanders is taken from Cannon's "Historical

[1] Partial account of the action fought in North Holland on the 2nd October 1799; by an officer engaged in Sir R. Abercromby's Division.—*Military Journal*, Vol. II.

Record":—"The 92nd Regiment was ordered to escort twenty pieces of artillery to the front along the sea-shore. In the performance of this duty it was attacked by a column of nearly 6000 men, when a most sanguinary conflict ensued, immediately under the eye of General Sir Ralph Abercromby.[1] Trusting to their superior numbers, the French advanced with resolution, and fairly met the bayonets of the regiment. Colonel the Marquis of Huntly was wounded in this memorable charge,[2] which completely overthrew the enemy and preserved the guns; the command then devolved upon Lieut.-Colonel Erskine.

"The French centre was supported by the town of Alkmaar,[3] and Sir Ralph Abercromby had passed Bergen, in order to turn the French position at Alkmaar, to which place the 92nd, immediately after the brilliant affair before recorded, advanced. As the men fought hand to hand, the furious conflict, which lasted 3½ hours, was signalised by many feats of individual bravery and devoted courage.

"It may be observed that this is one of the few instances on record of crossing bayonets by large bodies. Even the supernumerary rank of the 92nd on this occasion was bayoneted. Among the number Lieut. Gordon M'Hardy was killed, and Lieut. Donald M'Donald (who afterwards succeeded to the command of the regiment at Waterloo) received three bayonet wounds."

A private soldier of the regiment, in his journal published afterwards, mentions, as an instance of the fierceness of the fight, that when searching for the killed and wounded among the sandhills, they found a Highlander and a Frenchman fast locked together in the death grip, which was so firm that when they raised the Highlander, who was uppermost, the Frenchman's body was also raised from the ground. Several in the struggle were transfixed with each other's bayonets, clubbed firelocks were used, and sometimes the combatants were so close as only to have room to use their fists.

A French writer also says—"Les deux partis se chargèrent plusieurs fois à la baïonnette avec la plus grande fureur." "On remarqua le courage et l'adresse des montagnards Écossais qui combattaient dans les inondations, et franchissaient avec agilité tous les obstacles pour gagner le flanc des troupes qui leur étaient opposées."[4] And in the *Military Journal*, 1799, Vol. II. p. 146, in

[1] Sir Ralph's son served in the ranks of the Gordon Highlanders at Egmont-op-Zee as a volunteer.—*Letter from James Erskine of Cardross, dated October 26th*, 1799.

[2] The Prince of Wales presented the Marquis of Huntly with a Highland snuff-mull, set in gold with a Gaelic inscription, in memory of his conduct on this occasion.

[3] The country behind the sandhills near Alkmaar is flat and intersected by dykes, ditches, and canals, and part of it had been inundated by the French.

[4] "The two parties charged each other several times with the bayonet." "People remarked the courage and address of the Scottish mountaineers who fought in the inundations, and overcame with agility all obstacles, to gain the flank of the troops opposed to them."—Dumas' "Précis des événemens militaires," Vol. II.

describing the action of October 2nd, it is said :—" The courage and activity which were exhibited by the Scotch Highlanders [1] on this occasion drew the attention and excited the admiration of both sides. These brave fellows were seen up to their middles in water struggling to outflank the troops that were opposed to them, and daring with wonderful intrepidity not only the elements, but manfully and dexterously pushing forward in the midst of a severe and galling fire."

Had war correspondents accompanied our army then, no doubt many heroic actions would have been made public which were only heard of by the comrades of the actors, or in the glens and villages where they related them on their return.

Perhaps Private Norman Stewart summed up the performance of his comrades as well as a more official dispatch, when he told an English inquirer, " Ilka lad shot a shentleman to hersel."

The Gordon Highlanders had proved themselves as firm in fight as they were " polite " and " orderly " in quarters. They had won their first honour, the royal authority being given to bear the word " Egmont-op-Zee " on the regimental colour and appointments. They may have afterwards equalled, but they could never excel, the conduct displayed in this their first great battle.

An incident well known in the regiment may be here related.

Major-General Moore, having been severely wounded in the course of this eventful day, was lying on his face stunned and apparently dead ; as his senses were returning, he heard two soldiers of the Gordon Highlanders say, " Here is the general, let us take him to the doctor," and, having done so, they said, " We can do no more ; we must join the lads, for every man is wanted." On his recovery, the general inquired for these men, and offered a reward of £20, but no one claimed it. It was therefore supposed either that they were killed, or were prevented by the feeling common among Highlanders, that it is wrong to take money either for preserving the life of a friend, or for taking the life of an enemy.[2] General Moore remarked that " it was a noble trait of the regiment that no man in its ranks came forward to personate the parties and claim the reward." In any case, the circumstance says much for the honourable feeling of the men.

In 1804, General Moore, on being made a Knight of the Bath, wrote to the commanding officer,[3] that being as a knight entitled to supporters to his coat of arms, he had chosen a light infantry soldier for one, being colonel of a light infantry regiment, and a Highland soldier for the other, " in gratitude to, and in commemoration of " these two soldiers, adding, " I hope the 92nd will not have any

[1] The 79th and 92nd.

[2] See note on Moore's death at Corunna, p. 158.

[3] The letter is addressed to Lieut.-Colonel Napier of Blackstone, 92nd Regiment.

objections, as I have commanded them, and as they rendered me such a service," and he asked to have a correct drawing of the uniform.

That night the Gordons had to lament the loss of many a comrade who would see Lochaber no more. The regiment had 3 officers, 2 sergeants, and 65 rank and file killed in action; 12 officers were wounded (of whom Captain M'Lean was also taken prisoner), 208 n.c. officers and soldiers were wounded, of whom 29 died of their wounds, and 38 were taken prisoners of war, of whom some were wounded. So that out of the 33 officers (including the staff) and 713 n.c. officers and men who were fit for duty on the morning of October 1st, 15 officers and 313 n.c. officers and men were killed, wounded, or taken prisoner by the evening of the 2nd!

OFFICERS KILLED

Captain William M'Intosh.[1]
Lieutenant George Fraser.
,, Gordon M'Hardy.

OFFICERS WOUNDED

Colonel the Marquis of Huntly.
Captain John Cameron.
,, Alexander Gordon.
,, John M'Lean.
,, Peter Grant.
Lieutenant Norman M'Leod.
,, Charles Chad.
,, Donald M'Donald.
,, Charles Cameron.
,, John M'Pherson.
Ensign George William Holmes.
,, James Bent.

The following list of n.c. officers and soldiers killed or died of wounds is taken from the " Return of men of the 92nd Regiment (or Gordon Highlanders) killed, wounded, dead of wounds, and missing in consequence of the action which took place in Holland on the 2nd October 1799, as also those that were drowned at the landing on the 27th August 1799. Chelmsford, 15th November 1799."[2]

KILLED

RANK AND NAME.	PARISH.	COUNTY.
Sergeant William Gordon	Urquhart	Ross
,, Kenneth M'Intosh	Durness	Caithness

[1] Sergeant D. Nicol in his journal says that in Captain M'Intosh his company lost a friend who always took their part and who maintained that his men could do no wrong.

[2] Names of those drowned are not given here, but five of the fifteen were from Inverness-shire.

KILLED—*continued*

Rank and Name.		Parish.	County.
Corporal	Donald Fraser	Calder	Nairn
,,	John Hendry	Alloa	Clackmannan
,,	Hugh Porter	Ballyovey	Ireland
Private	James Anderson	Paisley	Renfrew
,,	Andrew Banner	Speymouth	Moray
,,	Duncan Cameron	Kilmallie	Inverness
,,	Roderick Chisholm	Kilmallie	Inverness
,,	James Dennie	Weymouth	England
,,	Alexander Dobbie	Bellie	Moray
,,	James Donald	Huntly	Aberdeen
,,	John Duncan	Dollar	Clackmannan
,,	Malcolm Ferguson	North Uist	Inverness
,,	James Fraser	Rhynie	Aberdeen
,,	John Fraser	Ardersier	Inverness
,,	John Geddes	Raffin	Banff
,,	Donald Gillies	Durinish	Inverness
,,	Alexander Galloway		Stirling
,,	John Grant	Kirkmichael	Banff
,,	William Hosack		Inverness
,,	James Jamieson	Fetteresso	Kincardine
,,	Gilbert Kemp		Aberdeen
,,	William Leslie	Achandoun	Banff
,,	Matthew Loches		Somerset
,,	Donald Lowe	Morven	Argyll
,,	Gilbert MacCormack	Kilmodan	Argyll
,,	Angus MacDonald	Boleskin	Inverness
,,	Ronald MacDonald	South Uist	Inverness
,,	Ewan M'Donald	Morar	Inverness
,,	Andrew MacFarlane	Gorbals	Lanark
,,	John MacKay	Killearnan	Ross
,,	John MacKinnon	Mull	Argyll
,,	Donald M'Lean	Kiltarlity	Inverness
,,	Charles MacLeay	Girvan	Ayr
,,	Archibald MacMaster	Ardnamurchan	Inverness
,,	James M'Neill	Barra	Inverness
,,	Roderick M'Neill	Barra	Inverness
,,	William MacPherson	Kingussie	Inverness
,,	James MacQueen	St Nicholas	Aberdeen
,,	Angus MacPhee	Kilmallie	Inverness
,,	Alexander MacPhee	Kilmallie	Inverness
,,	John Mathieson		Inverness
,,	John Micham		
,,	William Millar	Dunnet	Caithness
,,	Andrew Morrison	Muthill	Perth
,,	Ebenezer Monteith		Stirling
,,	Andrew Plamon (?)		Falkirk

KILLED—*continued*

RANK AND NAME.	PARISH.	COUNTY.
Private Duncan Rankin [1]	Kilmallie	Inverness
,, William Reid	Urquhart	Inverness
,, John Roxburgh	Newton-Stewart	Galloway
,, Peter Simpson	Old Machar	Aberdeen
,, Alexander Stuart	Kincardine	Inverness
,, James Stuart	Sleat	Inverness
,, James Stuart	Mortlach	Banff
,, James Sutherland	Thurso	Caithness
,, John Taylor	Forres	Moray
,, David Thomson		Falkirk
,, John Watson	Cromarty	Ross
,, Alexander Wood		Aberdeen

The above 60 names are all I find in the Return, but the "Historical Record" makes the number 65, and regimental record agrees.

DIED OF WOUNDS

RANK AND NAME.	PARISH.	COUNTY.
Sergeant Alexander Gordon	Cairney	Aberdeen
Corporal Donald MacPherson	Kingussie	Inverness
Private Alexander Barkly	Kinethmont	Aberdeen
,, Thomas Blain	Gifford	Haddington
,, John Bremner	Rothes	Moray
,, Norman Buchanan [2]	Skye	Inverness
,, Allan Cameron	Kilmallie	Inverness
,, Ewan Cameron (2nd)	Kilmallie	Inverness
,, Donald Cattanach	Kingussie	Inverness
,, Donald Catherwood		
,, John Duabble	Devon	England
,, Thomas Edmonston	Balfron	Stirling
,, George Elsworth	Dorset	England
,, Charles Ferguson		Armagh
,, James Forbes	Inveravon	Banff
,, Alexander Gow	Genvie	Banff
,, John Grant	Inveravon	Banff
,, Donald MacEachan	Strath	Inverness
,, John MacEachan	Morar	Inverness
,, Donald MacIntyre	South Uist	Inverness
,, John M'Pherson	Durinish	Inverness
,, John Martin	Innerat	Inverness
,, W. Nichol	Mortlach	Banff
,, James Rankin	Airdrie	Lanark
,, Alexander Reynolds		Falkirk
,, James Robertson		Stirling

[1] Not in Return, but mentioned by Captain Cameron as killed.
[2] N. Buchanan is mentioned later as d.o.w. prisoner.

DIED OF WOUNDS—*continued*

RANK AND NAME.	PARISH.	COUNTY.
Private John Stuart	Kirkmichael	Banff
,, Angus Sutherland	Lagie	Sutherland
,, James Wood [1]		Glasgow

In the Return they are given by companies, which are called the Colonel's (formerly Captain M'Intosh's), Lieut.-Colonel Erskine's, Major Napier's, Captain John Cameron's, Captain Ramsay's or Grenadiers, Captain Paton's or Light, Captain Gordon's, Captain M'Lean's, Captain MacDonald's (formerly Colonel's), Captain Grant's.

The troops passed the night on the ground they had won. The distance from Patten to Egmont is about fifteen miles, and the army had fought its way over ground sometimes of deep, yielding sand, and sometimes deep in water. They were worn out with fatigue, those on the sandhills by the sea had no water, and the state of the country was such that waggons for the wounded could not be brought forward till late next day.

Meanwhile, General Brune, finding himself outflanked by the British column under Sir Ralph Abercromby,[2] and finding his centre menaced, drew back his line to a position stronger than that he had left. On the 3rd October the order had just been given for the troops to send for their rations, when a report was made that the French were retiring from Egmont-op-Zee. Not a moment was to be lost. The troops stood to their arms, and marched forward without expressing a murmur, leaving their provisions on the ground. It was expected that a strongly armed battery would have to be carried by assault, but the French had retired two hours, and were not overtaken in a pursuit of three miles. The British entered Alkmaar and pushed their advanced posts forward, so as to stand parallel with those of the French-Dutch army. They were cantoned in barns and huts, but the remainder of the 92nd bivouacked that night on the ground near Egmont. The troops had suffered almost as much as human nature could endure, but their efforts had been crowned with success.

Sir Ralph Abercromby's Division did not suffer in the action of the 3rd, the brunt of it falling on the other British divisions and on the Russians. The French-Dutch suffered severely, and we took 500 prisoners.

On the 4th and 5th October the two armies rested on their arms, but on the 6th an attack was made by the Duke of York on the entire front of the enemy's line. In the beginning of the engage-

[1] Cannon's "History" gives the return of killed as 65, and wounded 208. (See Appendix IV).

[2] Major-General Moore being wounded on the 2nd, Major-General Knox, attached to the Russian column, was removed to command the 4th Brigade of Abercromby's Division.

ment the Anglo-Russian army made several successful impressions against the enemy, took Ackerslot, and advanced to Castricum ; but the action becoming general, General Brune, availing himself of a favourable opening, advanced at the head of his cavalry, and broke the Anglo-Russian line, and they were driven back with considerable loss. In this successful advance General Brune had two horses shot under him.

The engagement lasted till night, when the French-Dutch army returned to its original position, but the issue of the battle, though by no means decisive, was in their favour. They had rendered the British attack abortive, an attack on which depended not only the success of their enterprise, but their means of subsistence, as no provisions could be obtained in the situation of the Anglo-Russian army, whose position is described by the Duke of York in Orders of the 8th as one of " insupportable hardship."

A Council of War, assembled by the Duke of York, decided that the army could not maintain itself in its present position, and they accordingly retired behind Zype. Although in this new position the army was not more than six or seven miles from where they disembarked, its communications were rendered impracticable by the continual rains that had fallen, the breaking up of the roads, and overflowing of the dykes and canals.[1] The French, under Brune, entered Alkmaar on the 8th, and his right column, under General Daendels, entered Hoorn on the 9th. The Anglo-Russians retreated from Enkhuisen and Medemblick, after having destroyed the timber and dockyards, some ships of the Dutch East India Company, and most of the public stores ; but they were obliged to leave their wounded for want of conveyance.[2]

The rear-guard was attacked, and General Daendels threatened the left ; and there was constant skirmishing. On the 12th and 13th the Anglo-Russians occupied ground near Zype, before Patten, Warmenkuyzen, Dirkshoorn, and Winkel ; some in huts and some in tents covered with straw and branches. Here an armistice was proposed, and the terms were concluded at Alkmaar on the 18th. It was agreed that hostilities should cease, that the allied army should re-embark and leave Dutch territory by November 1st,[3] that 8000

[1] The men were employed in repairing the roads and dykes—for which they were allowed " working money "—" on account of the severity of the weather."

[2] It appears that, although the women were not allowed to accompany the troops, some had contrived to follow them, and those who had no children were given employment, if they wished it, as nurses for the wounded. The others were to return to England, and to receive two guineas each on arrival to enable them to get home.

[3] In General Orders, October 20th, H.R.H. " has the satisfaction to announce to the troops that it has been found advantageous to both armies to enter into an agreement and cessation of arms, the object of which, on our part, is the undisturbed evacuation of a country in which, from the untoward circumstances of the weather and lateness of the season, it is found impracticable any longer to carry on offensive operations, and on that of the enemy to

French and Dutch prisoners should be given up, among them Admiral de Winter, who had been taken prisoner at the naval battle of Camperdown in 1797.

General Brune's plan of defence, of which the inundation of the country by cutting the canals formed an important part, was so good that, notwithstanding the well-concerted manœuvres of the British generals and the valour of their troops, this victorious army was checked within six miles of the field of battle, and the enemy, though beaten, obtained a decisive superiority. Great Britain had, however, accomplished half her object in the capture of the fleet.[1]

The Gordon Highlanders, reduced as they were in numbers, took part in all the operations of the campaign without further loss in officers or men. The regiment received a draft of sixty-five volunteers from the Gordon Fencibles—ready made soldiers—who were posted to companies on October 9th.

At this time Lieut.-Colonel Erskine commanded the Brigade, Major-General Knox being otherwise employed.

The following letters from Captain John Cameron to his father at Fassiefern give a graphic account of part of this campaign as seen by a company officer, and serve incidentally to show the interest taken by the officers in their men.

Letter from Captain John Cameron to his father at Fassiefern :—

OUDESLUIS, *3rd September* 1799.

We set sail from the Downs on the 13th of August, and on the 15th we were in sight of Isle Showen, when such a violent tempest arose as drove us out to sea, where we tossed about at the mercy of the winds and waves till the 21st, when we again came in sight of land, and anchored that night in sight of the entrance of the Texel, in company with Admiral Duncan's fleet and the Russian fleet. On the morning of the 22nd we weighed and stood in towards the land, when a violent current had nearly driven us on the Hack, a dangerous shoal in the entrance of the Texel. We were immediately obliged to bear up and come to anchor. About an hour after, it blew a perfect hurricane from the north-west, which endangered the whole fleet. However, after the loss of about twenty anchors, we got to sea again, where the horrors of the scene were indescribable, and our situation became truly unpleasant ; but at last, after failing in another attempt to get inshore, on the evening of the 26th our fleet were all safely moored in a line extending from within four miles of Camperdown on the right to the battery on the Main at the entrance of the Texel on the left. Our gunboats and all the ships of war (as the enemy themselves afterwards acknowledged) were extremely judiciously placed, and moored so close inshore as effectually to protect our landing.

prevent the execution of strong measures of severity and destruction, which it appeared in our power to execute, but which are repugnant to British feeling and practice."

[1] Prize-money was afterwards given to the soldiers and sailors for the capture of the Dutch fleet and arsenals.

The situation of the country is very unfavourable to a landing if opposed by a spirited enemy. The beach, to be sure, is smooth and hard, but there is constantly a great swell upon it.

About fifty paces from the beach a double row of sand heights extend themselves from Camperdown on the right to a battery within two miles of the Helder on the left, forming the extreme point on one side, as Texel, a small village on the island of Texel, does on the other, to the entrance of the Texel or Zuyder-Zee. The entrance is so narrow that but one ship can pass at a time, and though it will admit the largest ships, yet were a 74 by any accident to turn crossways in entering, her stem and stern would be on the bank, whilst there might be twenty fathoms water under her waist. In this entrance the Dutch frigates were moored in line at a place called Vew Deep; about three miles further in lay the Dutch fleet under Admiral Hory. They were covered by a considerable battery at the extreme point of the entrance, being at the end of the sand height, and another betwixt that and the Helder, supposed to be the strongest in Holland, mounting 52 pieces of ordnance, 24 of them 24-pounders. Parallel to the sand heights lies one of the most immense plains I ever beheld—a dead flat, but partially divided from the heights by an ugly, though narrow, lake or marshy swamp, nearly opposite to the right of our fleet. There was a trifling kind of work with a few guns placed in the interval of the sand heights, called Pitten. The Dutch army was posted in front of Pitten, with the marsh in front, their left and rear covered by part of the Zuyder-Zee. A division of them occupied a post with a flag-staff on the sand heights above Pitten, and they had almost 2000 men in the battery at the point and Helder. About daybreak on the morning of the 27th we began to get into the boats, and at four the 3rd Brigade and a battalion of Guards and the reserve were engaged with the enemy in the interval betwixt the heights, below Pitten, on the seaside. I happened to be amongst the first of our regiment that landed, with two companies, and was ordered to march to the flag-staff on the heights above Pitten (where the battle then raged), there to remain till further orders. When we got to the heights, we found them hotly engaged about a quarter of a mile from us on our right. There and on the left, to which we were removed an hour or two after, we remained idle spectators of the whole action, which lasted with great fury on both sides till four in the afternoon. Sir Ralph exposed himself like a boy for a long time in the heat of action, which animated the troops in an extraordinary degree; but, notwithstanding, the victory was not immediately very complete, as General Daendilio retired without confusion, without interruption, and without being followed; but, indeed, they had a great train of artillery, and we had not got any landed, and our left was not engaged, being reserved for the attack of the batteries, which was the main object; but, to our astonishment, about eight at night they evacuated them without firing a shot, spiking their cannon and returning to their army, and we immediately took possession. Major Gordon was sent with my company and another to take possession of the Dutch arsenal at Newark, where we saw our fleet oblige the Dutch fleet to surrender, and where we met with more attention from the Dutch Admiral Boser then ever I met with in any part of the world. Never as long as I breathe will I forget the kindness of the honest B. and his beautiful Irish wife.

The Dutch army are returning towards Alkmaar, and our first line have advanced this far after them, about fourteen miles from Newark, which is about three miles from Helder. Our regiment compose the left of General Moore's Brigade, which is the left of the first line of the army, and opposed to the right of the Dutch army.

From the same to the same:—

FRIEDENHOF, 11*th September* 1799.

I wrote you very fully from Oudesluis, the place we were last in, about eighteen miles from this, giving as far as I could some account of our movements. I forgot to mention one very melancholy circumstance that took place as we landed in this country. A boat, carrying part of Captain Gore's company of ours, overset, by which means 1 sergeant and 14 privates of ours, a midshipman and 7 seamen were drowned. I am extremely sorry to say two of the men were Barra men, Donald M'Kinnon and Alexander M'Leod—poor fellows, their fate was really hard. We left Oudesluis on the 5th and marched to this place, from which Major Napier was advanced with two companies, one of which was mine, to the villages in front, and about eight miles distant from Alkmaar. There we had an opportunity of seeing the enemy daily, and of skirmishing with their advanced posts. On the 8th we were relieved and returned here. On the evening of the 9th I was sent out with a picket of reserve of eighty men, with some artillerymen and a howitzer, in rear of our former posts. At the village during the night the enemy were pretty quiet, except when they kept beating their drums and sounding bugles constantly. About daylight General Moore came up on his way to the advanced posts, and on my inquiring if I should return to my picket, he said, not this day. Soon after an extreme smart firing commenced on the right of the line of the army. In about half-an-hour the general returned and told me he perceived the enemy intended to attack the whole line. He had scarcely spoken when the firing became violent at the advanced posts at the villages, and the general told me to form my picket so as to cover the retreat of the advanced posts. The situation was an extremely advantageous one. Our men were placed on the reverse side of a pretty high dyke, just so far up as to be able to see well over it, with the howitzer on the top. The advanced posts began immediately to retire, and the enemy pursued with great activity, considering the place being completely intersected with canals, and the bridges all broken down. The general remained on the very spot where I was formed with my picket. The enemy had now forced all the advanced posts, and sharpshooters from their advance began to try their hand, when the Royals appeared and soon after formed. They had scarcely formed, when our regiment also came up, and my picket joined their companies and the action commenced. This battle was a severe one while it lasted, and was equally severe throughout the whole line. Three battalions of Grenadiers supported by field pieces attacked our post three different times, but were very soon repulsed, each time with great slaughter, and, which is very extraordinary, with little or no loss on our side (owing, I believe, to our being so well sheltered by the dyke), as we had only one man killed, and Captain Ramsay and six or seven men wounded. The man who was killed was, poor fellow, a Uist man of my company. I had also a man wounded, a Skye man. General Moore was wounded in the finger, and Captain Ramsay in the thigh, but the escapes on all sides were wonderful. The enemy took at least four or five hours retiring from the time they commenced their retreat, and filed off pretty regularly towards Alkmaar, which was at least ten miles distant. This action, I think, reflects great lustre on the British arms, and adds much to the character of Sir Ralph. General Moore's compliments to our regiment were, I think, beyond measure high and flattering. He said that to-day he was indeed proud of being a Scotchman, and that he never saw so much bravery joined to such a ready obedience of orders. Our conduct to the prisoners, upwards of fifty of whom we took, was truly meritorious, as we extended mercy to all, even to villains who were skulking behind ditches and in hop fields, and taking their deliberate aim hours after the action. One fellow in particular fired at one of our sergeants as he passed from one company to another, and that instant held up his hands for mercy, which was shown him. Another fellow fired at three of our officers and immediately surrendered.

From the same to the same:—

Deslyp, near St ——, North Holland,
October 4th, 1799.

It is with infinite concern and even reluctance that I find myself bound to write to you on the present occasion, as my letter must be one continued detail of the most tragical, though glorious, events. About three o'clock on the morning of the 2nd, the column of the allied army under Sir Ralph Abercromby, of which General Moore's Brigade (to which you know the 92nd belong) formed a part, commenced its march in order to attack the enemy by a flank movement. Our march was along the seaside with a range of very high sandhills on our left, which extends all along the coast. The passage is in some places so narrow that we could march but by half companies, and often only six or seven men abreast. The intention was that our column was to take Egmont-op-Zee, which lay about thirteen or fourteen miles distant on an angle of these sandhills, and by which the enemy's whole position would be exposed, as it flanks Alkmaar. During the march, and till we came in view of Egmont, the enemy were continually retreating before us and evacuating their posts on the hills with very little fighting, but then it was that the battle raged with great fury on both sides. Our situation at first was really a cruel one. We were a good deal exposed to their fire without having it in our power to return it, as they were screened by the sandhills, and several of our regiments were engaging them between us and them. During this period we had twenty men wounded, and poor Lord Huntly received an ugly wound in the shoulder. Our right wing was now ordered by Sir Ralph to charge them, which they did with the bayonet. However, our left was soon obliged to support our right, when, after a most obstinate resistance, we drove them from sandhill to sandhill about two miles back, but, God knows, not without considerable loss. The remains of the 92nd continued on that ground all night and last night also, and were this morning, I am told, to join the column in storming Egmont. You will see by the papers the loss we have sustained ; it is painful and horrible to me to think of it. There is, I hope in God, not much fear for the wounded officers and men, as any of them (Heavenly Creator, rest their souls !) that were severely wounded died from the coldness of the night in the field. Of the poor lads that left the country with me, as far as I can yet learn, there is killed—Alexander M'Phee and Angus M'Phee, and Duncan Rankin ; wounded—Sergeant M'Kinnon, Corporals Duncan and Richard Evans. Lieutenant D. M'Donald, Duncan's friend, is amongst the worst wounded of the officers that have survived. He has got two thrusts of a bayonet through his breast. Of my company I do not exactly know the loss, but when I left the field I could find but five or six of them that were not either killed or wounded. Charles Cameron, my only officer, was wounded half-an-hour before I was ; poor Hugh Achreachten and Captain Campbell, Duntroon,[1] were killed very early in the day within fifty paces of me, but alas! it was not in my power to see two such intimate friends put decently under ground. I am just at this moment told that there are thirty-six of my company still to the fore. Late in the evening I received a slight wound in the knee ; you may easily suppose it is a slight one, when I remained an hour with the regiment, charged once with them after I got it, and did not leave it till they forced me to the doctor. The clutching the doctor had at me to get out the ball was infinitely worse than the wound itself, and disabled me so much that I was obliged to be moved back to the cantonments.[2]

From the same to the same:—

YARMOUTH, *2nd November* 1799.

I wrote you immediately after the action of the 2nd October, mentioning some of the particulars of that dear-bought glory which the 92nd acquired that day, but I believe I forgot to mention Ewan Cameron, Glensuilach, being pretty severely wounded, though now doing well (as all the wounded are). His brother Allan, Glencoe's old servant, died of his wounds the day after the action ; little Dougald Cameron, from Dochanassie (Mr Ross's regimental weaver, who belonged to my company), fought like a lion that day ; though wounded in three different places, one being through and through his body, he refused to quit the field till the action was over. He was sent to England in one of the hospital ships.

A story was told me many years ago by an old man at Ballachulish of Ewen MacMillan, Captain Cameron's foster-brother, who was a soldier in his company and his servant. Ewen was a good shot and skirmisher, having been accustomed to deer-stalking[3] at home. One day on outlying picket in Holland, when the French riflemen were annoying them by firing from shelter, and at greater

[1] One of the 79th.

[2] He seems to have made light of this wound purposely, as it appears he was laid up a long time by it.

[3] In those days when strangers were seldom seen in the Highlands, and shootings or fishings were of no letting value, though the Game Laws were more severe than now, little notice was taken if the people took a deer or a salmon for their own sport and use, so long as they did not come for them too near the laird's residence. The feeling on this subject is embodied in the Gaelic saying, "Slat as a choille, Breac as an linnhe, Fiadh as an fhireach. Mèirle as nach do ghabh duine riabh nàire."—"A stick from the wood, a trout from the pool, a deer from the hill (sky-line), theft that a man was never ashamed of."

distances than the Highlanders' smooth-bores could reach, MacMillan, seeing one of the enemy take cover at a place where he thought he could get within reach unseen, proceeded without permission to leave his post and stalk the Frenchman, crawling along

ON THE SANDHILLS OF HOLLAND

the back of a wall, till, thinking himself sure of his aim and resting his musket on the wall, he was about to fire, but the Frenchman was too sharp for him and fired first, the bullet cutting off part of his ear. Not disconcerted by this near shave, he fired in return, wounded his adversary, and before the latter had time to reload,[1]

[1] Loading an old-fashioned rifle was a work of time; the bullet had to be rammed home, often with a mallet.

rushed in and bayoneted him ; then running back to his comrades, and pointing to his bleeding head, in an aggrieved tone said to his captain in his native tongue :—

"Am bheil sibh a faicinn gu'de a rinn Mac an Diabhul sin orm ?"

"Is math a thoill thu e, Eobhain, a choinn gun d'fhag thu d' aite," was the unsympathising rebuke.

"Cha dean e tuille e co-dhiu !"[1] said Ewen as he doubled back to his post.

[1] "Do you see what that devil's son did to me ?" "You well deserved it, Ewen, for leaving your place." "He won't do it again, whatever !" This story is also told in the "Memoir of Colonel John Cameron."

CHAPTER VI

ON the 28th October the regiment marched to Colenzugby, near the Helder, and embarked on board H.M. ships *Kent* and *Monarch*, landing at Yarmouth next day with an effective strength of 24 sergeants, 20 drummers, and 446 rank and file; and were billeted in the town. In Orders of the 29th, the following regiments are ordered to Ipswich—2nd Battalion Royals, 25th Regiment, 92nd Regiment (or Gordon Highlanders).

R.O., Yarmouth, 3rd November 1799.—The regiment will march to-morrow morning at half-past seven o'clock. The pipes will go round the town at seven.

No sooner, however, had they arrived at Ipswich than they were ordered to Chelmsford, a three days' march. Such of the convalescents among the wounded as were able walked under charge of a sergeant; the others went on waggons. Captain Cameron, in a letter dated November 8th, gives an amusing account of the journey of some of the wounded officers:—"Captain Grant with his arm broken, Lieutenant G. Fraser with a ball still lodged in his head, and your humble servant on crutches, set off from Yarmouth in a post-chaise together, with a stout fellow in his Highland dress mounted sometimes behind, sometimes before. You will readily conceive what an exhibition this appeared to honest John Bull. I every minute expected to be asked what was to be paid for the show." [1]

At Chelmsford they were again under the orders of Major-General Moore, to whom the commanding officer applied for leave to let a certain proportion of the men go on pass to Scotland for part of the winter. The journey was not so costly as might be thought, as the Carron Company ran packets from London to the Forth. These vessels were armed, for fear of privateers, and gave free passage to soldiers on furlough on their engaging to help to defend them.

While at Chelmsford, the bard MacKinnon, who was a corporal in the Gordon Highlanders, composed his well-known poem describing the battle of the 2nd October, which, with another on the Egyptian campaign, is given in Appendix V. They are considered among the finest and most spirited pieces in the Gaelic language.

Soldiers' letters signed by the commanding officer were allowed to go at a cheap rate of postage.[2]

[1] In a letter from Chelmsford, November 28th, "Captain M'Lean, who was wounded and taken prisoner, has been sent back, and has just joined us almost perfectly recovered. His accounts of the treatment he received are not much to the credit of the French." Captain Cameron goes on to mention the condition of such of the wounded as belonged to his father's neighbourhood, for the benefit of their friends, and as to money sent through him by one of them to his parents.

[2] The regular rate from England to Scotland was, for a single sheet 1s. 8d., for 1 oz. 3s.

On January 1st, 1800, the regiment wanted to complete 381 men.

R.O., January 9th.—Officers will appear in grey pantaloons, half-boots, bonnets, sash, and gorgets,[1] when the regiment is ordered to parade in marching order ; the grey pantaloons to have red binding on them.

By Horse Guards Order of February 3rd, 1800, hair-powder was ordered to be discontinued by His Majesty's command. This must have been an immense relief to the army, for although no gentleman of fashion was in those days considered properly turned out without it, it was very troublesome to those who did not aspire to that title. They, however, still wore false queues, " if their own hair does not admit of being tied in that form."

Here the regiment received canvas knapsacks painted yellow, having a circle of red in the centre, in which the crown and thistle, with the words " Gordon Highlanders," were inserted.

R.O.—The bonnets to be fifteen inches in height, reckoning from the bottom of the velvet to the top of the feathers. The heckle to be worn upright, and to be of an equal height with the black feathers.[2]

In February, a detachment of artillery with drivers, horses, and guns was attached to the regiment, and forty men were selected to learn the exercise of the guns, Ensign Campbell being put in charge of the whole. All the men were afterwards taught gun-drill.

During the spring a few volunteers from Fencibles were received, and also recruits. The colonel had power to discharge men, and " by Lord Huntly's order," a man is discharged " having found another man in his place."

R.O., February 22nd.—The men who have been fixed upon to be detached as riflemen will take with them their new clothing, and they will immediately set about cocking and making up their new bonnets. The officers will take care that they be neatly cocked.

R.O., February 24th.—The detachment of riflemen will march to-morrow at ten o'clock under command of Ensign Cameron. The major expects that the detachment will conduct themselves in such a manner as to do credit to the regiment they belong to, and that Ensign Cameron will so exert himself on the march, and after he has arrived at Horsham, that his detachment will appear as respectable in the corps they are to join, as the regiment has always done among other regiments.

The meaning of the above Order was that H.R.H. the Com-

[1] Gorgets were worn on duty till 1825.

[2] Pensioner Corporal Innes, Glenlivat, who served from 1794 till after Waterloo, told me that at one time they wore their feathers long " trailing ower oor shoulders," as he expressed it, while, he said, the 42nd wore them very short. Fashions change, as the 92nd certainly wore the feathers shorter at a later period.

mander-in-Chief had been impressed by the rapid movements and accurate fire of the enemy's riflemen, and by the execution they had done in Holland; he therefore determined to constitute a similar corps as part of the British establishment. The commanding officers of fourteen regiments of the line were directed to select from each 2 sergeants, 2 corporals, 30 privates, and 1 person qualified for a bugler, to compose a rifle corps; and to send in to the Commander-in-Chief the names of 1 captain, 1 lieutenant, and 1 ensign willing to volunteer for this service.[1] These detachments were assembled at Horsham, under Colonel Coote Manningham of the 41st Regiment, A.D.C. to His Majesty, and Colonel Stewart of the 67th,[2] as an "Experimental Corps." Three companies of these riflemen, made up of the detachments of the Royals, 23rd, 25th, 27th, 79th, and 92nd Regiments, under Lieut.-Colonel Stewart, embarked with the force under Sir J. M. Pulteney, and landed at Ferrol in Spain on the 25th August 1800; they covered the advance, and particularly distinguished themselves in the two skirmishes which took place near that fortress, the Spaniards being defeated in both. Colonel Stewart was dangerously wounded, also 2 captains, 1 subaltern, and 8 rank and file, of whom some died of wounds. This was the first day a British rifleman ever fired a shot at an enemy.[3]

The 79th and 92nd detachments formed the Highland Company, wearing their own dress. When the Rifle Corps was afterwards formed, the officers and men of the "Experimental Corps" were allowed to volunteer for it, but with two exceptions the men of the Gordon Highlanders elected to return to their own regiment. Lieutenant Alexander Clarke, Ensigns Charles Cameron and A. Cameron volunteered, however; but Lieutenant Clarke was killed in Egypt, and only the two ensigns donned the "green

1 *Military Journal*, February 1801.—Willoughby Verner's "Rifle Brigade"; Sir W. Cope's "History of the Rifle Brigade."

2 Colonel Coote Manningham, and Lieut.-Colonel the Hon. W. Stewart of the 67th (formerly of the 42nd) Regiment, fourth son of the Earl of Galloway, had addressed a letter to the Secretary for War urging "the importance of having in the British army a regiment armed with a rifled arm." The "Experimental Rifle Corps" being a success, it was determined to raise a "Corps of Riflemen." This corps had a great many men from the Scottish Fencibles; out of it grew the 95th Rifle Regiment, which became the Rifle Brigade.—W. Verner's "Rifle Brigade."

[The rifle used by the "Experimental Rifle Corps," 1800, was supplied by a London gunmaker named Ezekiel Baker; known as the "Baker Rifle." It was 2 feet 6 inches long in the barrel, 7 grooved, and rifled one quarter turn. The balls were 20 to the lb.; weight of arms 9½ lbs.; sighted to 100 yards, and by a folding sight to 200 yards; loaded with some difficulty, and small wooden mallets were supplied to assist in ramming down the ball. The corps carried a horn for powder and a pouch for bullets. In the stock of rifle was a brass box to contain the greased rag in which the ball was wrapped. A picker to clean the touch-hole and a brush were suspended by brass chains to the belts. A triangular sword-bayonet was fixed by a spring. The smooth bore bullets weighed 14 to the lb.—Sir J. Cope.]

3 There had previously been only foreigners in British pay armed with the rifle. The 60th were Germans.

jacket" as lieutenants. The latter[1] eventually became General Sir Alexander Cameron, K.C.B., of Inverailort, Colonel of the 74th Highland Regiment, and his son, Arthur Wellington Cameron, served in the 92nd from 1844 till 1876, and commanded the regiment in India.

R.O., March 24th, 1800.—As the issue of beer ceased to-day, and as in future the men are to receive 1d. a day in lieu thereof, that sum is to be added to their present weekly rate of subsistence.[2]

On April 10th, 1800, the regiment was ordered to the Isle of Wight, marching in three divisions on the 12th, 14th, and 15th, *via* London, Bagshot, and Winchester, and arriving at Newport on the 23rd, where they were quartered in billets, a detachment being at West Cowes. During its march, the regiment halted for a day at Islington and the men went sight-seeing all over London. Marching through Hyde Park they passed the King but did not recognise him till too late to salute. At Hounslow and Bagshot they saw bodies of highwaymen hung in chains. The soldiers greatly admired Winchester Cathedral.

A severe scarcity had followed the disastrous harvest of 1799, and provisions had risen almost to famine prices;[3] consequently, the commanding officer allowed married men with families to add to their means by working in the neighbourhood.

R.O., Newport, May 15th, 1800.—The lieut.-colonel begs officers commanding companies to impress on their men's minds that meat of every kind is procured at present with the greatest difficulty, and they cannot expect to receive it of as good quality as when it was in greater abundance. The soldiers, in looking about them and comparing their own comfortable position with that of the labouring class of people, the produce of whose labour does not admit of their buying meat at all, should submit with cheerfulness to bear, in common with the people in general, their share of the dearth and scarcity which at present unfortunately exists.

R.O., Newport, 17th May.—The regiment to be ready to embark

[1] In the eighteenth century, when Ensign Cameron joined, candidates for commissions were not required to pass difficult examinations, and young Highland gentlemen, instead of being sent south, were generally educated at home, and brought up among the country people, joining in their sports and also in the work of the estate or farm. Young Cameron was clipping a sheep when a letter was brought to him, then a rare event in the distant Highlands. It announced his commission in the Gordon Highlanders. "Cha rùisg mi caoraich tuilleadh." "I'll clip no more sheep," said he, tossing aside the shears, and left the Highlands, to return a general, with a "Sir" to his name. Lieutenant Alexander Stewart, an original officer of the 92nd, became captain in Rifle Corps.—See Appendix VI.

[2] The pay of the soldier, which in 1797 had been increased from 6d. to 1s. a day, with allowance of beer or wine, has been increased at various times, generally by waiving deductions and adding advantages, as better clothing, barrack and hospital accommodation. The 1d. beer money was given till 1867, when it stopped on an increase of money being given.

[3] Alison. In "Social Life in Scotland in the 18th Century," Graham states that in 1750 the price of ale was 2d. the Scots pint, equal to two English quarts.

at an hour's notice. Officers to have only one portmanteau, and no women or children to go, but they are to receive an allowance from Government.

Each company to carry a box for the new feathers which are to be served out on arrival at their destination.

The British Government had determined to send a secret expedition to France, of which the Gordon Highlanders were to form part. They embarked on ships of war on the 27th of May at Cowes, but did not sail till the 30th. In the interval they were joined on board ship by a large draft of volunteers from Highland Fencible Regiments, especially the Caithness, a regiment recruited chiefly in that county and Ross-shire.[1] Nineteen recruits had also joined from Scotland during the month, but the regiment was still below its establishment of 1000 rank and file, the embarkation strength being :—1 lieut.-colonel, 2 majors, 7 captains, 16 lieutenants, 6 ensigns, 6 staff, 36 sergeants, 22 drummers, and 600 rank and file. Several men had returned from French prisons, but 28 of those wounded in Holland, who had not yet recovered, and 8 who were still missing (supposed to have been killed or died of wounds) are struck off the strength. There were also recruiting parties in the north, and those attached to the "Rifle Corps."

Their destination proved to be against the forts in Quiberon Bay, where it was expected the Royalist party of that district would join them.

The troops, consisting of artillery, infantry, and royal marines, were in three divisions, commanded by Colonel the Earl of Dalhousie, Lieut.-Colonel Erskine, and Lieut.-Colonel Clephan. Erskine's Division consisted of the 1st Battalion 20th Regiment and his own regiment (the 92nd), with three six-pounders and two howitzers.

[1] Stewart's "Highlanders of Scotland." *Military Journal*, November 1800. H.M. thought fit to order the 21st, 71st, 72nd, 79th, and 92nd Regiments to be filled up by volunteers from Scottish Fencible Regiments in Ireland. Bounty 10 guineas, unlimited service, and the colonels could appoint an officer to an ensigncy in the line for every fifty men who volunteered. The numbers asked from each were :—

Aberdeenshire	100	Princess Charlotte of Wales' Highlanders	100
Angus	100	Argyle	100
Dumbarton	100	Breadalbane	150
Fraser	100	Clanalpine	150
Elgin	100	Fifeshire	150
Glengarry	100	Inverness	150
Lochaber	150	Reay	150
North Lowland	100	Rothesay and Caithness	200
Ross and Cromarty	100		
Tay	100		

The latter was called Rothesay and Caithness because the Prince of Wales had allowed his title as Duke of Rothesay to be added. The 200 men of this regiment went to the 79th and 92nd ; the 92nd also got men from the Clanalpine and Lochaber Fencibles, etc.

On the 7th of June the expedition disembarked at the Isle of Houat, in Quiberon Bay, where they encamped, and were constantly exercised in heavy marching order, recruits included, to prepare them for the expected "most active and honourable service." "It is not necessary," say the General Orders, "to ask the men to do their duty bravely, but whosoever is catched plundering or marauding after landing will suffer death."

On the 18th of June the regiment embarked on board H.M.S. *Terrible*, destined with others to attack the fortress of Belle-Isle, but the orders were countermanded on the 19th, and they returned to the Isle of Houat next day, without having landed on the mainland of France. "Embarking and disembarking, vessels running foul of each other, such is what we are engaged in," writes an officer ; as the song says—

At the siege of Belle-Isle
I was there all the while;
I was there all the while
At the siege of Belle-Isle.

The Gordons returned to camp at Isle of Houat, but they had little rest for the soles of their feet on land. On the 23rd they were on board H.M.S. *Diadem*, bound for Port Mahon, in the island of Minorca, where a large force was assembling.[1] They arrived on July 20th, but remained on board ship till the 7th August, when they landed and occupied barracks. Here they found the 42nd Royal Highlanders, who received them with the hospitality so characteristic of the Scot abroad. Wine flowed, the quaint old streets of Mahon re-echoed Highland toast and song, and no doubt many a Highland head ached next morning !

R.O., August 8th.—The lieut.-colonel will not take any particular notice of the irregularities which happened last night, on account of the men meeting so many of their friends, but he expects not to have anything more happening of the same nature.[2]

All officers of the 42nd and 92nd had two epaulettes, while officers of the other regiments in garrison under field rank had but one, so their men presented arms to the Highland subalterns, till cautioned in Garrison Orders how to distinguish them.

The men were not so healthy as they had been at sea,[3] which was attributed to excess in wine and in the delicious fruits with which

[1] General Sir Ralph Abercromby took the command on August 4th.

[2] The influx of so many troops seems to have caused scarcity of fresh meat in Minorca. The magistrates, with the approval of the Commander-in-Chief, fixed the price. The naval and military hospitals were first served, then what was absolutely necessary was allowed to the sick of the town, one-fifth of the remainder was set apart for the supply of the town—no more than ½ lb. to any one family, while three-fifths was apportioned to the army and navy.

[3] Very great care was taken to keep the lower decks clean and well aired, and the bedding was brought on deck in dry weather.

Minorca abounds; but, though they indulged occasionally too freely in the juice of the grape, the Highlanders did not forget the old folks at home. When there were neither savings banks nor postal orders it was not easy for a soldier to send money. According to Sergeant Duncan Robertson's journal, most of the Gordon Highlanders belonged to the estates of their officers, and they seem to have entrusted to them, as their natural guardians and friends, the money they wished to send to their parents. In Captain Cameron's letters I find constant reference to this practice, mentioning the sums belonging to each man, with his name or the by-name by which he was distinguished at home from others of the same name—as "Ewen dubh Tailear," "Ewen dubh Coul," both men of the name of Ewen Cameron, etc.

The regiment embarked on the 30th August, forming part of the greatest armament which had left Britain during the war,[1] which was collected in the Straits of Gibraltar, menacing the coasts of Spain. The united fleet consisted of twenty sail of the line, twenty-seven frigates, and eighty-four transports, having on board about 20,000 foot soldiers. This formidable force appeared off Cadiz on the 5th October, but, finding that yellow fever was raging in the city, the British commanders, dreading the contagion among the troops which would result if the city were taken, countermanded the orders to land, decided to withdraw from the enterprise, and returned to Gibraltar.

By Regimental Orders of September 23rd, on board the *Stately*, the men of the regiment were ordered to have "half gaiters" made; hitherto they seem only to have had their hose.

October 31st.—The discharge of Sergeant Donald MacKinnon, of the recruiting company, is mentioned, which is the only reference to any kind of depôt.

After remaining a short time on board ship at Gibraltar, where they lost Sergeant Lachlan MacIntosh, who was drowned in the bay, the regiment proceeded to Minorca,[2] and, after a fortnight's stay, sailed again on the 21st November, and anchored at Malta on December 1st.[3] Here the troops were landed as often as possible for exercise. The detachment of riflemen rejoined, and the regiment received a draft of recruits and volunteers from the Fencibles. Officers are desired to make most minute inquiries into the wants

[1] Alison.

[2] Minorca, which had been off and on in British possession for about a hundred years, was finally given up to Spain by treaty in 1803. There are still many remains of the British occupation in the island. English expressions and names are adopted into the language of the people. There is also a family of Frasers owning an estate, descended from a Highland officer who married the heiress.

[3] Malta, having been closely blockaded by the British for two years, had surrendered in September 1800, the French being permitted to march out with the honours of war.

of these men, and to provide them with everything that can be procured. There was evidently some difficulty in fulfilling these orders, and old bonnets and jackets had to be made up for them ; "and collect all the yellow cloth they can, so that at all events they may have yellow cuffs and collars."

At Malta the regiment furnished four carpenters, two masons, and one blacksmith to be employed under the chief engineer.

At this time 6 officers, 7 sergeants, 1 drummer, and 15 privates were employed recruiting in Scotland.[1]

Meanwhile the French army which had been left by Napoleon in Egypt, consisting of about 30,000 men, with about 15,000 Copts, Greeks, and Arabs as allies, was in possession of that country, and the British Government determined to wrest it from them.

It was intended that three armies should co-operate. The Turks were to march across the desert from Asia Minor ; a body of British troops under Sir David Baird was to be brought from India by the Red Sea ; and the main army of the British was to be landed on the opposite shores. This last and most important body was part of that which had been harassing the coasts of France and Spain, and which was now assembled at Malta, about 15,000 strong, under the command of Sir Ralph Abercromby.[2]

The 92nd numbered 730 n.c. officers and rank and file on January 1st, 1801.

The regiment sailed for Marmorice Bay, on the coast of Asia Minor, and arrived at that fine harbour early in January. The expedition remained there in expectation of Turkish reinforcements and horses for the cavalry. There was much sickness in the army, which, with the exception of short intervals, had been at sea, often in very bad weather, since May 1800. All the sick were encamped on shore, and regiments were successively encamped while their ships were being cleaned ; the rest landed daily, practising disembarking, and for exercise and bathing. Great attention was paid to their health, every man being obliged to wear flannel next his skin, etc. At this place they experienced a fearful storm, which continued three days and nights, the hailstones being as large as walnuts. Many ships were drifted on shore, tents were blown away, and horses broke loose, causing great confusion and suffering to the invalids.[3] A good market was established, where sheep cost a dollar each, and plenty of fruit and vegetables could be had at cheap rates, the good conduct and fair dealing of the soldiers giving confidence to the inhabitants.

[1] These seem to have composed the Recruiting or 11th Company.

[2] At Malta officers and men of the "Experimental Rifle Corps" rejoined their regiments. Sir R. Abercromby took with him only the troops for unlimited service ; those enlisted for service limited to Europe he left.

[3] R. T. Wilson's "History of the Expedition to Egypt."

In February, Volunteer Mungo Macpherson, of the 92nd, was promoted ensign in the 42nd.

A curious story was told of two officers of the 92nd, who, while at Marmorice, met during a walk a very magnificently dressed Turk, followed by a number of retainers. One of the officers, with British contempt for this display of Oriental grandeur, and thinking it possible the grandee might understand English, but certainly would be ignorant of Gaelic, said to his companion, "Co a ghalla is mathair d'on chu leisg so?" What was his astonishment when the Turk answered, "Seadh a' ille agus gu'de an seorsa mathair dh'araich thusa mar chuilean."[1]

After mutual apologies and explanations, the Celtic Turk dined on board with the officers, and afterwards sent boatloads of fruit and vegetables for all the men. His name was Campbell, and, having in a quarrel killed a school-fellow at Fort-William, he had fled the country, entered the Turkish service, and had risen to high position.

At last Sir Ralph, tired of waiting for the Turks, determined to attack the French with the force under his immediate command, and set sail on the 22nd February, the voyage being enlivened by the capture of several French ships laden with luxuries for their army, which afforded a seasonable supply to the British. On the 2nd March[2] this magnificent array of nearly 200 ships cast anchor in the Bay of Aboukir, near Alexandria, at the spot where Nelson had, three years before, defeated the French fleet.

On the 1st March, at sea off Alexandria, the 92nd had only 496 rank and file present and fit for duty, 129 present being sick, 38 in hospital, and others left in various places.

The General Orders as to a landing express the satisfaction of the Commander-in-Chief with the behaviour of the troops in their transactions with the inhabitants at Marmorice Bay; he trusts the same exemplary conduct will be continued, and the manners, customs, and religious opinions of the people most rigidly respected. The regimental preparations for landing included putting leather peaks to their bonnets, and giving into store their greatcoats, fatigue jackets, and trousers, taking with them two shirts, a blanket, razor, and cleaning materials.

The landing, which had been delayed by stormy weather, was

[1] Dr Clark, in the "Memoirs of Colonel J. Cameron," gives it translated more politely than it was told to me in the original:—"Do you see the fellow with the tail? It is easy telling who his mother was, the lazy dog!" "Ay, my lad, and what sort of mother may own you for her whelp?"

[2] On the 2nd March a French frigate, which turned out to be the *Régénerée*, was seen standing into Alexandria. She had found herself during the night unexpectedly among the British fleet, and actually continued her course with it unsuspected till she hoisted French colours as she stood into Alexandria.—*R. T. Wilson.*

effected on the morning of the 8th. The troops, with their muskets unloaded, took their places silently in the boats, each boat containing fifty men, while armed vessels covered the flanks as the line rowed vigorously towards the shore. The French, confident in the strength of their position on the heights above the beach,[1] had not thought it necessary to oppose so difficult an operation with a large force; but it was composed of artillery, cavalry, and infantry, while the guns on the fort of Aboukir also commanded the landing.

They allowed the invaders to approach within easy range, then opened so heavy a fire of shot, shell, grape, and musketry that the water was ploughed up by it, and the foam was like surf rolling on breakers. Several boats were sunk, and the loss among the crowded crews was severe, causing momentary confusion; but the sailors pulled hard, the soldiers sat silent and steady, their arms in their hands, anxious for the moment to use them, and the line pressed forward with such precision that the bows of almost all the first division struck the sand at the same time.[2] The troops instantly jumped into the water, which in many places reached their middles, loading and forming as quickly as they could. The French cavalry charged, and advancing into the water, made havoc among the British, till they, overcoming all difficulties and falling into line, with loud cheers drove the horsemen before them. The 23rd and 40th rushed up the heights without firing a shot, but charging with the bayonet two battalions which crowned it, and taking three pieces of cannon.

The 42nd formed as on parade, then mounted the position notwithstanding the fire from two guns and a battalion of infantry. They were then attacked by cavalry, which they quickly repulsed. The 28th also distinguished themselves. The Guards had hardly got out of their boats when they were suddenly charged by cavalry from behind the sandhills, but the 58th, already formed on their right, checked the cavalry by their fire, giving the Guards time to present a front, when the cavalry again retreated with loss. The 54th and Royals landed at the instant that 600 infantry were advancing through a hollow against the left flank of the Guards, but on seeing the fresh arrival they fired a volley and retreated.

The exaltation at this moment cannot be described. The French, finding the heights carried, and General Coote advancing with his brigade, ran from all points of their position, but main-

[1] The sandhills rose in one part to a height of 180 feet. The French had 2000 men and twelve guns.—*R. T. Wilson.*

[2] Alison. The landing was effected by the Reserve under Major-General Moore, the Brigade of Guards (Coldstream and 3rd Guards) under Major-General Hon. J. Ludlow, and part (including the 92nd) of Major-General Coote's Brigade, who commanded the whole, about 5500, all infantry. The soldiers carried sixty rounds of ammunition and two spare flints each.

tained a retreating fight for about an hour, when the whole of the British troops were established on the heights, though weaker by 500 men killed and wounded. The enemy had lost 300 men and 8 pieces of artillery.

The Gordon Highlanders did not meet with such determined opposition as did the regiments on their right, but they did considerable execution, and pursued the enemy for some distance, when they were ordered to halt and pile arms.

This brilliant opening had most important effects on the fate of the campaign ; the gallant conduct of the troops, and the rapidity of their success in sight of the whole fleet, filled the soldiers, many of whom were young and inexperienced, with confidence in their own prowess, even against these veterans of France, of whom some regiments, from their constant career of victory in Italy, were called "The Invincibles."[1]

G.O., Aboukir Heights, March 9th, 1801.—The gallant behaviour of the troops in the action of yesterday claims from the Commander-in-Chief the warmest praise that he can bestow. It was with particular satisfaction that he observed that conduct marked equally by ardent bravery, and by coolness, regularity, and order.

After thanking Major-Generals Coote, Ludlow, and Moore, and Brigadier-General Oakes, who commanded the troops engaged, he goes on to acknowledge "the effective assistance received from the navy on this occasion, in consequence of the judicious arrangements made by Admiral Lord Keith."

No doubt many feats of individual bravery were performed, and long remembered. It is still told in Brae Lochaber how Donald Cameron, known as Donald Mor Og, a man remarkable for his great strength, was on this occasion attacked by a French dragoon, when Donald, parrying the Frenchman's blow, transfixed him with his bayonet, and lifting him from the saddle threw him over his shoulder among his comrades, crying, "Sin agibh fhearaibh, spéic a dh' Abercromby ! "[2]

[1] Alison.

[2] "There, men, is a blow for Abercromby ! " Donald, when he had left the regiment, was often visited by the Duke of Gordon (who, as Marquis of Huntly, had been his colonel) when passing through his Lochaber estates, and would press on his old commander the best his cottage afforded, treating him with the respectful familiarity which is characteristic of the old-fashioned Highlander. Donald was a good deer-stalker ; shootings were not let then, and his friend the Duke did not allow him to be interfered with. Donald Mor Og came of a fighting family. His father, Donald Mor Cameron, carried Lochiel's standard at Prestonpans. In crossing the moss then existing there, at the beginning of the action, the men got out of order, and Lochiel ordered them to halt and dress their ranks, when Donald Mor cried, "An Diabhul 'halt' na 'dress' bhios an so an diugh ; leigibh leis na daoine dol air an aghairt f'had s'tha iad blath ! " "Gum beannachadh Dia thu," answered Lochiel, "biodh mar a tha thu agradh." "The devil a 'halt' or 'dress' will there be to-day ; let the men go on while their blood is up." "God bless you, let it be as you say,"—and the clan rushed on to victory.—*Told by the Rev. A. Stewart, Nether Lochaber.*

When the Gordons halted, the heat of the day, and the powder getting into their mouths in biting off heads of cartridges, had made them very thirsty, when, to their relief, they found in the huts which had been occupied by the enemy not only plenty of water, but camp kettles on the fires, containing mutton, poultry, and everything in preparation for a good dinner, which the French had intended to enjoy after they had driven the British back to their ships. Though not eaten by those for whom it was cooked, the feast was not wasted; the Highlanders did it ample justice, and cracked many a joke on the excellence of the Frenchman's *cuisine*, and the mortification they would feel at losing so good a meal.[1]

They passed the night on the sandhills (where Sir Ralph had taken up a position about three miles forward, his right on the sea and his left on Lake Maadieh), a bitter wind blowing on them from which they had no protection but their ordinary clothes, and stood to their arms an hour before daylight, when the Commander-in-Chief, noticing their cold condition, ordered a gill of rum to each man. The 92nd were then ordered to advance with a party of cavalry, under Lieut.-Colonel Erskine and Major R. Wilson. After an hour's march they arrived at a battery on an eminence which they were to attack, but the enemy had anticipated them and carried off their guns. Here they had a fine view of Alexandria, and halted to refresh.

Meanwhile every exertion was made to complete the disembarkation of the cavalry, artillery, and stores. The castle of Aboukir was invested, and entrenchments thrown up round the camp. The sick and wounded were sent on board ship.

The army remained here till the 12th; the advanced posts skirmished, however, occasionally, and a surgeon and twenty men of the Corsican Rangers were taken by a sudden dash of the enemy's cavalry.

G.O., Aboukir Heights, March 11th, 1801.—The main body of the army is considered for the present as forming three lines, as follows:—1st line, Brigade of Guards, Major-General Coote's Brigade and Major-General Finch's; 2nd line, Major-General Cradock's and Major-General Lord Cavan's; 3rd line, Major-General Stewart's and Major-General Doyle's.

"The army will advance to-morrow morning at eight o'clock according to the above order. The Brigade of Guards, marching from their right, will head the first column; they will proceed along the road near the sea-beach, leaving the Redoubt of Mandora to their left, and will be directed on the flag-staff about two miles in front of it. Fifty dragoons from General Finch's Brigade will join the Guards. The mounted men of General Finch's, followed by

[1] Sergeant Robertson.

General Coote's, marching also from the right, will lead the second column. They will proceed along the Lake, leaving the Redoubt of Mandora to their right. They will march on a conspicuous green hill in front of Mandora. Lieut.-Colonel Murray will conduct them. The brigades on the right of the second and third lines will follow the Guards, and the brigades on the left will follow General Coote's," etc. etc.

The 92nd were in Major-General Coote's Brigade, along with the 1st Royals and the 1st and 2nd battalions 54th Regiment.

During the first two days at Aboukir the men had suffered greatly from thirst, till they found springs by digging in the sand near the sea. Sergeant D. Nicol in his journal records that his comrades had the idea that the Bible says there is no rain in Egypt, and that as they had showers at Aboukir the Bible must be wrong; after discussion it was decided that the Bible meant that the country did not depend on rain as others do, but on the inundation of the Nile. Apropos of religious discussions, another soldier of the 92nd mentions, that in hospital at this time, he had long arguments with a comrade who tried to convince him of the superior claims of the Scottish Episcopal Church.

The army moved forward on the evening of the 12th, and came in sight of the enemy strongly posted on high ground in front of an old Roman camp, with his right to the canal of Alexandria, and his left towards the sea ; though weaker in infantry than the British, he had more cavalry and artillery.[1] During this forward movement the 92nd were extended as skirmishers in advance of the left, and exchanged shots with the enemy's cavalry, who skirmished with them all the way till they formed line within cannon shot of the French position, and were allowed to rest and refresh ; but they were quickly compelled to retire to a safer distance by round shot pitching among them, which, however, did no great execution, though one poor fellow had his head carried away while sitting eating his supper.[2]

The Gordons at this time had more than 200 sick, and on the morning of the 13th had not quite 350 rank and file on parade. They had been all the preceding night on picket in front of the enemy, and had not been allowed to lie down or take off their knap-

[1] The British army, including 1000 sick, 500 Maltese, and all kinds and descriptions of people attached to an army, amounted to 15,330 men ; but of these so many were left on board ship for care of sick, stores, etc., that not 12,000 fighting men landed, and these were afterwards reduced by the effects of service. They were weak in cavalry, which consisted of detachments of the 11th Light Dragoons and Hempach's Hussars, the 12th and 26th Regiments of Light Dragoons (of whom only a few had horses, the plan of mounting them at Marmorice not having been very successful). The few guns were dragged through the sand by sailors. The French army in Egypt amounted to about 32,180 men, exclusive of the Coptic battalions.—R. T. Wilson's "History of the Expedition to Egypt."

[2] Sergeant Robertson.

sacks, and were very glad of the allowance of rum which was served out at daylight. They were also ordered to leave their packs in charge of weakly men, one from each company.[1]

The enemy had received a reinforcement of two half-brigades of infantry and a regiment of cavalry from Cairo, and other corps from Rosetta, making their total force about 6000 men, of which 600 were cavalry, with between 20 and 30 pieces of cannon. Their position was on very commanding ground, the approach to which formed a glacis for the whole range of fire from their numerous artillery.[2]

As Sir Ralph Abercromby had determined to turn their right, the British army marched in two lines and column of regiments from the left, the reserve covering the movement on the right, and keeping parallel with the first line.

It had not advanced out of the wood of date-trees in front of Mandora Tower before the enemy left the heights and moved down by their right, commencing a heavy fire of musketry, and from all their cannon, on the 92nd Regiment, which formed the advanced guard of the left column. At the same time the cavalry, under General Bron, charged down on the 90th Regiment, forming the advanced guard of the right column. This regiment firmly maintained its ground, and, allowing the cavalry to approach, fired such a volley as completely altered their direction ; they then passed on to the 92nd, which regiment, with steady precision, fired a volley with terrible effect. Wounded horses were galloping about, some without riders, some with their wounded masters, their feet entangled in the stirrups, dragged along by the maddened steeds, while the bayonet did its work among those of the dismounted who refused to surrender, and among the few dragoons who reached the ranks of the regiment.[3]

" The discipline and steadiness of the 90th Regiment were most honourable and praiseworthy." " The conduct of the 92nd had been no less meritorious. Opposed to a tremendous fire, and suffering severely from the French line, they never receded a foot, but maintained the contest alone until the marines and the rest of the line came to their support." [4]

On the approach of the supports, the 92nd were ordered to lie down and allow the 17th and 79th to pass over them, the army pushing on with the greatest vigour, but with the strictest regularity. The French kept up a constant fire of musketry and artillery, but did not again oppose in line, but only as *tirailleurs*. The Gordons

[1] Captain J. Cameron's letters and Sergeant Robertson.
[2] R. T. Wilson.
[3] Sergeant D. Robertson.
[4] R. T. Wilson.

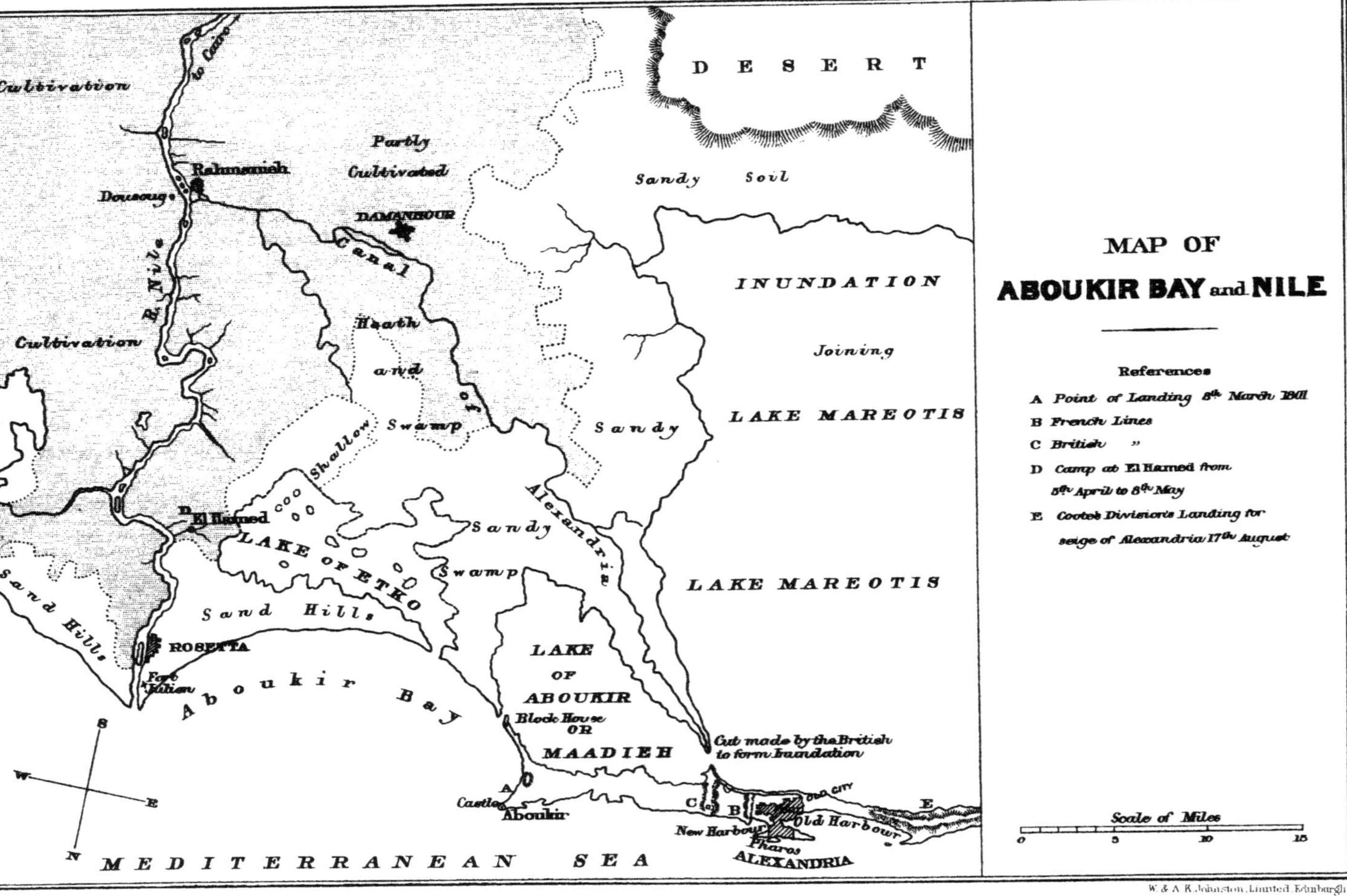

W. & A. K. Johnston, Limited, Edinburgh

now proceeded to the left, where the Dillon Regiment was closely engaged near the old tower of Mandora, where the French had two guns protected by infantry and cavalry, which, in spite of their efforts, were captured and turned against themselves.

Sergeant D. Nicol in his journal says that at one time at Mandora the 92nd was in danger of being surrounded. Ammunition for the two guns failed and the guns were sent to the rear. Five companies were in line, the others being extended to the left amongst bushes towards the lake. The Colonel ordered the men not to fire till they could see their enemy's feet as they advanced from a hollow in front. When the order to fire was given, "like magic it dispelled the gloom from our countenances. Every man did his duty manfully. We encouraged each other, firing, and at the same time praying, for soldiers do pray, and that fervently, on such occasions."

Sir Ralph, in his dispatch dated "Camp before Alexandria, March 16th, 1801," in describing the action of March 13th, mentions the 90th and 92nd as having "behaved in such a manner as to merit the praise both of courage and discipline"; no other regiment is named. The 90th and 92nd and Dillon's Regiment were the only ones which afterwards received Royal authority to bear the word "Mandora" on their colours and appointments. Dillon's, which was composed of foreigners, was afterwards reduced. The Cameronians (Scottish Rifles) and the Gordon Highlanders are therefore the only two corps in the army which have Mandora among their honours.

On the heights before Alexandria, Sir Ralph, wishing to follow up his success and to carry the position to which the French had now retired, advanced across the plain. General Hutchinson with the second line moved forward towards the left to secure a rising ground known as the Green Hill; General Moore was directed to the right, that both flanks might be assaulted at once, and the first line remained in the plain rather to the right. During this movement the 90th, 92nd, and Dillon's Regiments, still keeping to the left, found the enemy inclined to make a stand on the Green Hill, but with the assistance of the 44th, they compelled them, after a warm contest with the bayonet, to abandon their position and retire to the works in front of Alexandria, from which the enemy began to play from all their field artillery and heavy ordnance. The army continued under the most terrible and destructive fire for a long time, whilst Sir Ralph reconnoitred the enemy's position, but they patiently endured this exterminating fire. "If a word was heard, it only contained a wish to be led to the assault." [1]

At length Sir Ralph decided it would not be prudent to attempt

[1] R. T. Wilson.

forcing the heights, which appeared to be commanded by Fort Crétin; and for want of artillery he could not have maintained them. The army was therefore withdrawn at sunset, still marching as if on parade, and occupied that position from which it had driven its opponents, and which was soon to be the theatre of its glory and revenge. Sergeant Robertson, in his journal, thus describes the situation of the Gordons after the contest on the Green Hill:—"We received orders to rest ourselves, but were not permitted to remain quiet, for the French commenced a heavy cannonade upon us (we being pretty close to the works), by which many were killed, and which compelled us to retire as far as the heights the French quitted in the morning. During this retrograde movement, I observed a ball carry off both the feet of one poor fellow, and another who, imagining a ball was spent, in attempting to stop it had his leg carried away above the knee. The latter was a tailor in the 92nd."

The regiment bivouacked in sight of Alexandria and of the objects of historic renown which surround it. The nights are very cold in Egypt in March, which is the more trying from the contrast with the heat during the day. The men had only a light blanket, and many of them made holes in the soft dry sand, covering themselves with it. The evening was spent in the melancholy duty of collecting the wounded and burying the dead, friend and foe alike. Our men noticed that nearly every Frenchman had a pack of cards on him. The loss of the British army was about 1100 men killed and wounded. The French from their position did not suffer so much, losing 500 in killed and wounded; four field-pieces were taken and a great quantity of ammunition. Sir Ralph Abercromby had a horse shot under him.

G.O., *14th March* 1801.—"The Commander-in-Chief has the greatest satisfaction in thanking the troops for their soldierlike and intrepid conduct in the action of yesterday. He feels it is peculiarly incumbent on him to express his most perfect satisfaction with the steady and gallant behaviour of Major-General Cradock's Brigade, and he desires that Major-General Cradock will assure the officers and men of the 90th Regiment that their meritorious conduct commands his admiration.

To the 92nd Regiment and Regiment of Dillon's an equal share of praise is due, and where it has been so well earned, the Commander-in-Chief has the greatest pleasure in bestowing it."

The Gordon Highlanders had 11 officers wounded (of whom 4 died of their wounds), 22 n.c. officers and soldiers were killed, and 77 n.c. officers and men were wounded.[1]

In a letter dated Egypt, March 15th, 1801, "from a hole in the sand," Captain Cameron writes:—"We had 11 officers, 8 sergeants,

[1] Cannon's "Historical Record."

FRENCH SOLDIERS, 1801, EGYPT

and 119 rank and file killed and wounded." Captain Cameron mentions among other wounded who are doing well, "a poor little boy not quite fourteen, son to our Quartermaster Wilkie, and ensign in the regiment." "As to the lads that left the country with me they have been very lucky this time, as only two of them have been wounded, and those not dangerously. Ewen Coul's son is one, and Alexander Kennedy the other."

Lieut.-Colonel Erskine was wounded early in the day by a grape-shot, which mangled his thigh. At first hopes were entertained of his recovery, but he sank after the amputation of the limb and died in a few days. He had asked his brother officers to let him carry to the grave a gold locket which he wore round his neck, containing a lock of his sister's hair and of the lady to whom he was engaged.

In 1894 a soldier of the garrison of Alexandria, while employed on some excavations in the sand, found the skeleton of a man, having only one leg and with a gold locket on the neck. He wrote mentioning the circumstance to the *Times* newspaper, and Colonel Erskine's story being well known both to his regiment and his family, there was no doubt the remains were his. They were re-interred in consecrated ground at Alexandria, and a suitable monument to his memory erected at the mutual cost of the officers of the Gordon Highlanders and Mr Erskine of Cardross, the locket being kept by the latter as an heirloom.[1]

"In him the Service lost one of its best officers," writes an officer of his regiment, and there can be no doubt that his thorough knowledge of his profession, his high sense of duty, the impartiality with which he carried on discipline, combined with his kindness and attention to the welfare of his men, had the best effect on the character of the regiment, with which he had served since it was raised. His portrait still hangs in the officers' mess-room of the 2nd Battalion Gordon Highlanders.

On the death of Lieut.-Colonel Erskine, the command devolved upon Major Napier.

[1] The last letter ever written by Sir Ralph Abercromby to a private individual was to Colonel Erskine's father, telling him of his son's wound.—"Prospect of Private Life of the Gentry, 1784 to 1804."

CHAPTER VII

THE army remained stationary from the 13th till the 20th March. Tents were brought, which protected the men from the sun by day and the dew by night; a market was established where fresh provisions could be bought, sheep being sold for half a dollar, or about 2s. 3d. each. Ostriches also were sold at a dollar, these being principally bought by the Highland soldiers, in order to add extra feathers to their bonnets, in which they took great pride.[1]

In addition to the severe military duties necessary in a position so close to the enemy, the army was constantly employed in constructing batteries, bringing up the guns, provisions, and firewood, consisting of date palms, which would scarcely burn when it was obtained; the palm branches were, however, useful to make shelters from the sun. The want of horses or camels was distressingly felt, as all had to be dragged or rolled through the sand by soldiers and sailors.

At first they had no water, but every regiment was ordered to dig for it, and the 13th found a copious supply by hitting on an ancient aqueduct arched over, the existence of which was unknown to the natives.

Several skirmishes took place at the outposts, a soldier of the 92nd being wounded on the 17th, and in one of them our cavalry lost 2 officers wounded and 3 taken, 7 men killed, 6 wounded and 12 taken, and 42 horses killed or wounded—a great loss to an army very ill provided with that arm.

On the 20th a column of infantry and cavalry was seen to enter Alexandria, and an Arab chief gave warning of the arrival of General Menou, the French Commander-in-Chief, with a large army, and of his intention to surprise the British camp.

The wounded Gordons had been sent on board ship, and some to Rhodes, and the regiment was so reduced by sickness, especially ophthalmia, and by the casualties of the 13th March, that it was, in General Orders of the 20th, ordered to march (the Commander-in-Chief's guard excepted) to Aboukir[2] at four o'clock next morning, there to rest and recruit its strength.

B.O., March 20th, 1801.—"Major-General Coote feels extreme concern that the 92nd Regiment is ordered to march to Aboukir; he hopes they will soon return to the army, and cannot part with that corps without requesting that the officers and men will accept his best thanks for their attention to their duty since they have been under his command."

On the morning of the 21st the army, as usual, was under arms

[1] Sergeant Robertson.

[2] The siege of the Fort of Aboukir had been delayed on account of the scarcity of ammunition. It surrendered on the 18th.

at three o'clock.[1] All was quiet for half an hour, when the report of a musket was heard on the extreme left, then a cannon and scattered musketry. A thick mist hung over the plain, through which the dawn of day was hardly perceptible. The 92nd, mustering little more than 150 effective men,[2] had just begun their march to Aboukir, when these shots were heard, and the regiment was halted.

The men openly expressed their wish to be placed in line, but the Commander-in-Chief, who was near, called the officers round him, told them the firing was only an affair of outposts, and that they must proceed on their route ; but as the firing extended to the right, the boom of cannon, added to the rattle of musketry, followed by loud shouts, showed the real attack to be in that direction. Sir Ralph mounted his horse, and was about to ride to the scene of alarm, when, at the loud calls from the ranks to permit them to take their part in the fray, he consented to their request, and Major Napier immediately countermarched the regiment in order to resume his former station in the line.[3] As the Gordons arrived, they had barely time to form between the Royals and the 1st and 2nd Battalions of the 54th, when they heard through the mist the French drums beating the charge. The brigade was on rising ground, and their general, cautioning them to be cool and steady, waited to receive the attack.

When the enemy came near enough for the muskets to tell, the brigade poured in a volley which staggered them, but they bravely continued to advance till the British prepared to charge. At the flash of the steel the French fell back, but advanced again, and this was repeated more than once, while they still kept up a musketry fire. Finding the red-coats could not be forced from their position, the enemy brought up two field-pieces loaded with grape, which had murderous effect on the British brigade. The latter threw out skirmishers as the French retired, the rest of the men being ordered to lie down to avoid the grape-shot which flew like hail, till one of the guns burst, or the ammunition exploded, killing the gunners, whose bodies could be seen blown into the air. The other gun was then drawn off, their infantry retiring with cavalry protecting their rear. Fortunately for them there was no artillery in that part of the British line to give them a parting salute.

The brigade in which the 92nd was placed had to their left rear the 8th, 18th, 90th, and 13th Regiments ; on their immediate right were the Guards, to the right and rear of the Guards were Derrol's

[1] They paraded every morning before daybreak. "Any officer late to be put under arrest ; any man found with accoutrements off during the night to be tried by a drum-head Court-martial."—*R.O.*

[2] Captain J. Cameron.

[3] R. T. Wilson and Sergeant Robertson.

and Dillon's Regiments, and on the right front of these were Stuart's,[1] the 42nd and 28th Regiments. The whole of the above formed for a time the front line of the British army.

Meanwhile the battle was raging with still greater fury on the British right. The enemy had advanced against the Roman ruins, but were met by so heavy a fire that they fell back in confusion, their general being killed. A strong fresh column, however, at once advanced and, carrying on the broken troops, renewed the attack in greater force on the front and flank of the British who were defending the ruins. Menou (the French Commander-in-Chief) supported this attack by a grand charge with all his cavalry on the right and right centre. The 42nd was suddenly charged in flank by the French horse and broken, but the Highlanders formed in little groups, standing back to back, and bravely resisted the dragoons. The 28th were maintaining the fight to their front, and hearing French shouts behind them, the rear rank had just time to face about when it was assailed by a French regiment, which had got round under cover of the mist, and these gallant men, standing back to back, maintained this extraordinary contest for a considerable time. In memory of this exploit the 28th wear their badge at the back as well as on the front of their head-dress.

Sir Ralph, who had proceeded to the right on the first alarm, had sent his aides-de-camp with orders to different brigades, and whilst thus alone he was attacked by some French dragoons and thrown from his horse. An officer then rode at him and attempted to cut him down, but the veteran general seized the sword and wrested it from his hand, and at that moment the officer was bayoneted by a soldier of the 42nd.

A singular incident happened soon after, an aide-de-camp to General Cradock, whilst going with orders, having his horse killed. Seeing Sir Sidney Smith,[2] he asked to mount his orderly's horse, when just as Sir Sidney was turning to tell the man to give it, a cannon ball struck off the dragoon's head!

Sir Ralph was wounded in the thigh, but only mentioned pain from a blow on his breast, supposed to have been received in the scuffle with the French dragoons, and when the cavalry were repulsed, he walked to the redoubt on the right of the Guards, from which he could view the whole field of battle.

A second charge of cavalry was attempted, but completely failed. The guns in the British battery had exhausted their ammunition, and some of the infantry on both sides were in the same condition; in one instance they actually pelted stones at each other, and a

[1] Derrol's, Dillon's, and Stuart's were all foreigners serving in the British army.

[2] Captain Sir Sidney Smith, R.N., was in command of the sailors who were landed from the fleet.

sergeant of the 28th was killed by one of these unusual weapons at the redoubt.

While this was going on to the right, a column of Grenadiers had advanced as soon as day dawned, supported by a heavy line of infantry, to assault the centre, and tried to turn the flank of the Guards; but the officer commanding the latter wheeled back some companies and checked this movement, and the advance of General Coote's Brigade compelled them to retreat, and terminated the important operations of this eventful day. General Menou, finding that every one of his movements had failed, and that no serious impression had been made upon the British, retired in very good order to the heights of Nicopolis in his rear, under cover of the cannon placed on that formidable position. The want of cavalry, artillery, and ammunition on the part of the British was fortunate for the French, whose loss would otherwise have been very much greater. By ten o'clock a.m. the action was over.

Sir Ralph had remained in the battery, where several times he had been nearly killed by cannon shot. He had continued walking about, paying no attention to his wound. Officers came and received his orders without knowing from his appearance that he was wounded, unless when they saw the blood trickling down his clothes. It was only when exertion was no longer necessary that his spirit yielded to nature; he became faint, and was carried to a boat by a party of the Gordon Highlanders who formed his permanent guard, cheered by the feeling expressions and blessings of the soldiers as he passed. He was taken on board Admiral Lord Keith's ship, the *Foudroyant*, where he died on the 28th March, leaving a name enshrined in the grateful recollection of his country.[1] He was succeeded in the command of the army by Major-General Hutchinson, but before his death he desired that officer to thank the whole of the troops "for their conspicuous and brilliant conduct in the action of the 21st, which has dismayed an insolent enemy, has raised the glory of their country, and established their reputation for ever."

The ultimate effects of the battle of Alexandria, though not immediately apparent, were very great and far-reaching. It increased the confidence of the British soldiers, and revived the military spirit of the nation, which supported it through the arduous conflict awaiting it, which lasted almost without intermission till it culminated in the battle of Waterloo. It taught Continental nations that the army of Britain was to be respected as well as the navy; it delivered Egypt from the French yoke, and decided to some extent the fate of the civilised world. The prophetic eye of

[1] Sir Ralph was brother to Colonel Abercromby, who raised the 75th Highland Regiment, now the 1st Battalion Gordon Highlanders.

to Alexandria
Position of French before and after action
MEDITERRANEAN
Shores of Lake Mareotis passable in many parts both for Cavalry & Artillery
W
S
N
E
Canal of Alexandria
French advancing in Columns
Roize
Lanusse
Regnier
Friant
Rampon
Cavalry
British Gunboats
Stuarts
C
Corsicans
D
A
Guards
Queens
SEA
Dromedary Corps
B
Armed boats British
Scale of One Mile
0
1/4
1/2
3/4
1
Artillery Depot
LAKE ABOUKIR
Called by the French
LAKE MAHADIE
From Aboukir
BATTLE OF ALEXANDRIA
Cavalry
Red British
Blue French
British 1st Posts without colour
A Ruins of Nicopolis Roman ruins
B Redoubt taken by French
C Two guns abandoned by French
D French column taken prisoners
W. & A. K. Johnston Limited, Edinburgh

Napoleon at once discerned the magnitude of its consequences, and he received the intelligence of the disaster at Alexandria with a degree of anguish only equalled by that experienced from the shock of Trafalgar.[1]

In this battle the British army had an effective strength of less than 10,000 men, including 300 cavalry. They had two 24-pounders and 34 field-pieces defending their position. The French General Reynier puts the number of their army at 9700, including 1500 cavalry, with 46 pieces of cannon; they had also a dromedary corps, all picked men, which was used in the same way as our camel corps. After the action, when the Gordons were resting, Sir Ralph Abercromby was carried past, and when it became known that he was dangerously wounded, "the interest excited was so great that everyone ran to get a sight of him whom we all loved." [2]

The men were at once employed to carry the wounded to the boats; in this humane work French and British were treated with equal care, and these poor fellows, their passion cooled, were to be seen making friends and helping one another.

G.O., March 21st, Camp near Alexandria.—"The works on the left to be completed in the course of this day and night. As it is possible the enemy may be desperate enough to make a night attack, Major-General Hutchinson is under the necessity of requesting that the troops may remain with their accoutrements, and lie in their blankets in the position they occupy. General officers are to take care not to throw away fire during the dark, but to use the bayonet as much as possible," etc.

The loss of the British was 1500 killed and wounded. 1040 Frenchmen were buried on the field, and it was estimated that they had lost in killed, wounded, or prisoners about 4000 men, among them most of their principal officers. Two pieces of cannon were taken, and 400 horses were lying in the field; a standard was also taken, belonging to a regiment of the "Invincibles," emblazoned with the names of French victories in Italy.[3]

The Gordon Highlanders lost 3 rank and file killed, and 2 officers and 44 rank and file [4] wounded. Among the wounded was M'Kinnon the bard. Sergeant MacLean, a friend and admirer of the poet, found him insensible, but seeing that he still breathed, had him conveyed on board ship, where he recovered and composed two poems, one describing the landing in Egypt, the other the battles

[1] Alison. [2] Sergeant Robertson.

[3] Wilson. The standard was first taken by Major Stirling of the 42nd. It was given by him to the charge of Sergeant Sinclair, who, in a subsequent charge of cavalry, was wounded, and the standard recovered by the French, and afterwards again taken by a soldier of the Regiment of Minorca or Stuart, named Anthony Lutz, who received the regulated reward, 20 dollars.

[4] Cannon's "Historical Record."

of the 13th and 21st of March, which are still the admiration of the lovers of Gaelic poetry.[1] The latter poem is given in Appendix V. On the 22nd the Gordons could only muster about 150 rank and file fit for duty. All hands were employed in bringing up guns, even boat carronades being utilised to strengthen the position.

G.O., March 22nd.—"The troops to sleep in their tents this night as usual. In case the enemy does not attack to-morrow morning, the 92nd Regiment will march to Aboukir at six o'clock to relieve the detachment of dismounted dragoons who are in the fort."

Accordingly the regiment marched to the Fort of Aboukir. Soon after their arrival the Capitan Pasha, with 6000 Turkish allies, landed at Aboukir, and encamped near the spot where the French had buried 4000 of his countrymen two years before, and where the remains were barely covered by the drifting sand.

On April 1st, Mr Archibald Campbell (paymaster, from lieut. 91st Highlanders) writes from Camp near Aboukir Fort, for Captain J. Cameron, to tell that he had been wounded on the 21st, and had just rejoined from ship-board, the regiment being fifteen miles from the army to recruit its strength, having been reduced on the 21st to about 150 effective men, the rest either killed, wounded, or sick of a slight fever which they brought with them, and of which they are now fast recovering. He mentions that they have plenty of good water, good market for meat, fish, and vegetables, etc. ; that Lieutenant Clarke from Badenoch is dead of his wound, and that Captain Cameron desires him to say that " all the young lads that came from your country escaped in the different actions, except those he mentioned in his last letter, and also a Corporal Alexander Cameron from Clunes, who was wounded on the 21st. They are all doing well."

On April 1st they had 244 rank and file fit for duty, 250 sick and wounded present in camp, and 103 on board ship ; 89 sick or on duty at Rhodes, Lisbon, recruiting in Scotland, etc.—total, 686 rank and file.

Officers commanding companies were desired to see that all convalescents had fresh provisions, and that all the men had the option, instead of their ration of 1 lb. of salt pork, of taking 4d. a day to supply themselves with fresh meat.

[1] Corporal Alexander M'Kinnon the bard was born in 1770 in Morar, on the west coast of Inverness-shire (where his father was tacksman of a farm), and enlisted in the Gordon Highlanders 1794. He was 5 feet 10 inches, and a man of great strength, of amiable disposition, and a very good Gaelic scholar. When he composed a song or poem he would repeat it to his comrades for their approval of his description of the incidents. (" Beauties of Gaelic Poetry, or Sar obair nam Bard.") *N.B.*—No doubt his songs were sung by the Gordons in many a bivouac and barrack, and did much to keep up the traditions and *esprit de corps*. They are still repeated in the West Highlands.

The regiment remained here till the 2nd May, when, being sufficiently recovered, it marched, carrying three days' provisions through the desert, where it suffered from want of water, to El Hamed, near Rosetta, which had been taken in April, when the Castle of St Julien had also surrendered to the combined British and Turkish forces.[1]

General Hutchinson had cut the isthmus which separated Lake Maadieh from the dried bed of Lake Mareotis, and by filling it, to a great degree isolated Alexandria from the rest of Egypt.

The British and Turkish troops advanced along the banks of the Nile ; at the same time, a flotilla of armed vessels and transport boats ascending, captured an important convoy descending the river for the support of the garrison of Alexandria.

The evacuation of the entrenched camp at Ramalieh and its occupation by the British cut off the supplies from Alexandria, and prevented co-operation between the French there and their army at Cairo.

Meanwhile the Grand Vizier, encouraged by the defeat of the French, crossed the desert which separates Syria from Egypt. At his approach the French drew back to Cairo, where their forces were increased by the arrival of the troops from Ramalieh.[2] General Belliard moved at the head of 6000 chosen troops to meet the Turkish force at El Hanka, but the movements of the Grand Vizier were now directed by British officers, and after an indecisive action of five hours, Belliard retreated to the capital ; a result so different from any that had previously attended their warfare with the French, that it raised the courage of the Ottomans, and disposed them to trust to the guidance of the British for the future.

General Hutchinson advanced to Cairo with part of the English army, and on the 20th of May invested the city on the left bank of the Nile, while the Grand Vizier did so on the right bank ; with the result that on the 27th of May General Belliard capitulated, on condition that his troops should be conveyed to France with their arms, field-artillery, and baggage. They numbered 13,672, besides the civil servants ; and they left in the hands of the British 320 pieces of heavy ordnance, besides the field-pieces they carried away with them.

During the advance to Cairo the Gordon Highlanders were brigaded with the 1st Royals, 50th and 30th Regiments, under General Doyle. Starting from El Hamed on May 8th, they at once got into the cultivated country, and those who have had a similar

[1] R. T. Wilson.

[2] Ramalieh capitulated with small loss to the British, but the garrison of 4000 infantry, 800 cavalry, and 33 guns escaped towards Cairo, except about 100 men who were too drunk to accompany them.—*R. T. Wilson.*

experience will understand the joy and delight they felt, after being so long scorched in the sandy desert, when their eyes once more rested on corn-fields and green grass. Their first camp was on the banks of the Nile, where they luxuriated in the delicious water-melons and vegetables which abound there. Among these the tobacco plants, looking like magnificent cabbages, which none of the Highlanders had seen before, at once caught the eye of the cooks, who thought they would delight their comrades with real kail broth, and accordingly put what they supposed to be succulent "kail blades" in the camp kettles. The results, physical or moral, were hardly what they expected!

The 92nd was employed in the affair of Ramalieh, where a sergeant (James Clark) was wounded. The soldiers were amused at the Turks who were in advance; when one man was wounded, six or seven of his comrades would assist in carrying him to the rear, which practice our men thought might account for their being so constantly beaten by the French two years before.

The march was enlivened by frequent skirmishes, and after one of these the kindly Scots were horrified at finding a party of Turkish troops engaged in barbarously hacking and killing a number of French prisoners they had taken. The Gordons quickly put a stop to this inhuman proceeding, and expressed their detestation of it by soundly thrashing every one of their Oriental allies they could catch.

On the 17th May, the men were bathing in the river near Algam, when the alarm sounded. General Doyle's Brigade instantly fell in, and marched off into the desert in pursuit of a large convoy of which an Arab had brought intelligence. It was marching from Alexandria to Cairo, escorted by a considerable body of troops, who moved by the river at night and kept to the desert by day. General Doyle, with 250 of the 12th and 26th Dragoons, led the advance. There was a burning sun above them, and when they got among the loose hot sand, reaching sometimes above the ankle, the progress of the infantry was slow and very fatiguing. Many of the men had not had time to fill their water-bottles, and the rations had not been served out when they marched; their throats were parched with intolerable thirst, which drove some nearly mad, whilst others tantalised their imaginations by thinking of the clear streams of their own country; but they plodded on for about seven miles till they heard firing behind a sandhill. The cavalry had found the enemy halted for the day; a corps of infantry formed his front line, three divisions of the Dromedary Corps and heavy Dragoons in echelon protected the flanks; on the right flank was a piece of artillery, and in the centre were the baggage camels.

The British infantry were far behind when the small force

of cavalry approached the French square. They halted at some distance, while Major Wilson, waving a white pocket-handkerchief, rode up and asked for the officer in command. Colonel Cavalier, for that was his name, came forward, when Major Wilson told him he was sent to offer terms of surrender on condition that his troops should lay down their arms and be sent back directly to France. The colonel angrily ordered him to retire. The major answered him that it was the humanity of his general which induced him to offer these terms, and reminded Colonel Cavalier of the responsibility which now attached to him. To this the colonel paid no attention, and Major Wilson was returning to General Doyle when an aide-de-camp galloped after him, and begged him to go back to Colonel Cavalier, who, on his repeating the proposed conditions, requested time to consult his officers. An evident sensation of joy was to be observed on the faces of the French soldiers, who, though they were veterans of many campaigns, were demoralised by the unaccustomed reverses of the 13th and 21st March. They were sick of Egypt now that they had to play a losing game, and, as they afterwards allowed, when they heard Major Wilson say that they would be at once sent back to France, that word had such an irresistible effect that Colonel Cavalier, after trying for easier terms, was constrained to surrender. They were marched, escorted by the 250 dragoons, afterwards joined by the infantry, to a field close to headquarters at Algam, where they grounded their arms, and our tired and thirsty soldiers were dismissed to their well-earned meal, washed down, no doubt, by a long drink of rum and water, for tea was as yet an unknown luxury. The convoy was composed of 569 men—infantry, cavalry, and artillery—including 120 of the Dromedary Corps, who were the picked men of the army, and excited universal admiration among the British.[1] One gun and 550 camels with their Arab drivers became the property of the captors, as also the horses, which were sold to the commissariat, where they were much needed. There were also a number of asses, which were taken by the men to carry their knapsacks.

On the 18th they had the first view of the Pyramids, and on the 19th they encountered one of those hot winds called Sirocco, which darkened the atmosphere with sand, making it so difficult to breathe that several horses and camels died; the ground was heated like the floor of a furnace, everything metallic became burning hot, and the thermometer registered 120 in the shade.[2] The army halted and pitched their tents, using blankets also as a defence against what one of the Gordons described as "a Highland snowstorm, only sand and heat in place of snow and cold."

On the 22nd May the regiment encamped opposite Cairo,

[1] R. T. Wilson. [2] *Ibid.*

which surrendered on the 27th. Some time after the capitulation of that city, the army of Major-General Baird arrived, consisting of 3600 British and 2800 Sepoys. They had sailed from Bombay in December, but were delayed by contrary winds. They were accompanied by the usual Indian camp-followers, and the army from Britain were astonished at their admirable arrangements and the comparative comfort in which they lived.

The regiment remained here till July, and took part in a review before the Grand Vizier. "The British troops, notwithstanding their rags, formed a very martial parade, and the Scotch regiments, from their being 'sans-culottes,' particularly excited his wonder." [1] The men made excursions to the Pyramids, visiting the celebrated Sphinx, which was afterwards granted as a crest to the regiment in honourable memory of its gallant conduct in the campaign. They had also freedom to enjoy the sights and amusements of Cairo, where many strange Oriental customs then existed, including a market where, among other commodities, women were sold by auction!

The Gordons formed part of the escort of the French army of Cairo, who were marched as prisoners of war to the place of embarkation on the Nile, and afterwards the regiment proceeded to Rosetta, on the march to which three popular regimental characters died, of whom one was known as "Strong Ale Rob."

G.O., Headquarters Camp near Imbebo, July 14th, 1801.—"Lieut.-General Sir John Hely-Hutchinson, K.B.,[2] has received His Majesty's orders to return the generals, officers and soldiers of the army his thanks for the brilliant services they have rendered their country; and for the manner in which they have sustained and increased the honour of the British name and the glory of the British arms. You landed in Egypt to attack an enemy superior in numbers, provided with a formidable body of cavalry and artillery, accustomed to the climate, flushed with former victory, and animated by a consciousness of hard and well-earned renown. Notwithstanding these advantages, you have constantly seen a warlike and victorious enemy fly before you, and you are now in possession of the capital. Such are the effects of order, discipline, and obedience, without which even courage itself must be unavailing, and success can be but momentary. Such also are the incitements which ought to induce you to persevere in a conduct which has led you to victory, has acquired you the applause of your Sovereign, the thanks of Parliament, and the gratitude of your country.

To such high authorities it would be superfluous in me to add my testimony; but be assured your services and conduct have made

[1] R. T. Wilson.

[2] General Hutchinson had lately been made a Knight of the Bath.

the deepest impression on my heart, and never can be eradicated from my memory. During the course of this arduous undertaking, you have suffered some privations which you have borne with the firmness of men and the spirit of soldiers. On such painful occasions no one has ever felt more sensibly than I have done. But you yourselves must know that they are the natural consequences and effects of war, which no human prudence can obviate; every exertion has been made to diminish their extent and duration; they have ceased, and I hope are never likely to return. Nothing now remains to terminate your glorious career but the final expulsion of the French from Egypt, an event which your country anticipates, and a service which, to such troops as you are, can neither be doubtful nor difficult. The prevalence of contrary winds has prevented the arrival of ships from England with money; your pay has been in arrear; but this inconvenience is now at an end, and everything that is due shall be put in course of payment, and discharged as soon as possible."

The following Order was also promulgated to the army at large:—

Horse Guards, May 18th, 1801.—"The recent events which have occurred in Egypt have induced His Majesty to lay his most gracious commands on H.R.H. the Commander-in-Chief to convey to the troops employed in that country His Majesty's highest approval of their conduct, and, at the same time, His Majesty has deemed it expedient that these, his gracious sentiments, should be communicated to every part of his army, not doubting that all ranks will thereby be inspired with an honourable spirit of emulation. Under the blessing of Divine Providence, His Majesty ascribes the successes that have attended the exertions of his troops in Egypt to that determined bravery which is inherent in Britons; but His Majesty desires it may be most solemnly impressed on the consideration of every part of the army that it has been a strict observance of order, discipline, and military system which has given its full energy to the native valour of the troops," etc.

His Royal Highness having thus obeyed His Majesty's command, added his own remarks on the example set to the whole British army by Sir Ralph Abercromby and the troops under his command. "The boldness of the approach to the coast of Aboukir, in defiance of a powerful and well-directed artillery; the orderly formation upon the beach under the heaviest fire of grape and musketry; the reception and repulse of the enemy's cavalry and infantry; the subsequent charge of our troops which decided the victory, and established a footing on the shores of Egypt, are circumstances of glory never surpassed in the military annals of the world. The advance of the army on the 13th of March presents a spectacle of a movement of infantry through an open country, who, being attacked upon their march, formed and repulsed the enemy ; then advanced in line for three miles, engaged along their whole front, until they drove the enemy to seek his safety under the protection of his entrenched position ; such had been the order and regularity of the advance. Upon the 21st of March, the united force of the French attacked the position of the British army. An attack begun an hour before daylight could derive no advantage over the vigilance of an army ever ready to receive it ; the enemy's most vigorous and repeated efforts were directed against the right and centre. Our infantry fought in the plain greatly inferior in the number of their artillery, and unaided by cavalry. They relied upon their discipline and courage. The desperate attacks of a veteran cavalry joined to those of numerous infantry, which had vainly styled itself *invincible*, were everywhere repulsed ; and a conflict the most severe, terminated in one of the most signal victories which ever adorned the annals of the British nation. In bringing forward these details, the Commander-in-Chief does not call upon the army merely to admire, but to emulate such conduct," etc.

G.O., 18th July—Gazette of 23rd May 1801—*92nd Foot.*—Quartermaster-Sergeant Donald M'Barnet to be ensign *vice* Wilkie, promoted.

On the same date (18th July) the thanks of both Houses of Parliament were communicated to the regiment, in common with the other troops in Egypt.

The regiment left Rosetta on August 7th, and marched by the heights of Aboukir, arriving at the camp near Alexandria on the 9th. Here it found a draft of recruits, who were regularly drilled morning and evening.

R.O., Camp before Alexandria.—In consequence of the reduction of the recruiting company,[1] Captain Peter Gordon's company will be called Captain Grant's.

On the 11th Brigadier-General Hope's Brigade, and on the 13th Major-General Moore's Brigade arrived, as did Lieut.-General Sir J. H. Hutchinson on the 15th ; he immediately determined to besiege Alexandria on the eastern and western fronts, having an army of 16,000 men under his command.

On the evening of the 16th, General Coote's Division embarked in boats on the inundation for the purpose of attacking on the west, and the same evening Major-General Doyle's Brigade, the 30th, 50th, and 92nd Regiments, were held in readiness to make a diversion in favour of General Coote, and to gain ground necessary for the progress of the siege on the western front. General Moore was also ordered with a corps of light troops to act on the right.

[1] In monthly returns this company is alluded to as the 11th Company.

The left column were to storm the Green Hill on the right of the French line, and the right column to occupy the Nole Hill in front of the French left.

At two o'clock on the morning of the 17th August, the troops were under arms ; the 30th Regiment directed itself against the Green Hill, the 50th on their right, and the 92nd remained in reserve between them, ready to support either. The light company of the 30th leading was challenged by the advanced French sentry, who, receiving no answer, fired and wounded the leading officer, but was himself immediately shot. Very little opposition, however, was made, and our troops took possession of the works. General Moore also occupied the Nole Hill with the Corsican Rangers,[1] after slight resistance, from which he was able to reconnoitre ; but finding it too advanced to maintain, withdrew some distance, when the enemy reoccupied the hill.

The French, finding their pickets attacked, and fearing a general assault, beat to arms, and began a heavy fire from their works, which continued for three hours ; but, our troops being covered by the inequalities of the ground, little execution was done. At seven o'clock a strong body of the enemy unexpectedly appeared, and advanced rapidly against the Green Hill, and the post particularly possessed by the 30th. That corps had been ordered to shelter themselves from the cannonade in the ditches of the works, and behind ridges of the hill ; thus they were scattered when the French began to ascend, supported by a heavy fire from all the batteries. The assembly was beat, 170 rank and file were collected, who instantly charged with the bayonet and routed the French. The 50th and 92nd at once advanced to the charge, and the enemy was driven back to the walls of Alexandria. The 30th lost 28 men killed and wounded ; the 92nd suffered comparatively little, but 7 men were wounded, of whom Private G. Tod died of his wounds.

The siege lasted till the 26th, when an armistice was agreed to, and on 1st September articles of capitulation were signed by General Menou, after a defence which his country had reason to approve. His troops were so short of provisions that, when General Hope dined with him after the capitulation was signed, the bill of fare consisted entirely of horse flesh.

The 92nd, accompanied by a few dragoons, was one of the first regiments that entered the place, and pitched their tents between the ancient and modern walls of Alexandria.

General Morning Order.—The 92nd Regiment is for the future to be attached to Brigadier-General Hope's Brigade.[2]

[1] This corps had been in the British army since our occupation of Corsica, and by their conduct and appearance did honour to the country of Napoleon.—*R. T. Wilson.*

[2] General Hope afterwards became their colonel.

R.O., September 16th.—The Marquis of Huntly has requested Lieut.-Colonel Napier to inform the regiment that he was much pleased with the accounts he had of their conduct, and that no person could be more gratified by the credit the 92nd acquired in Egypt than himself. His Lordship has also enclosed a letter from the Duke of York to be put in Regimental Orders.

HORSE GUARDS, *30th May* 1801.

MY LORD,—I have to acknowledge the receipt of your Lordship's letter of the 24th inst., and I need not assure you how sincerely I unite with you in regretting the loss of so deserving an officer as Lieut.-Colonel Erskine of the 92nd Regiment. I have ever entertained too high a sense of the gallant services of that corps not to have recommended on this occasion that the step should go in the regiment, of which His Majesty has been pleased to approve.

I am, My Lord,
Yours,
FREDERICK,
Commander-in-Chief.

Major-General the Marquis of Huntly,
etc. etc.

Accordingly, Major Napier had been promoted lieut.-colonel, and Captain John Cameron major.

The French had lost in the campaign in all 32,180 of their military establishment, of whom 3000 were killed in the different actions and 1500 had died of disease since the British landed, the rest having been taken in battle, or in the garrisons, convoys, or detachments which had surrendered. They also lost 600 deserters and 768 of the civil establishment. The British and Turkish troops had captured 1003 pieces of cannon, besides about 500 that were unserviceable.

The loss of the British in killed and wounded (of whom, however, many died) was :—

Officers killed, including quartermasters, 23.
,, wounded, 169 ; missing, 8.
Sergeants killed, 20 ; wounded, 149 ; missing, 2.
Rank and file killed, 505 ; wounded, 2728 ; missing, 73.

Of that number the Gordon Highlanders had 13 officers wounded, of whom 4 died of their wounds ; 23 n.c. officers and soldiers were killed in action, and 128 n.c. officers and soldiers, including Sergeant-major M'Intosh, were wounded, of whom 36 died of their wounds.[1]

The names of the officers wounded are as follows:—

On March 13th—

Lieut.-Colonel Charles Erskine, died of wounds.
Captain the Hon. John Ramsay.
,, Archibald MacDonald.

[1] One man was killed in action whose name cannot be found. Surgeon Findlay and 63 n.c. officers and soldiers died of disease.

Lieutenant Norman MacLeod.[1]
,, Charles Dowle, died of wounds.
,, Donald MacDonald.
,, Tomlin Campbell, died of wounds.
,, Alexander Clarke, died of wounds.
,, Ronald MacDonald.
Ensign Peter Wilkie.
,, Alexander Cameron.

On March 21st—
Captain John Cameron.
Lieutenant James Stewart Mathieson.

Killed in Action

Corporal W. Cunningham.
,, John MacDonald.
Private John Beaton.
,, Donald Beaton.
,, Adam Boyd.
,, George Donaldson.
,, Donald Downie.
,, Duncan Gordon.
,, Andrew Hawley.
,, James Horne.
,, Thomas Hutchison.
Private Robert Innes.
,, Stephen Low.
,, James MacCallum.
,, George MacBeath.
,, Alexander MacDonald.
,, Duncan MacDonald.
,, John MacLennan.
,, Peter Montague.
,, John Nichol.
,, Alexander Ross.
,, Montgomerie Wilson.

Died of their Wounds

Sergeant Charles MacHardy.
,, Thomas Macpherson.
Corporal Hugh Crauford.
,, Duncan MacKinlay.
,, Lachlan Milne.
,, Alexander Steuart.
Private W. Alston.
,, George Anderson.
,, Alexander Bean.
,, William Brown.
,, John Cameron.
,, Charles Campbell.
,, Alexander Cattanach.
,, John Cunningham.
,, William Dallas.
,, W. Fraser.
,, William Fyfe.
,, W. Gibson.
Private Cosmo Gordon.
,, Alexander Gordon.
,, William Gunn.
,, George Hogg.
,, William Knox.
,, Robert Johnstone.
,, John MacDonald.
,, D. MacKinlay.
,, James Maclean.
,, Henry MacPhee.
,, Alexander Murdoch.
,, David Sime.
,, J. Skeen.
,, Alexander Smith.
,, William Swan.
,, Arthur Thomson.
,, John Tiplin.
,, George Tod.

[1] Lieutenant MacLeod also died from the effects of his wound some time after. He was a son of MacLeod of Eyre in Skye.—M'Innes' "Brave Sons of Skye," p. 172.

Before the garrison of Alexandria were embarked for France, the men of the Gordons got on very friendly terms with the Frenchmen, some of whom could speak English. They shared their provisions and exchanged articles with each other, and both used to laugh heartily at the slovenly habits of the Turkish troops, who when on sentry would sit down and smoke their long pipes, giving themselves little trouble about their duty.

From the Regimental Orders of the 5th October, it appears that the men had generously expressed a wish that the money due to them for providing themselves with fresh meat in lieu of salt pork, which they had not used, should be given to the widows and orphans of the regiment, who had been made so by the campaign. The lieut.-colonel informs them that it amounts to £150, which will be paid to them ; that he will be happy to put himself at the head of any subscription for so charitable and proper a purpose ; and that the men can subscribe what they wish individually, so that each shall be free to give or not as he likes.

G.O., Camp before Alexandria, October 5th, 1801.—The 23rd, 42nd, and 92nd Regiments, with detachment Royal Artillery, will march to-morrow morning at six o'clock to Aboukir for embarkation.

The regiment embarked on H.M.S. *Renommée* and *Modeste* Frigate on the 6th October. Since the landing on the 8th of March it had been engaged in three battles, had taken part in the investment of Cairo and the siege of Alexandria, besides several minor affairs and skirmishes.

The 92nd, in common with the other corps employed in the campaign, afterwards received the Royal authority to bear on their colours and appointments the *Sphinx* and the word *Egypt.* They had also won the exceptional honour *Mandora,*[1] now only possessed by the Scottish Rifles and the Gordon Highlanders.

They had, however, suffered much, and tales of the sore eyes, the flies, the thirst and the sand-storms of the deserts of Egypt are repeated to this day by their descendants on the banks of the Spey.

The Grand Seignior of Turkey presented gold medals to the officers, which varied in size according to rank.

R.O., H.M.S. Renommée, October 8th, 1801.—The lieut.-colonel was happy to receive the following letter from Captain Probyn, one of the Commander-in-Chief's aides-de-camp.

HEADQUARTERS, *6th October* 1801.

SIR,—As the guard of the 92nd who have been doing duty at headquarters are about to join their corps, the Commander-in-Chief has directed

[1] *Egypt* and the *Sphinx* were given June 12th, 1803, and *Mandora* on February 23rd, 1813. In the officers' mess there was, and still is, preserved the horn of an Egyptian ration ox, made into a snuff-mull.

me to express to you his entire approbation of the exemplary conduct of the guard in general, and of Sergeant Mark in particular.[1]

I have the honour to be, Sir,
Your most obedient and humble servant,
(Signed) J. Probyn, A.D.C.

To the Commanding Officer
of the 92nd Regiment.

The ships arrived at Malta on the 19th October, and the regiment remained there till the 15th November, part only being landed.

The men's Highland clothing seems to have been entirely worn out, and no tartan being available to renew it, pantaloons of various sorts and colours were served out to them. The officers, who appear to have been in the habit of going ashore in the Highland dress, are desired in Orders not to do so till the men have it, "but they may wear their bonnets." While at Malta, Sergeants Allardyce and M'Arthur were suspended for four months respectively from "rank and pay" by sentence of court-martial.

On December 25th Orders are dated "At sea off Carthagena," and on January 2nd Lieut.-Colonel Napier is sorry to have to acquaint them that he is informed by Captain M'Kellar (of the *Renommée*), that from the uncertainty of the wind, it is absolutely necessary to stop one-third of the allowance of bread and spirits, which they both lament extremely, but hope in a few days communication may be had with the shore. They got supplies at Malaga on the 5th January.

[1] Sir Ralph Abercromby had selected the regiment to furnish the guard at his headquarters from his first landing in Egypt, and at his death Lieut.-General the Hon. J. H. Hutchinson had continued them. Sergeant Mark belonged to the parish of Glass.

CHAPTER VIII

AFTER touching at Gibraltar, they arrived at the Cove of Cork on the 30th January, without further adventure, but remained in quarantine there till February 12th, when they marched for Kilkenny. The effective strength consisted of 1 lieut.-colonel, 2 majors, 4 captains, 15 lieutenants, 5 ensigns, 6 staff, 42 sergeants, 22 drummers, and 559 rank and file.

Lieut.-Colonel Napier in Orders "trusts that the regiment will behave with the greatest order, whether on the march or in quarters," and that they will "remember that the credit they have already gained will make their conduct the more observed," and cautions them against the temptations of a garrison town. The Regimental Orders show that their conduct on the march to Kilkenny was excellent, but it is not surprising to find that after a year and eight months, spent almost entirely at sea or on active service, they should have celebrated their return to the land of fair women and poteen whisky by a carouse, which the commanding officer excuses under the circumstances, but "positively assures them" that he will not do so again.

Sergeant D. Nicol in his journal says that on the march to Kilkenny each company was entertained by the corresponding company of the Lancashire Fencibles. "A merry evening but all done in good order and with sobriety." At Kilkenny a girl was accidentally shot by one of the men. The regiment subscribed £50 for her parents. There were at this time two brothers named Young in the regiment who had nine other brothers in the army, eleven in all. The Queen gave their mother a pension.

Here the men received their new clothing and the "plaid and hose tartan"[1] due to them. The bonnets, however, had stood the rough work, and only required to have the feathers taken down and washed, and new ribbons put on. The regiment was medically inspected by a staff surgeon on February 25th, when many of the sick and wounded were convalescent, but 4 sergeants, 2 corporals, 1 drummer, and 92 privates were discharged as unfit for service. There is also a list of 14 drummers and 37 privates "very stout men and fit for service, but under size," whom Major-General Sir Charles Ross, Bart., who inspected the regiment on the 25th, recommended to be kept. The regiment received prize-money for Egypt, but the amount does not appear.

R.O., Kilkenny, 5th April.—The following letter is published:—

ROYAL HOSPITAL, DUBLIN.

SIR,—I am commanded by the Commander of the Forces to acquaint you that the definitive treaty of peace was signed at Amiens at four o'clock in the

[1] The tartan was for many years called *plaid* tartan, though intended for kilts as well as plaids, owing to the two garments having been originally called a plaid.

afternoon of the 27th ult. by the Plenipotentiary of His Majesty and the Plenipotentiaries of France, Spain, and the Batavian Republic, on which happy occasion the troops under your command will fire a *feu-de-joie* as usual.

I have the honour to be, Sir,
Your most obedient humble servant,
(Signed) S. BECKWITH.

To M.-General Ross,
Clonmell.

The principal features of the Treaty of Amiens were, that Great Britain restored all her conquests during the war except Trinidad and Ceylon, which were ceded to her by Spain and by the Batavian Republic.[1] Portugal ceded some possessions in Guiana to France. The territories of the Ottoman Porte were maintained in their integrity. The Ionian Republic was recognised, and Malta was to be restored to the Knights of St John. The French agreed to evacuate the Neapolitan and Roman States, and Great Britain the ports she held in the Adriatic and Mediterranean.

In March a recruiting party [2] was sent to Athlone, probably to get volunteers from the Fencible Corps quartered there ; it was at any rate withdrawn soon after.

The men received an allowance in money in lieu of the clothing which they had not got at the proper time, which at this period was Christmas Day. Recruits had blue pantaloons for fatigue.

The regiment was ordered to Dublin, to march in three divisions, under command of Major Gordon.

R.O., Kilkenny, April 9th, 1802.—" From the general good conduct of the regiment, particularly on the last march, and the high character the men get in every quarter, Major Gordon is sensible the men will feel it incumbent on themselves to maintain that character, and nothing marks a good soldier more than regularity and sobriety on the march."

There were still a good many sick, who were left at Kilkenny in the general hospital, with Sergeant Alexander Cameron to take care of them. *En route*, the destination of the regiment was changed to Belfast, where they arrived on the 28th ; some companies were quartered in barracks, the others were billeted in the town.

B.O., Belfast, 23rd April 1802.—" Brigadier-General Drummond makes known to the officers, n.c. officers, and privates who served in Egypt, the thanks of both Houses of Parliament, and also to the officers, petty officers, n.c. officers, and men of the Navy and Marines. Commanding officers of ships and corps are respectfully desired to thank those under their command for their gallant and

[1] Holland.
[2] Lieutenant Phipps, Sergeant MacLean, Corporal Walker, and John MacPherson, piper.

exemplary behaviour. The following is from the Speaker of the House of Commons :—

In transmitting these resolutions to Your Royal Highness, I have at the same time the sincerest satisfaction in bearing testimony to the sentiments of gratitude and admiration with which the House expressed its sense of those services which have raised our naval and military glory to a height unknown in former times, and have displayed so many brilliant and successful examples of valour and enterprise for after ages to admire and emulate.

I have the honour to be, etc.,

(Signed) CHARLES ABBOTT,
Speaker.

To Field-Marshal
H.R.H. The Duke of York,
Commander-in-Chief of His Majesty's Forces.

On May 27th, 70 men were discharged, 68 of them to out-pensions ; and the regiment wanted 390 men to complete establishment, the total strength being 610.

On May 29th, the regiment and a battery of artillery fired a *feu-de-joie* in memory of the restoration of King Charles II.

Early in June the Gordon Highlanders embarked for Glasgow,[1] this being their first visit to Scotland since they were raised eight years before. Sergeant D. Nicol in his journal says that the regiment was happy and comfortable at Glasgow, officers and men knowing each other well, and the new barracks in the Gallowgate being the best in Europe. Sergeant Nicol walked to Edinburgh on furlough, 55 miles, in 24 hours.

The Orders from May 1802 till July 1804 are missing, but from the monthly returns and other documents a certain amount of information can be gathered. It appears that it was expected that the army would be reduced at the peace, and that the regiment would be disbanded. Sir John Moore writes in September to Major Cameron at Glasgow Barracks, " I am sorry to find it is determined to reduce the 92nd. Their gallant services entitled them to a better fate." There were great rejoicings throughout the country at the peace, but it was soon felt that it could not last ; Bonaparte evidently intended to exclude England from all Continental influence or commerce, and her statesmen did not think it safe to make any considerable reduction in the national establishments. Consequently, the idea of reducing the Gordon Highlanders was aban-

[1] *G.O., Dublin, April 12th,* 1802.—On a regiment arriving in Ireland, it is to be continued on the British establishment up to the end (*i.e.* 24th inclusive) of the military month in which it shall land, and the paymaster is to settle his accounts with the English War Office accordingly, the officers and men receiving their pay in English money for the whole period of the pay list ; and on a regiment embarking from Ireland, it shall continue in like manner upon the Irish establishment and be paid in Irish money, the paymaster settling his accounts with the Government of Ireland.

doned, and they were retained in the service of their country. To some of the men this was a disappointment; they had expected to be discharged at the end of the war, as had been the case after former wars with several of the Highland regiments; many of those enlisted on the estates of the officers so understood the terms of their enlistment. Their experience of soldiering had been of the roughest; they had the old Highland idea of a fight, and then home again to tend the cattle, and they demanded their discharge in fulfilment of what they believed to be their engagement. Lord Huntly, as colonel, arrived in Glasgow and inquired into the matter, which ended in a few of the men being found to be entitled to their discharge, which they received.

At Glasgow on the 24th October 1802, Archibald Campbell, Esq., paymaster in the 92nd Regiment, was appointed "our true and lawful attorney for us, and in our name and for our use and benefit to ask, demand, recover and receive," etc., "all sums of money that are, or may be, due to us in name of prize-money, bounty, or reward, on account of the capture of the Dutch shipping lying in the Zuyder-Zee in the year 1799, and the conquests in Holland made by the British in that year," etc., signed by 28 sergeants, 25 corporals, 19 drummers, and 297 privates.

Recruiting was now actively carried on. The standard height was lowered to 5 feet 5 inches; the levy money was £6 6s., of which the recruit received in money and necessaries, £5 5s.; the other payments being—for attesting, 1s.; surgeon's fee, 2s. 6d.; postage and paper, 3s. 6d.; reward to party, 10s. 6d.; and for conducting to place of approval, 3s. 6d. Recruits were enlisted at headquarters, and parties were sent to various places, such towns being chosen as were frequented by young men from the north in search of employment, from which they visited the fairs and feeing markets throughout the neighbouring districts.[1] They also received volunteers from the Highland Fencibles, among them twenty-eight men from the "Regiment of the Isles."[2]

In the spring of 1803, it became known that extensive military preparations were making in the ports of France and Holland, which

[1]

Recruiting Stations.	Rank and Name of Officers commanding Parties.
Fochabers	Major John Gordon.
Dundee	Captain the Hon. John Ramsay.
Aberdeen	„ Peter Grant.
Huntly	„ P. Gordon.
Paisley	Lieutenant James Mitchell.
Stirling	„ Donald MacDonald.
Inverness	„ W. M'Kay.
Perth	„ W. Phipps.
Fort-William	Ensign Donald M'Barnet.

[2] Letter from Lieutenant-General R. Vyse to officer commanding 92nd.

it was pretended were destined for the French colonies ; but King George III, in a message to Parliament, adverted to the necessity of precautionary measures for the security of the country. This message was received with applause alike by Whig and Tory, who were on this occasion unanimous in the determination to prevent the despotism of Bonaparte from becoming universal throughout Europe. The French demanded the expulsion from England of their exiled countrymen, who had availed themselves of her hospitality since the Revolution, and insisted on the immediate surrender of Malta, which had been retained by the British till security should be found against a fresh seizure of the island by the French. Other causes of irritation existed, and the patriotic feeling in both countries was inflamed to the highest pitch. A spark only was needed to light the blaze of war ; it was supplied by Bonaparte, who publicly insulted the British Ambassador in Paris. Thus after a short and feverish peace, the two countries were again plunged into war,[1] and a contest was begun which was only finally decided on the field of Waterloo. Britain stood alone to resist Bonaparte's career of conquest, and he resolved to strike the first blow. He assembled an army of 100,000 men at Boulogne for the invasion of Britain, and a host of flat-bottomed boats for their conveyance, while every effort was made to hasten the preparations. "Fifteen millions," he is reported to have said, "must give way to forty millions."[2] "Let us be masters of the Channel for six hours, and we are masters of the world."

It therefore became necessary for Britain to make proportionate exertions to meet the emergency. The Government devoted all their energies to the task, in which they were nobly seconded by the patriotism and spirit of both Parliament and people.

The whole force of the United Kingdom was immediately mobilised, and stationed principally on the eastern and southern coasts. The military force of the United Kingdom, at home and abroad, consisted of 150,950 men, of which 98,314 were the regular army, the remainder militia.[3] This number was so infinitely disproportioned to the formidable army expected to be brought against us,[4] that the Government felt it their duty to increase it by every

[1] War was declared on the 18th May, and at once Bonaparte ordered all English travellers and residents in France, to the number of 10,000, to be arrested and thrown into prison. This unusual and barbarous act produced a strong feeling of animosity against him among all classes in Britain.

[2] Alluding to the population of the United Kingdom and of France. "Short History of the English People" (J. R. Green).

[3] Military Transactions of the British Empire for 1803-7, inclusive, Commander-in-Chief's Office (J. W. Gordon, Secretary to Commander-in-Chief).

[4] The army of France, without counting the subsidiary forces of Holland, Switzerland, and the Italian States subject to its command, amounted to 427,000 men, besides the National Guards and Coast Guards, which were above 200,000.—*Alison.*

means the country could afford. The Militia, 80,000 strong, were called out on the 25th of March; on June 28th, the House of Commons agreed to the unusual step of raising 50,000 additional men by conscription, under the denomination of the "Army Reserve Act"; a certain quota of men from each county were chosen by ballot, in the proportion of 34,000 for England, 10,000 for Ireland, and 6000 for Scotland, on condition of their serving only in Great Britain and Ireland, but with liberty to volunteer for general service with bounty.[1] The men raised by this Act in the counties of Nairn, Inverness, Moray, Banff, and Aberdeen were sent to form a 2nd Battalion of the Gordon Highlanders, which was made up by men recruited in the usual manner. This battalion became a nursery of good recruits to supply the casualties consequent on the more active duties of the 1st Battalion, till the peace of 1814.[2] The 2nd Battalion 92nd Regiment (or Gordon Highlanders) was placed on the establishment on July 9th, 1803.[3] Lieut.-Colonel J. Willoughby Gordon, from the Staff of the Army, was appointed Lieut.-Colonel of the 2nd Battalion, 4th August 1804.

Meanwhile volunteer corps were enrolled in the different parts of the kingdom, and such was the enthusiasm, that in a few weeks their number amounted to 300,000.[4] Among them were to be found men of all ranks, professions, and classes, from the Prince to the peasant.[5] The voice of political faction was stilled, and Whig and Tory stood as comrades in the ranks, united in one great effort for the preservation of their country.

On June 6th the Gordon Highlanders marched from Glasgow, embarked at Leith on the 9th, landed at Harwich on the 24th, and proceeded to Colchester. On the 1st July they marched to Weely Cross in Essex, where they encamped, soon after going into Weely Barracks. The 2nd Battalion was formed there on November 24th; officers and n.c. officers, with a proportion of old soldiers, being posted to it from the 1st Battalion.

[1] The balloted men were allowed to find substitutes, and as it was inconvenient to many business men and others to serve personally, they often paid large sums to men who wished to join the army to become their substitutes, who afterwards added to the money thus earned the bounty for volunteering. Many young men materially increased the comfort of their parents by these means. A Peninsular veteran told me, when I admired his comfortable house, that he and his brothers had built it for their parents with their bounties, adding: "They have ruined Scotland, sir, with their Poor-Law! Lads lippen to the law to keep their mothers."

[2] Stewart's "Highlanders," p. 299.

[3] Cannon's "Historical Record."

[4] Alison.

[5] Among others, in Edinburgh, was a corps composed of lawyers. Their instructor having ordered them always to step off with the left foot, "Why not the right?" asked an advocate, keen for argument. "Because it's His Majesty's order," said the sergeant, adding, "I would rather drill ten fools than one philosopher!"

The members of an Angus corps found great difficulty in remembering which was the left and which the right foot, till their sergeant hit on the expedient of marking the left shoes with chalk, and ordered "Caulkit fut foremost."

	1st Battalion.	2nd Battalion.
Colonel	1	0
Lieut.-Colonel	1	1
Major	2	2
Captains	10	10
Lieutenants	12	12
Ensigns	8	8
Paymaster	1	0
Adjutant	1	1
Quartermaster	1	1
Surgeon	1	1
Assistant Surgeons	2	2
Sergeant-Major	1	1
Quartermaster-Sergeant	1	1
Paymaster-Sergeant	1	0
Armourer-Sergeant	1	1
Sergeants	50	50
Corporals	50	50
Drummers	20	20
Fifers	2	2
Privates	950	950
Total	1116	1113

In 1803, field-officers ceased to hold companies.

At this time flank companies had each two lieutenants, and battalion companies one lieutenant and one ensign.

Both battalions were commanded by Lieut.-Colonel Napier. On the 15th October the 1st Battalion had 58 English, 632 Scots, 25 Irish, and 1 foreigner[1]—total, 716 n.c. officers and men.

The levy money was raised to £10, 10s., of which the recruit got in cash and necessaries £7, 12s. 6d., the remainder going to the party and costs. In November 62 recruits and volunteers for general service (*i.e.* from the reserve) joined the 1st Battalion. In the returns of the early months of 1804, considerable numbers of recruits are mentioned as approved at headquarters, and also as being with the parties, and " approved by Lord Huntly," who was then a major-general on the staff of North Britain, but who took an active interest in getting the right stamp of men for his regiment. A great many reserve men[2] volunteered into the 1st Battalion for general service, and on March 1st the whole of the n.c. officers, drummers, and rank and file of the battalion were for unlimited service, except one drummer.

1 The " foreigner " seems to have been Donald Cameron, born at Charlestown, U.S.A.

2 235 up to the 15th May 1804, of whom " 16 are from the reserve attached to the 42nd Regiment "—Effective Size and Description Roll, 1st Battalion 92nd, January 1st, 1805. In this roll there are also many men who had served in—Argyll Militia, Ross-shire Militia, Shropshire Militia, Glengarry Fencibles, Isles Fencibles, Lochaber Fencibles, Independent

Scotland was then divided into four military districts: Northern, headquarters at Aberdeen; Central, Dundee; Western, Glasgow; and Southern, Musselburgh and West Barns. Major-General the Marquis of Huntly commanded the Northern District.

In May, Napoleon Bonaparte was crowned Emperor of the French.

R.O., July 23rd, 1804.—The regiment is to march to-morrow morning for Colchester, etc. etc. One subaltern from the regiment, and 1 sergeant, 2 corporals, and 12 privates from each battalion for baggage guard.

On August 1st the total strength of the 1st Battalion was 845 men.

R.O., Colchester, July 29th.—Officers of the 2nd Battalion are not to carry walking-sticks at church parade.

Battalion Order.—Sergeant Charles MacGregor is sentenced by Battalion Court-martial to be "suspended from the rank and pay of a sergeant to those of a private sentinel, for one month from the date hereof," for not having his company muster-roll ready on the 24th inst.[1] On account of his good character and the recommendation of the Court, Lieut.-Colonel Napier pardons him, but takes the opportunity of informing officers and n.c. officers that no excuse will be taken for neglect of public duties, and that nothing can be so unmilitary as the circumstance of trusting public documents upon a march to a common carrier, which appears to have been the case with Sergeant MacGregor.

At Colchester, where there was a large garrison, the two battalions 92nd Gordon Highlanders formed, with the two battalions 42nd Royal Highlanders, a brigade under Major-General the Hon. John Hope. The brigade marched and encamped on Lexden Heath on the 6th August, only light baggage being allowed; the 2nd Battalion 92nd was commanded by Major (Brevet Lieut.-Colonel) Morris, who was soon afterwards removed to a veteran battalion.

Here instruction was carried on in outpost and field duties of all sorts, light infantry and target practice; but during harvest a number of old soldiers were allowed to take work with the farmers. It appears the men sometimes did jobbing work where they were quartered, though it was not encouraged, and sometimes forbidden.

About this time pressing application was made for reinforce-

Company of Gordon Fencibles, MacLeod Fencibles, Grant's Fencibles, Clan Alpine Fencibles, North (or Gordon) Fencibles, Caithness Fencibles, Caithness Legion, Reay Fencibles, North Lowland Fencibles.

[1] Suspension from rank for a short period seems in those days to have been a common sentence of courts-martial for slight offences.

ments both in the East and West Indies, which, combined with the hostile aspect of affairs in Europe, made it necessary to augment the army. The Government therefore brought forward a measure for giving an addition to the army of 78,951 men, and an immediate levy of fifty battalions for home service, so as to set the regular army at liberty for disposal abroad. The "Additional Force Act" was passed in July, and continued in force till June 1806, when it had produced to the army only 15,771 men.

In October the men of the 14th Battalion of Reserve, commanded by Lieut.-Colonel Grant, who had volunteered for general service, were allowed to choose the 42nd, 91st, and 92nd Regiments.

On the 24th October, 1 subaltern officer, 4 sergeants, 4 corporals, and 1 drummer from the 2nd Battalion marched to London, *en route* for Elgin, where they were detached under instructions of the Inspector-General for the purpose of receiving the men to be raised under the "Additional Force Act."

On the 26th October both battalions returned to Weely Barracks, and before the camp broke up, Lieut.-General Sir J. Craig, commanding the Colchester District, published an order complimenting the troops on their good conduct, in which he states "that he has not received one complaint of irregular or disorderly conduct during the three months they have been on the ground."

G.O., *Weely Barracks*, *13th November* 1804.—"Officers commanding battalions will be so good as recollect that the original formation of their corps will always be in three ranks, unless the ten divisions fall in short of twelve complete files each.

Bat men and bat horses to be paraded daily with all they are intended to carry on service ; also the carts for entrenching tools, and all are to be exercised under the quartermaster of the day."

R.O.—The greatcoats of the men when in marching order are to be folded under the flap of the knapsack.

On the 25th October 1804, the 2nd Battalion was held in readiness to proceed by the Inland Navigation to Liverpool, and there to embark for Ireland. It accordingly marched on the 29th, taking only twelve rounds of ammunition and one flint each man,[1] returning the rest to store, and leaving its bat horses to the 1st Battalion.

The 1st Battalion was now brigaded with the 1st 42nd (the 2nd 42nd being also gone to Ireland) and the 1st 91st. The men were all regularly practised in gun-drill—1 sergeant and 14 gunners from each battalion "whenever ye weather permits."

R.O.—"Ten men if possible to be in a mess, and a n.c. officer at the head of each mess, and to be ye responsible person. The names of every man and his comrade to be posted at the foot of each berth." (Two men slept in each berth.)

[1] Sixty rounds and two flints was usual for each man.

The officers, at any rate, seem to have returned to the use of hair-powder, which was still fashionable in civil life.

R.O., December 22nd, 1804.—" Major Cameron directs that no officer will presume to appear on parade without having his hair tied according to regulation and properly powdered." And by a later Order, officers' hair to be tied two inches from the head, and the tail five inches in length.

R.O., February 7th, 1805.—The men not to wear greatcoats except on duty and never on fatigue.

R.O., May 21st.—Officers to wear the Highland dress on all duties, and when off duty white cassimere or cloth pantaloons and half-boots.

Regimental mounted officers' horse appointments—saddle-cloth of the regimental facing, with double row of regimental lace round it. Holsters covered with black bear-skin, the fronting of the bridle covered with ribbon the colour of the saddle-cloth.

R.O., February 11th.—The barrack guard to turn out once a day to the commanding officer of the regiment with presented arms, after which it turns out with ordered arms. To field officers of the regiment it turns out once a day with ordered arms, after which it only stands to arms.

At this time a number of men were discharged unfit for service, and were sent home to Scotland, and others, fit only for home duty, were sent to a veteran battalion in the Isle of Wight.

Clothing was now issued in April instead of at Christmas, and the battalion received it, with one pair of shoes per man, on April 16th. " Hose, not being now an article of clothing, will be paid for."[1] At this period the *Sphinx* was by Regimental Order adopted as the crest to be worn on the bonnets in place of a regimental button as heretofore.

On the 17th April the battalion was inspected. Field officers and staff in white breeches and boots, the other officers in Highland dress ; and Major-General Hope, in Orders of the 18th, expressed the highest approbation of the " complete and efficient appearance of the 92nd Regiment yesterday, as well as the precision with which the battalion moved," and he " requests Lieut.-Colonel Napier to express to the officers and men his sense of the diligence they have bestowed on the several field duties, and of the proficiency to which they have attained."[2]

Sergeants were sometimes allowed to resign rather than stand trial.

R.O., May 21st.—" Sergeant Lewis MacGregor, having been

[1] There is still only three yards and a half of plaid tartan in each kilt.

[2] This was the more satisfactory, as General Hope is by no means so complimentary in his Order to a regiment he had inspected the previous day.

guilty of an irregularity, Lieut.-Colonel Napier has accepted his resignation."

In May six officers were employed in receiving volunteers from the Scottish Militia regiments in England. Three volunteers were also received from the Oxfordshire, and six from the Irish Militia.

At Weely the married men were allowed to live in huts belonging to Government outside barracks, but their wives sometimes abused this privilege by selling liquor and allowing disorderly conduct. "Major Cameron was perfectly shocked at the infamous scene of gambling he witnessed to-day in rear of the centre huts of the 92nd Regiment," etc. "On account of the irregularities allowed by Private M'Kean, his hut is to be burned down by the pioneers, and if Mrs M'Kean continues these disgraceful scenes, she will be drummed out of the regiment."

On the 4th of June, being the King's birthday, the battalion fired three volleys, and drill was dispensed with in honour of the occasion.

R.O.—"It is to be observed that when the open column marches past in parade order the drummers on the flanks of divisions are not to roll ; the pipers may, however, be allowed to play." [1]

At this time a few carpenters and masons were allowed to volunteer from the regiment to the corps of Royal Military Artificers with ten guineas bounty.

R.O., July 10th.—Complaints having been made in an irregular manner by Captain Watts' company, Lieut.-Colonel Napier admonishes the men, "who know perfectly the proper mode to be adopted when they feel themselves to be aggrieved. It appearing, however, that some negligence has taken place on the part of Captain Watt, Lieut.-Colonel Napier has reprimanded him in presence of the field-officers and captains of the battalion."

In August a certain number of men of good character and large families are allowed to assist at harvest, but are not to solicit work from the farmers, who are to apply to the commanding officer if they want men.

At a time when the price of bread varied, according to the season and district, from 7¼d. to 10¼d. for the 4-lb. loaf,[2] it was a great help to men with children to be able in this way to add to their means of support.

On September 2nd the 1st Battalion, along with the 1st 42nd, 91st and 95th (Rifles), marched to Colchester, leaving a detachment

[1] It was the custom in the regiment till after the Crimean War, when marching past in slow time was done away with, for the piper of each company to play "Failte am Prionsa" or some other salute on the flank of his company as it passed the General, the band all the time playing the "Garb of Old Gaul."

On May 11th, in Regimental Orders, "Alexander Cameron the piper is to be taken on the strength of the Grenadiers as drummer from the 25th of last month" (probably to get him drummer's pay, to which, as piper, he was not entitled).

[2] Monthly commissariat prices in Orders. In March 1805 it was 10¼d.

in charge of the heavy baggage at Weely, but on the 4th it was ordered to hold itself in readiness for embarkation, and returned to Weely on the 6th. On the 8th October it was again at Colchester for exercise and field days.[1] While there, it was reviewed at Broxted by H.R.H. the Commander-in-Chief, who expressed his approval of the troops' "soldier-like appearance and correct discipline." On the 18th the Highland regiments returned to Weely, and on the 29th the 1st Battalion marched to Tilbury Fort, *en route* for Ospringe Barracks, where it arrived November 6th.

R.O., *October* 28*th*, 1805, *Weely*.—In leaving the district and quarters so long occupied by the regiment, Lieut.-Colonel Napier "hopes and trusts that the regiment will conduct themselves in a manner worthy of the high character they hold in the army, and that they will not, by drunkenness or otherwise, give people reason to suppose that their present character has been established merely from their appearance in the field or at a review."[2]

In a letter to the commanding officer in November, Messrs Greenwood and Cox, the army agents, inform him that each married woman left behind on embarkation is to be allowed £1 1s., and 5s. for each child. In a letter from Ospringe Barracks, Faversham, Kent, Major Cameron writes—"We are waiting here till the transports are ready for us.[3] We are all of us this morning in the most extraordinary mixture of joy and sorrow, for the loss, in the midst of such glorious achievements, of that greatest of men—Nelson." Again, on the 25th December 1805, "to the astonishment of every person here we are still without a word of moving; the 2nd Battalion of the 79th and we form the brigade of the Honourable Major-General John Hope, commanded *ad interim* by his brother Charles."

Meanwhile Napoleon's plan by which the British fleet was to be divided, and the whole French and Spanish fleets concentrated to crush the British squadron in the Channel, and protect the vast armament at Boulogne in crossing to the shores of Britain, had been frustrated by the death of the Admiral destined to command it, and by the vigilance of the British Admirals. Austria, Russia, and Sweden had formed an alliance to wrest Italy and the Low Countries from the French Emperor, who had reluctantly abandoned for the time his dream of the invasion of Britain ; and marching the "Army

[1] At this period Sunday could hardly be called a day of rest. It was the custom to march past in review order after church, and "evening parade is to be formed with arms at same hour and in same manner as on week-days."

[2] I do not find once in General or Brigade Orders any fault found with the 92nd by general officers, which is by no means the case with every regiment in the district.

[3] They had probably been intended to form part of the expedition which embarked in November under Lord Cathcart, landed at Cuxhaven and occupied a position on the River Weser, for the purpose of joining the Russian and Swedish forces in causing a diversion on the side of Holland ; but the French victory of Austerlitz in December put an end to hopes for its success, and Lord Cathcart returned to England.

of England," as it was called (now amounting to 150,000 men), towards the Danube, he forced an Austrian army to capitulate at Ulm, three days before Nelson's victory of the 21st October at Trafalgar.

"England has saved herself by her courage," said Pitt, in what were destined to be his last public words. "She will save Europe by her example."[1]

The 1st Battalion 92nd marched to Canterbury on the 26th November, there to be quartered; and, instead of being sent to the Continent, the battalion, by order from the Commander-in-Chief, marched in three divisions to London, on the 2nd, 3rd, and 4th of January 1806, in order to attend the funeral of Admiral Viscount Nelson on the 9th.

They were to return after the solemnity, and to leave everything in their present quarters, except "only such articles as are necessary to their appearance on parade on the day of the funeral."[2] There was evidently some difficulty as to the "appearance" they were to make. The men carried new hose and new jackets in their packs, but many of the latter were not ready, and the tailors had a hard time of it, and no "Sunday out" in London; they were ordered to "bring as many jackets in the waggon as they can finish on Sunday."

On the morning of January 9th, the 92nd marched from their quarters in the outskirts of London to the Horse Guards Parade, where the troops, consisting of three brigades of infantry, with cavalry and artillery, were formed for the funeral procession, which was led by a detachment of cavalry and four companies of light infantry, under Major Cameron of the 92nd, immediately followed by the battalion companies of the 92nd and 1st Battalion 79th, the other troops following, and four Grenadier companies bringing up the rear. "The music of the 79th to be in front of the brigade, and occasionally play a few bars of solemn and appropriate music."

The four light companies under Major Cameron, and an officer and twenty men of the 2nd Dragoons, were formed inside the railings, to the door of St. Paul's, while the Grenadier companies lined the aisle from the door to the place where the remains of the hero were deposited. The rest of the troops passed on to Moorfields, where they fired three volleys in the air, and where refreshments were provided for them; they then returned over London Bridge to their quarters.

"Though the honour was great," says Major Cameron, "the trouble was by no means slight, as we were under arms from six in the morning till seven at night."

[1] Green's "Short History of the English People."

[2] White leather gloves had lately been introduced for the men.

While the Gordon Highlanders were in London, it was announced that their colonel, Major-General the Marquis of Huntly, had been removed to the 42nd Royal Highland Regiment, and Major-General the Hon. Sir John Hope (afterwards the Earl of Hopetoun), from the 60th Regiment, was appointed by His Majesty King George III to be colonel of the 92nd. Although the official connection was thus severed, their first colonel by no means gave up his interest in the regiment. The officers were always welcome guests at Gordon Castle; several spent the evening of their days as tacksmen on the Gordon Estates in Lochaber and Badenoch, where numbers of the veteran soldiers also found comfortable homes.

Up to this period the monthly returns were always headed "92nd Regiment of Foot (or Gordon Highlanders)"; the last return so headed is that of July 1st, 1807. The title used afterwards is generally "92nd (Highland) Regiment of Foot," sometimes simply "92nd Regiment of Foot." Still "Gordon Highlanders" was used colloquially in the regiment, and in the Highlands "Reismaid Gordonach" continued to be the usual appellation of the 92nd.[1] It is remarkable that by some inadvertence there was, for many years after its formation, no reference in the Army List to the nationality of the regiment.

Instead of returning to Canterbury as they expected, where all their belongings were left, the battalion marched to Colchester in two divisions, leaving London on Saturday the 11th and Monday the 13th of January. A sergeant of each company had to march *via* Gravesend to Canterbury, to collect and take particular care of the men's necessaries, and give over the barracks. It is no wonder that Lieut.-Colonel Napier tells them "that the duty they are sent upon requires great attention, and may be attended with much trouble." The officers and men must have been patient indeed, if they did not use strong language, at such a want of forethought and arrangement on the part of the authorities! However, recruits, women, and children, with all their possessions, reached Colchester, where the battalion arrived on the 15th, and was inspected on the 19th by Major-General Milner, in "their new clothing and plumes." Here they were brigaded with the 1st and 2nd Battalions of the 79th, under Brigadier-General Dunlop.

On the 1st of February, the total strength of the 1st Battalion was 945. Soon after, a number of men were sent to the 9th Veteran Battalion at Edinburgh,[2] and others were discharged.

[1] Also on recruiting placards at various periods; and the *Sphinx* with the words "Gordon Highlanders" was the crest on the regimental writing-paper in 1850. The title is also used by Stewart and other writers in 1815.

[2] Veteran Battalions are expressly for the reception of meritorious soldiers, who, by wounds, infirmity, or age, are become unfit for the more active duties of the line, but who retain sufficient strength for the less laborious duty of garrison.—*Military Journal.*

While at Colchester, great attention was paid to elementary drill, and the position of the soldier in marching, etc.; to the health and comfort of the men, and to carrying on the whole system and duty of the regiment with alacrity.

Men of the 2nd Battalion were encouraged to extend their limited service, and a number of volunteers were received from it in June by the 1st Battalion.

In July the garrison was reviewed by Lieut.-General the Earl of Chatham.

It was ordered that soldiers' letters should only be charged 1d. instead of the usual high rate of postage, which was, from Colchester to Scotland, 3s. for 1 oz. weight.

At this time the 2nd Battalion 79th was affected by ophthalmia to an alarming extent, and steps were taken to prevent the disease spreading, the affected battalion being sent to encamp.

Lieut.-Colonel Napier keeps officers and men up to their work, and rebukes the former if the men's rooms and clothing are not in the best possible order.

2nd August.—Men are allowed to go harvesting, particularly married men.[1]

On August 26th, the Highland Brigade (1st and 2nd 79th and 1st 92nd) marched to Weely, and Sergeants Gunn, Alexander Cameron, and Donald MacDonald were sent recruiting to Chelmsford, the Ross-shire and other Militia Regiments being in that district.

Return giving the country of officers, n.c. officers, drummers, and privates of the 1st Battalion 92nd Regiment or Gordon Highlanders, September 2nd, 1806.

	Officers.	N.C. Officers, Drummers, and Privates.
English	1	63
Scots	28	938
Irish	4	33
Foreigners	1[2]	0
Total	34	1034

On October 3rd the Highland Brigade marched to Colchester, and were reviewed by H.R.H. the Duke of York, Commander-in-Chief, who expressed his "highest approbation of their appearance in the field this day," and desired his Order to that effect to be

[1] The price of bread had fallen to 7½d. the 4-lb. loaf in Essex and 6¾d. in Norfolk; good beef and mutton, 7¼d. per lb. In 1806 the pay of regimental officers and n.c. officers was increased, and corporals and privates of infantry were allowed additional pay at 1d. a day after seven years' service, and 2d. after fourteen years' service; two years' service in the East or West Indies counted as three years elsewhere. According to the Regulations of the 7th October 1806, infantry could engage for three distinct periods of seven years each, cavalry for three periods of ten, seven, and seven years, and artillery for three periods of twelve, five, and five years. No man could be drafted without his consent. (Fortescue.)

[2] William Charles Grant.

read to the men at evening roll-call. Next day they returned to Weely.

Horse Guards, October 27th, 1806.—Sergeants to have collars and cuffs of the regimental facings on their greatcoats.

B.O., October 21st.—Soldiers of the Highland Brigade to be allowed two months' furlough to go to Scotland, to the extent of twenty men per battalion.

In November the men were exercised in firing ball, fifty rounds a man. Corporals to have brushes and prickers for the arms.[1]

In February 1807, the levy money for infantry recruits was £18, 12s. 6d., of which the recruit received in necessaries and cash £11, 11s. The remainder went to the party and for incidental expenses.[2]

Ophthalmia now existed in the 92nd as well as in the 79th, and, with the view of checking this complaint, seven companies of the 1st 79th and the 1st 92nd marched on 3rd February to Harwich, the 92nd leaving all their ophthalmia cases behind at Weely, and every precaution being taken to prevent the disease spreading by infection, all suspicious cases being at once separated till seen by the doctor.

The men were desired to scour all their old clothing, and to wear their new kilts and waistcoats till it was dry, and also to wash their hair ; " they will not be expected to wear their pads at every parade, but to tie their hair neatly with a string."

R.O., 3rd March 1807, *Harwich Barracks.*—The commanding officer having observed several of the officers appearing with handkerchiefs tied with large knots in front, also the shirt appearing very much above the handkerchiefs, he particularly forbids the practice, except at evening parties, otherwise he will be under the necessity of ordering them to appear again in their leathern stocks. Officers are requested not to leave the barracks unless dressed in every respect regimentally.

On the 4th March the battalion was inspected by Brigadier-General Warde, who expressed his approbation.

The following is the general return of their country, height, age, and time of service :—

Country.	Sergeants.	Corporals.	Drummers.	Privates.
English	0	2	1	59
Scots	52	48	19	802
Irish	2	0	2	31
Foreigners	0	0	0	0
	54	50	22	892

[1] To clear the pan and vent.

[2] The levy money for cavalry recruits was £15, 4s. 6d., and for boys £10, 15s. Boys' pay had lately been increased from 8d. to 10d. a day.

Height.	Sergeants.	Corporals.	Drummers.	Privates.
6 feet 2 and upwards	0	0	0	8
6 feet and upwards	3	1	0	31
5 feet 11 ,,	4	2	0	26
5 feet 10 ,,	10	1	0	35
5 feet 9 ,,	7	9	2	41
5 feet 8 ,,	9	15	2	105
5 feet 7 ,,	7	5	5	128
5 feet 6 ,,	8	8	3	184
5 feet 5 ,,	4	9	5	163
Under 5 feet 5	2	0	5	170
Boy	0	0	0	1
	54	50	22	892
Age.				
55 years and upwards	0	0	0	1
50 ,,	1	0	0	7
45 ,,	1	2	0	21
40 ,,	4	3	0	53
35 ,,	9	1	0	81
30 ,,	23	14	1	164
25 ,,	15	22	1	214
20 ,,	1	8	10	284
18 ,,	0	0	5	55
Under 18	0	0	5	2
Not particularised	...	...	...	10
	54	50	22	892
Service.				
30 years and upwards	0	0	0	1
25 ,,	1	0	0	0
21 ,,	1	0	0	3
18 ,,	0	0	0	2
14 ,,	0	0	1	7
12 ,,	31	13	5	203
10 ,,	4	6	1	26
8 ,,	5	5	2	70
7 ,,	3	8	1	47
6 ,,	1	4	2	61
4 ,,	2	5	0	88
3 ,,	2	5	2	220
2 ,,	1	4	7	89
1 ,,	1	0	0	40
Under 1 year	0	0	0	12
Not particularised	2	...	1	23
	54	50	22	892

Total n.c. officers, drummers, and privates, 1018.

On March 5th the battalion marched to Colchester, where the officers "are on all occasions to wear the bonnet, with the exception of mounted officers."

They were inspected by Major-General Grosvenor on the 8th, returning to Harwich a few days later, under the command of Lieut.-Colonel Lamont, and on the 27th April the battalion returned to Weely, where new bonnets were issued to the men, who were ordered to take down the feathers and make them up on the new bonnets, and to wear the old bonnets, without feathers, as foraging caps. In May a draft of 32 rank and file, described as of good appearance, was received from the 2nd Battalion.

By letters from the Adjutant-General, H.R.H. the Commander-in-Chief directs "that the sergeants must be taken from the regiment generally, without reference to the battalion in which the vacancy may occur, and exclusively from the men who are enlisted without limitation. This is not, however, to have a retrospect to men engaged for a limited service who have already been appointed sergeants or corporals," etc. "H.R.H. is pleased to approve of your leaving it to the discretion of the officer commanding the 2nd Battalion to appoint a certain number of men of limited service to be corporals." "To Major-General the Hon. Sir John Hope, or officer commanding 1st Battalion 92nd Regiment."

The corporal punishments in the army and navy were at this time very severe, especially for desertion, absence, and making away with regimental necessaries. It seems almost incredible that it should have been necessary for the King to "express his opinion," as he did in December 1806, "that no corporal punishment should exceed 1000 lashes"! Corporal punishment appears, however, to have been comparatively rare in the 92nd.

CHAPTER IX

WHILE the Gordon Highlanders were thus quietly preparing themselves at home for the future service of their country abroad, Napoleon, who had defeated a Russian and Austrian army at Austerlitz, taken possession of Vienna, and conquered Prussia, had in the summer of 1807 made peace with Russia at Tilsit, and forced Sweden, the only ally England retained on the Continent, to renounce her alliance. He had made one brother King of Holland and another King of Naples, and now having command of the whole of Western Europe, determined on the entire exclusion of British commerce from the Continent, trusting to defeat her by the ruin of her manufactures. Notwithstanding the precautions taken by the Emperors of France and Russia to conceal their designs, the British Government had been made aware of their determination to seize the fleets of Denmark and Portugal, and thus destroy the maritime supremacy of Great Britain. To prevent this, an expedition was at once secretly prepared to demand the surrender of the Danish fleet, with a promise that if it was given up without resistance, it would be returned at the close of the war.

In this expedition the 1st Battalion of the Gordon Highlanders was selected to take part. A draft was received from the 2nd Battalion, consisting of 103 rank and file—good-looking young men.

Brigade Orders, *Weely Barracks*, *July 25th*, 1807.—The 1st Battalions 79th and 92nd Regiments are to march to-morrow morning at five o'clock for embarkation at Harwich. They will be formed in the barrack-field at Harwich for the inspection of Sir David Baird, after which they will immediately embark on board ships which have been named for each regiment, etc. Men to have sixty rounds ball cartridge and two flints each. Regiments to have a sufficient number of camp colours, each regiment a different colour, so that each boat may have one on disembarking before an enemy (92nd flags red, 79th red and green). Field officers allowed 2 cwt. of baggage, captains and staff 1½ cwt., subalterns 1 cwt.

The strength embarked was 3 field officers, 10 captains, 20 subalterns, 5 staff, 48 sergeants, 19 drummers, 981 rank and file.

R.O.—Lieut.-Colonel Napier desires that officers commanding different transports on which the 92nd are embarked will pay particular attention to the cleanliness of the ships ; they will keep both men, women, and children on deck during the day when weather permits ; bedding to be brought on deck in fine weather, and men to dine on deck. Ships to be fumigated three times a week.

Officers had an embarkation allowance in money, and a drawback allowed to the merchant on wine purchased by them.

Water was precious on board ship in those days. A sentry was put on it, no person to take any without leave from the officer of the watch, except cooks or officers' servants, who may take drinking water to the cabin. Soldiers to drink at the cask, or if they want to cook anything they must show it on a dish to the officer of the watch before they can get water. No basins of water to wash allowed.

Major-General Burrard was in command of the expedition till the arrival of Lord Cathcart. The troops were divided into :—

The 1st Division, Lieut.-General G. Ludlow, consisting of Brigade of Guards, the Hon. Major-General Finch ; the 28th and 79th Regiments, Brigadier-General Ward.

2nd Division, Lieut.-General Sir D. Baird, consisting of the 4th and 23rd Regiments, Major-General Grosvenor ; the 30th, 50th, and 82nd Regiments, Major-General Spencer.

The Reserve, under Major-General Sir Arthur Wellesley, and Colonel Stewart acting Brigadier-General, consisted of the 43rd, 52nd, 92nd, and 95th Regiments.[1] "The two battalions of the German Legion to be under the command of their own Major-General if they have one with them ; if not, under the senior officer till further orders." [2]

On August 12th, a curious Brigade Order explains that the troops may be ordered to land with or without knapsacks, and that if it is determined to land without them, the men "must wear their trousers over their breeches and leggings," and any other articles in their greatcoats or haversacks. This, of course, did not apply to the Highlanders.

R.O., August 12th.—The men will land in their new clothing, and officers in the Highland dress.

R.O., August 14th.—The plaids to be immediately made into wide-mouthed trousers, as light as possible ; should the quantity be too small for a large man, they can make the head bands of their old trousers. The trousers to be folded with the greatcoats when they have on their packs, and in the packs when they have on their greatcoats.

R.O., H.M.S. "Goliath," August 12th, 1807.—The ordinary formation of the Reserve under the command of the Right Hon. Sir Arthur Wellesley is as follows :—The 1st Battalion 95th on the right, the 1st 43rd, then the 2nd 52nd, then the 1st 92nd and 2nd 95th on the left. Circumstances may render necessary a deviation from this formation, of which due notice will be given.

G.O., H.M.S. "Prince of Wales," August 15th, 1807.—Notice is thus early given to the army that the Danish territory is to be considered as that of a Power with which war has not been declared, and that all officers in command will be held most strictly responsible

[1] This was the beginning of the regiment's service under Wellington, which continued with little intermission till the occupation of Paris, 1815.

[2] There was also cavalry and artillery.

that the persons and properties of individuals, subjects of His Danish Majesty, shall be protected as strictly as those of His Majesty's, or as those of any foreign Power in alliance with Great Britain where the seat of war might happen to be. Force and military operations are to be directed against those who oppose us in arms, and to the enforcement of such restrictions as it may be found necessary to impose upon the inhabitants. At the same time, it is to be understood that hostility and vigorous resistance may be expected on the part of the Danish troops. These, of course, are to be repelled in the most vigorous manner, and, as usual, whenever it is practicable, by the bayonet in preference to fire.

Any waggons, horses, forage, cattle, or provisions that may be pressed or taken are to be reported in the first instance to the officer commanding the brigade, in writing, who will be called upon to transmit the same.

The fleet, consisting of 25 sail of the line and convoying transports with 27,000 troops,[1] after passing Elsinore, anchored inside of Croningsberg[2] Castle ; and the Danish Government was summoned to surrender its fleet into the keeping of Britain, to be held as a deposit, and restored at a general peace. The offer being refused, the island on which Copenhagen stands was invested by the ships, and on Sunday morning, the 16th August, at five o'clock, a landing was effected by the troops near the village of Welbeck.

The Light Company of the 92nd was the first that touched Danish ground, but except a party of dragoons, who galloped off as soon as the Highlanders jumped from their boats, there was no enemy to be seen.[3] The regiment advanced for some distance through a wooded country, and halted in the open beyond till the three divisions were landed. Here Private James Duncan was accidentally shot.[4] He was a fine young soldier, and his untimely death caused great grief to his comrades.

When the troops were ordered to advance, the reserve moved on the right by Marum, having Baidle on the left, by Lyngby. Sir G. Ludlow's Division in the centre marched towards the Hermitage, Sir D. Baird's towards Charlottenburg, etc. "The right of the army will be supported upon the lakes of Lyngby, the left by the sea. Pickets of 2nd line to be placed in rear, the chain of posts fronting towards Croningsberg[2] Castle. The Reserve will also place pickets to cover the right of the army."

Reserve Order, *August 16th*, 1807.—Position of Frywgt. Two companies of the 43rd, 52nd, and 82nd Regiments are for picket this day.

[1] Mahan's "Influence of Sea Power upon the French Revolution and Empire."
[2] Cronborg.
[3] Sergeant Robertson.
[4] Regimental Return and Sergeant Robertson.

R.O., Husum, August 18th.—The 1st and 2nd companies will march this day at eleven o'clock, the 1st to Herlem to occupy it, the 2nd to Bredding for the same purpose. They will post pickets at night to prevent surprise. Immediately on arrival of the 1st company at Herlem, Captain Seton will detach a subaltern and twenty men to Gladsan, there to remain in the same manner.

Other villages were occupied by cavalry, artillery, and infantry; great care was taken that the inhabitants should be well treated, and that complaints should be immediately attended to. Batteries were constructed at the windmill to the left of the line, and working parties of 200 men from each of the 43rd, 52nd, and 92nd Regiments, with arms and accoutrements, were employed, being relieved by others in the middle of the day, and so on night and day. The troops were ordered to be formed till further orders two deep instead of three deep, as was usual. All supplies abandoned by the proprietors were taken possession of by the Commissariat, and payment made when the persons entitled to it could be discovered.

General Orders of August 23rd give detailed instructions for an attack which is to be made on the enemy the next day, and the 92nd is one of the regiments named to take part in it. On the morning of the 24th, cavalry, infantry, and artillery were under arms before daybreak, and advanced towards Copenhagen, halting near the suburbs ; but the plan of attack was not carried out, and the regiment returned with the reserve to its cantonments. In the course of the day information was received that some Danish troops were at a neighbouring village, and the 92nd were sent to dislodge them. The advanced guard, entering the street very quietly, found no troops, but only a sentry, who had fallen asleep at his post, and was frightened out of his wits when he woke to find himself in the hands of a Highland sergeant ! The detachment to which he belonged had gone, having neglected to relieve him. While in these cantonments, Colonel Napier, when taking a ride, fell in with a number of waggons laden with ammunition, escorted by a small body of soldiers. He told them that our troops were close at hand, on which they surrendered to him, and he marched them in as prisoners.[1]

Reserve Orders, August 25th.—Corps to march to-morrow morning, etc. Infantry, cavalry, and artillery to take up quarters at various places, "the 92nd at Kleitöfte."

R.O., August 26th, 1807.—The men will immediately cook two days' provisions, and one day's rum will be served out. Officers to make every possible inquiry for maps of the country at all times.

They marched on the morning of the 27th, and for once the conduct of the regiment gave pain to its commander.

[1] Sergeant Robertson.

R.O., August 27th, near Roskilde.—Lieut.-Colonel Napier has observed with extreme concern the bad conduct of great part of the regiment in regard to plundering the houses on the road. The regiment used to be a pattern to others on all marches, but they have now shown a very different one. A great deal has been done through mere mischief and wantonness, and he is sorry to be obliged to say the officers have not done their duty in preventing it. He desires that on a march, officers and n.c. officers may keep their sections perfectly formed, and though they may march easy and unconstrained, the regiment must be kept in that formation. It is strictly forbidden to any man to go into a house on the march, or when in cantonments into any house but his own. If the officers are attentive they can prevent this, and are to report any offender, who will be punished immediately when the regiment halts. The necessity of the case, and the wanton conduct of the men make it necessary for Lieut.-Colonel Napier to adopt, reluctantly, this line of conduct.

Poultry were evidently an attraction, and had been found with some of the soldiers. And another Order says, "The goose and fowles found yesterday are to be divided among some poor people, viz. the woman who has been defrauded by some men, and another poor woman in the house with Captain Campbell."

The misconduct was not confined to one regiment, for "Sir Arthur Wellesley regrets the necessity of resorting to measures of severity, but he is so fully convinced that soldiers guilty of marauding are unfit for any service, that he is determined to put a stop to their disgraceful practices." It is requested that requisitions upon the country may be confined solely to articles necessary for the public service, and not extended to those private conveniences which it is always intended should be paid for by individuals. Receipts to be given for all articles received on requisition, and every receipt signed by the commanding officer of the regiment.

Reserve Order, August 28th, Roskilde Knoe.—The troops will march to-morrow morning by the left in the same order as yesterday morning. The 92nd to furnish a company for the rear guard. Men's packs to be carried in waggons.

On the 29th, Sir Arthur Wellesley, having sent round Baron Linsingen's Brigade to fall on the enemy's left flank, himself advanced to attack the Danish army drawn up near the town of Kioge, with their artillery on a rising ground, and cavalry on the flank. They had entrenchments in their rear, and between them and the British ran a high road through the plain. The 92nd was on the road, the 43rd to its right, the 52nd on the left, their front covered by skirmishers of the 95th Rifles and the Light Company of the 92nd, and also by the fire of our artillery. The

Light Company of the 92nd, being on the left of the Rifles, came in contact with the Danish cavalry, but they galloped off on the first fire of our guns. The line advanced, the 92nd leading the attack; the Danes retired to their entrenchments, which the Highlanders charged and carried successfully; they pursued the enemy into the town of Kioge, the Danes retiring and firing on them, but fortunately shooting high. The 92nd returned their fire in the street with effect. The enemy had placed a gun at the end of the street, but just before the gunner applied his match to it, he was knocked over by one of the Gordons, and they immediately, as they did at Mandora, turned it on the enemy, whose retreat became a rout. The regiment took a general and many other officers, the general calling out, "Old Nelson, old Nelson," to conciliate his captors.[1]

The 92nd had two privates, William Dallas and James Grant, killed, and one private wounded.

The following is an extract from Major-General Sir Arthur Wellesley's dispatch to Lieut.-General Lord Cathcart, dated Kioge, August 29th, 1807:—

"I therefore thought it proper to make the attack in an echelon of battalions from the left, the whole covered by the 1st Battalion 95th Regiment, and by the fire of our artillery. It fell to the lot of the 92nd Regiment to lead this attack, and they performed their part in the most exemplary manner, and were equally well supported by the 52nd and 43rd. The enemy soon retired to an entrenchment which they had formed in front of a camp on the north side of Kioge, and they made a disposition of their cavalry upon the sands to charge the 92nd in flank, while they should attack the entrenchment. This disposition obliged me to move Colonel Redins' Hussars from the right to the left flank, and to throw the 43rd into a second line, and then the 92nd carried the entrenchment, and forced the enemy to retreat into the town in disorder. They were followed immediately in the most gallant style by Colonel Redins and his Hussars, and by the 1st Battalion 95th Regiment, and by the whole of the infantry of my corps. I cannot close this letter without expressing to your Lordship my sense of the good conduct of the troops; all conducted themselves with the utmost steadiness. But I cannot avoid to mention particularly the 92nd Regiment, under the command of Lieut.-Colonel Napier, the 1st Battalion 95th, under the command of Lieut.-Colonel Beckwith, the British Artillery under command of Captain Newhouse, the Hanoverian Hussars under Colonel Redins, and the Hanoverian Light Artillery under Captain Sympter, as corps that had particular opportunities for distinguishing themselves."

Reserve Order, Headquarters, Kioge, August 30th, 1807, *six o'clock a.m.*—Major-General Sir Arthur Wellesley has much pleasure in expressing his satisfaction with the conduct of the troops in the action of yesterday. All behaved in a manner deserving the appro-

[1] Sergeant Robertson.

bation of their superior officers. And the major-general has not failed to report their conduct in the most favourable manner to the Commander-in-Chief. But it is particularly incumbent on him to notice and report on individuals who had opportunities of distinguishing themselves. Among these are the 1st Battalion 95th Regiment under Lieut.-Colonel Beckwith, and the British Artillery under Captain Newhouse, in covering the advance of the infantry, and the former in the attack on the town of Kioge; the 92nd Regiment under Lieut.-Colonel Napier, in the attack of the enemy in their camp, and the regular and orderly manner in which they marched through Kioge, and formed beyond it; the Hussars under Colonel Redins throughout the day, and particularly in the charge at the town of Kioge; and the Light Artillery of the Legion under Captain Sympter, in covering the advance of the infantry, and supporting the charge of cavalry into the town.

The Order goes on to mention the Honourable Captain Blacquier and Brigade-Major Campbell, and to thank Major-General Linsingen and Brigadier-General Stewart, for the assistance he received from them in the formation and execution of a plan by which the enemy's force in the field has been defeated and dispersed, and is signed:—W. Cotton, A.A. General.

Extract from dispatch from Sir A. Wellesley to Lord Cathcart, 29th August, half-past 10 p.m.

After remarks as to the prisoners—" Our loss has been very small. We have lost no English officer, and but few men of the 92nd and 95th. Some officers of Hussars have been wounded. The rout of the enemy appears complete. We have prisoners coming in every moment."[1]

G.O., Headquarters, Hilinaup, 30th August 1807.—The Commander-in-Chief congratulates the army on the brilliant success which has attended Major-General Sir Arthur Wellesley and the detachment under his command in attacking and totally defeating a large body of the enemy collected at Kioge. The enemy's cannon, a general officer, and a great number of prisoners and military stores were taken, and the pursuit is still continued.

G.O., September 1st, Headquarters, Hilinaup.—The Commander-in-Chief of the Forces desires to express in the strongest manner his thanks to Major-General Sir Arthur Wellesley, and the officers and men under his command, for the judgment, valour, and discipline exhibited in the two attacks made upon the enemy in the general action of the 29th ult. at Kioge. The detail of this affair as reported by the major-general will be laid before His Majesty at the earliest possible opportunity.

[1] Wellington's "Dispatches," 1st Edition, Vol. IV.

In the action of Kioge the enemy, who numbered over 12,000 men, lost several hundred in killed and wounded, and 1200 prisoners, besides artillery, standards, waggons, and horses. It was the only general action connected with the fall of Copenhagen. On September 1st the place was summoned, and the same terms offered as before—viz. that the fleet must be surrendered to the keeping of Great Britain, to be returned at a general peace. This offer being again rejected, the bombardment began on the 2nd, both by sea and land; the inhabitants sustained it with heroic resolution for three days, when, the town having caught fire, and much of it being burnt down, a flag of truce was sent to the British outposts on the 5th to treat for a capitulation. But the time for equal negotiation was past, and the British would agree to nothing short of the unconditional surrender of the whole fleet, with the artillery and naval stores contained in the place, which terms were agreed to on September 7th.[1]

The magnificent prize consisted of 18 ships of the line, 15 frigates, 6 brigs, and 25 gunboats, besides 2 sail of the line and 3 frigates which had been destroyed; 3500 pieces of artillery were taken and quantities of stores. The prize-money due to the troops engaged was estimated at £960,000.[2] The Island of Heligoland was also captured, and became a depôt for English goods to be smuggled into the Continent.

The regiment was cantoned at Blessingberg and at Osted; the men received a daily ration of 1½ lb. of wheaten bread or 2 lb. of rye-bread; officers of companies made their own arrangements for supplying their men with meat, "the men to be charged as cheap a rate for it as possible."

Sir Arthur Wellesley requested the troops to be careful not to injure the harvest, and the soldiers seem to have made great friends with the farmers and others in whose houses they were cantoned. The Danes and Scots had many tastes in common, especially the love of liquor and music.[3] There was a good deal of drunkenness, which Colonel Napier attributes to the practice of the women bringing brandy and wine from Roskeld, "and Mrs Semple of the 1st Company having been found in the act, her provisions are to be stopped"; and in the case of another, whose conduct had been still worse, "it was the intention of Lieut.-Colonel Napier that she

[1] "The rage of Napoleon at this intelligence was terrific."—"The Student's Hume," London, 1865.

NOTE.—No honours were given to regiments for this campaign. The Berkshire (49th) and Rifle Brigade bear Copenhagen, having been on Nelson's fleet in 1801.

[2] Alison, chap. li.

[3] Sergeant Robertson mentions that the farmer in whose house he lived was giving them a tune on the violin the night on which Copenhagen was partly burned; on seeing the flames he threw down his instrument, bewailing the conflagration of the capital of his country.

should be drummed through the quarters of the regiment, but out of respect to the character of her husband, the lieut.-colonel will be satisfied with her disappearance for ever—and he gives her forty-eight hours to do so."

The Grenadiers seem to have given the commanding officer nearly as much trouble as the ladies. The parish clergyman's ducks having been found in their cantonments, the corporal and twelve men, in whose apartment the ducks were discovered, have to furnish a guard and prevent any Grenadier passing out till the culprit who stole the poultry is given up. And the officers of the company are not to leave their cantonments.

G.O., September 7th, 1807.—The Commander of the Forces congratulates the army on the capitulation of Copenhagen, which includes the capitulation of the Danish fleet. The Grenadier detachments of the army will march into the citadel at four o'clock this evening. A detachment will also embark at the same hour to occupy the dockyard, etc. All hostilities will cease.

G.O.—Officers are requested to discontinue the practice of shooting deer in the woods of His Danish Majesty. Danish officers are to be complimented by our guards and sentries as officers of allied armies.

On September 18th Sir Arthur Wellesley left for England, and his command was taken over by Brigadier-General Stewart.

The troops were drilled to a uniform field exercise and firelock exercise, "no deviation such as may have been ordered by general officers commanding particular districts." The 43rd, 52nd, and 92nd "are permitted to act as light infantry according to the rules and instructions approved and practised by light infantry."

All the outposts were now called in, and on the 23rd September the 2nd Battalion 52nd, and 1st Battalion 92nd marched to Roskilde Kroe, and next day joined the army before Copenhagen.

Reserve Order, Blessingberg, September 22nd.—Prize lists to be made up as soon as possible.

On the 28th September a dispatch from Viscount Castlereagh, His Majesty's principal Secretary of State for the War Department, was published to the army expressing His Majesty's approbation of the conduct of his troops, both British and Hanoverian, and of the harmony and zealous co-operation which had prevailed in all departments of the naval and military service. The Commander of the Forces thanked the general and other Staff officers and the regiments for their "patience, discipline, and exertions, to which, under Providence, he is indebted for the complete success of the whole expedition."

G.O., October 16th, 1807.—The baggage of the 92nd Regiment is to be embarked at the jetty head near the citadel this afternoon at three o'clock.

Distribution on board ships :—

Minotaur	336
Neptunus	600
Iris	200
	1136

The regiment embarked on the morning of the 17th, and sailed on the 21st. The crews of the English men-of-war were divided among the prizes, the soldiers assisting in working the ships. The *Neptunus*, a fine new Danish 84-gun ship, by a mistake of the pilot, ran on a sandbank and was wrecked. Some merchantmen, under convoy of the *Sybil* frigate, passing, came to their aid ; the stores were got out of the ship, and the men were landed on the island of Hewan, belonging to Sweden, where they remained fourteen days, when ships came from England to take them off. The weather was intensely cold, and being unprepared for it, the men suffered severely. However, they landed safely at Chatham, on the 24th November, and marched to join the rest of the regiment at Weely Barracks.

By Horse Guards Order of 8th December all drafts from 2nd Battalions are to be made under the inspection of a general officer, in order that a fair and equal draft may be made in such a manner as to keep both battalions efficient for foreign service.

R.O.—Officers commanding companies will immediately get their men to sign a power of attorney for Archibald Campbell, Esq., late paymaster of the regiment, to empower him to draw their prize-money. Men going on furlough to sign before they go.

At this time the battalion subscribed for a Garrison Lying-in Hospital for soldiers' wives.

R.O., *Weely Barracks*, *January 23rd*, 1808.—Corporal James Gray of the Light Company is appointed sergeant in the same company *vice* Dugald Cameron, promoted to a commission in the Royal East Middlesex Regiment of Militia.

In March the quartermaster is desired to serve out 3½ yards of kilt tartan for each man ; the men to have their kilts neatly made up, but are not to make their old kilts into pantaloons, but are to wear them till further orders. From this it appears they sometimes had tartan and sometimes grey cloth fatigue trousers.

The Highland Brigade, consisting of the 1st Battalions 79th and 92nd, marched to Colchester on the 22nd March, leaving the 2nd 79th at Weely.

Hair-dressing was still a serious business:—

" The hair two inches from the head queued within one inch of the end, of which one inch of hair is to be below the lace of the neck of the coat. The double knot of ribbon to be one inch in length, and the single ends to be two inches. It is also wished that the side locks be set back with a little pomatum, as well as that of the forehead, but in no case to be stiffened with soap."

As it was expected that the troops might be ordered on active service at the shortest notice, they were regularly exercised in marching short distances of from six to twelve miles. "Officers are desired to see that the men have good dinners, and that everyone has a knife, fork, and spoon." Up to this time dinner was the only meal of which official notice was taken, but a breakfast mess was now recommended to be introduced as a salutary regulation, to which officers commanding battalions are requested "to turn their most anxious attention," and the Order gives a recipe for a pottage made of either milk or small beer with oatmeal, suet, and molasses, which is highly recommended, to cost 7d. for six men.

The sentence of court-martial on Lieut.-General Whitelock, commander of the expedition for the reduction of the province of Buenos Ayres, was published, "for shamefully abandoning and delivering up to the enemy the strong fortress of Monte Video, which was committed to his charge," and which was "sufficiently garrisoned and provisioned"; for which he was sentenced "to be cashiered, and declared totally unfit and unworthy to serve His Majesty in any military capacity whatever." This was read at the head of every regiment, as were the sentences for various causes on several officers commanding battalions, and on other officers for drunkenness, but none of them belonged to the Gordon Highlanders.

The regiment had not much time for rest or enjoyment of their prize-money, for on the 19th April they were held in readiness for immediate embarkation. Ensign Hector Innes, in a letter to his family at Loanhead, Cullen, says :—

COLCHESTER, 24*th April* 1808.

We received our route yesterday to march to Harwich for immediate embarkation, our destination supposed to be Norway in the first instance. The town is a scene of dissipation, the soldiers having a great deal of money. I could live but indifferently in England on an ensign's pay ; living is extremely expensive, 14s. a week for dinner without beer ; no supper, tea, etc. England seems a delightful country in some ways, but I would rather live in Scotland.

R.O.—The men who have not fired ball, including a draft lately joined from the 2nd Battalion,[1] are to fire fifteen rounds.

They were to take with them on service sixty rounds and three flints a man.

Six women per company to embark with their children.

[1] In 1808, recruits might enlist for a limited or unlimited period at their option.

On 28th April 1808, the 1st Battalion marched to Harwich. "They will march in old clothing and new bonnets and kilts, and carry new coats and trousers in the packs."

The Gordons embarked the same day with a strength of 3 field-officers, 9 captains, 22 subalterns, 4 staff, 50 sergeants, 22 drummers, and 934 rank and file, and sailed on May 4th for Yarmouth, the place of rendezvous appointed for the troops destined to form the armament under Lieut.-General Sir John Moore, consisting of 10,000 men intended to assist Sweden in maintaining her independence, which was menaced by France and Russia. At Yarmouth the expedition was joined by Major-General the Hon. Sir John Hope (Colonel of the 92nd) as second in command.

On May 18th Mr Innes writes from Gottenburgh Roads:—

We arrived here yesterday in great health and spirits after eight days from Yarmouth; we are to go up the Baltic to take possession of the islands of Zealand and Boresholm; we are in fine spirits and impatient for a landing. We have about 15,000 men, and will have 50,000 Swedes co-operating with us. Subs. can live better abroad than in England, and we have an allowance of wine. On our voyage we drove a Danish Indiaman ashore and captured some small craft.

June 2nd.—Lieut.-Colonel Murray, A.G., was sent with dispatches to the King at Stockholm, and on his return was ordered immediately to England. The Commander of the Forces and Staff have lodgings in Gottenburgh, and have landed their horses, but few officers are allowed to land.

During their stay the troops were exercised in disembarking and embarking in boats, and the men had plenty of fresh fish to vary the salt pork and biscuit.

Negotiations were carried on between Sir John Moore and the King of Sweden, for the purpose of concerting operations; but as it appeared that the views of the Swedish monarch as to the disposal of the forces differed greatly from the intentions of the British Government, the expedition was recalled, to be employed in the operations in the Spanish Peninsula.

The fleet accordingly sailed on July 3rd, with orders to rendezvous at Yarmouth, but were met by a dispatch vessel with orders which changed their destination to the Downs; where they arrived after a rough passage on July 20th, and proceeded to Spithead, where a draft of seventy rank and file of good appearance joined from the 2nd Battalion. Here they took in provisions and water for six weeks; neither officers nor soldiers were allowed to land. They had been nearly three months at sea, generally on salt provisions, and there were several cases of scurvy among the men, these being sent ashore to Gosport Hospital.

At this period political interest was centred in Spain and Portugal. In October 1807, France and Spain had agreed to divide Portugal between them ; the Portuguese Royal Family had fled for refuge to Brazil, and a French army under Junot had entered Lisbon. No sooner was Napoleon in possession of that country, than he induced the weak Charles IV, King of Spain, to meet him at Bayonne, where he extorted from him, and from his son Ferdinand, a renunciation of the Spanish throne in his own favour.

It was declared that the Spanish Bourbons had ceased to reign. Joseph Bonaparte, who had been crowned King of Naples, was removed to Spain, whilst Napoleon's brother-in-law, Murat, took his place at Naples.[1] A French army at once invaded Spain, and Joseph Bonaparte entered Madrid on July 20th, 1808. The Spaniards immediately rose against the French usurper, established a " supreme Junta of Spain and the Indies " at Seville, and proclaimed Ferdinand VII as King. The news of the rising was received throughout Britain with joy ; Whig and Tory agreed in the determination to support the Spaniards and Portuguese, and to strike a bold stroke to rescue the world from the power of Napoleon. Supplies in profusion were sent to the Spanish Loyalists, and a small army of about 10,000 men, under Sir Arthur Wellesley, was sent to the Peninsula. The first efforts of the patriots were successful ; in July a French force which had invaded Andalusia surrendered to the Spaniards at Baylen.[2] Joseph Bonaparte was driven out of his new capital, and had retired to Vittoria, when the British army, reinforced by General Spencer from Cadiz, and numbering about 14,000 men, landed near Figueras, in Mondego Bay, on the 1st of August. On the 15th, the first British blood which flowed in the Peninsular War was drawn in a skirmish at Obidos. Wellesley defeated Laborde in the combat of Roliça on the 17th, and in the battle of Vimiera[3] on the 21st had forced Junot to retreat, and was about to follow up his success, when Sir Harry Burrard, who had left the fleet on which Sir John Moore's troops were embarked, and had arrived that morning, but had generously declined to take the command from Sir Arthur during the battle, gave orders for the army to halt and remain in position at Vimiera, till the reinforcements under Sir John Moore should arrive. On August 22nd Sir Hew Dalrymple arrived from Gibraltar and assumed the command of the army, so that in thirty hours a battle had been fought, and three generals had commanded the forces.

The 1st Battalion 92nd, along with the troops under Sir John

[1] Napoleon had also made his brother Louis King of Holland, and a third brother, Jerome, became King of Westphalia.

[2] The bravery and steadiness of the veteran Swiss and Walloon Guards in the Spanish service conduced greatly to the defeat of the French.

[3] It was at Vimiera that Clarke, a piper in the 71st, severely wounded in the leg, and unable to walk, continued to play, saying, " Deil hae me an' the lads want music." A monument was afterwards erected by his regiment over Clarke's grave at Fort-George.

Moore, after a calm and slow passage, arrived at Mondego Bay on the 21st of August, and were to land there on Sunday the 22nd, when an express from Sir Arthur Wellesley arrived, and the transports proceeded to Maceira[1] Bay, where the 1st Battalion landed on the 27th, with great difficulty, owing to the swell on the rocky coast ; several boats were upset, but none of the regiment were drowned. They at once marched inland, and pitched their tents close to the battlefield of Vimiera, which exhibited a horrible spectacle, strewn as it was with dead bodies, which the inhabitants were burning, according to the custom of the country.[2] They were told that very large sums of money were found in the knapsacks of the French soldiers, who had pillaged the country in all directions.

Next day they marched twelve miles to Ramalhal, "half-roasted" by the sun, and in the night were "half-drowned by torrents of rain"; their route lay through a country of bleak hills and fertile valleys, where the superabundance of fruit sent some to the sick list. They were at a small village near Torres Vedras on 30th August, on which day the famous Convention of Cintra was signed, by which it was arranged that the French should evacuate the whole kingdom of Portugal, and be conveyed to France with their field artillery, and sixty rounds of ammunition for each gun, all other artillery and ammunition to be delivered up to the British; they were to carry with them cavalry horses and private property; their sick and wounded to be taken care of till fit to be sent to France, and the fortresses of Elvas, Almeida, Peniche, and Palmela to be delivered up as soon as British detachments could take possession of them. Also the Russian fleet in Lisbon Harbour, which had been blockaded by Admiral Sir Charles Cotton, to be conveyed to Britain with all its stores, and to remain in deposit till six months after a general peace, but their officers and crews to be returned at once to Russia ; also the Spanish troops in custody of the French armies to be liberated.[3]

This convention was received with indignation both in the Peninsula and in the British Islands. The Portuguese, who had been in no hurry to confront the invader in the field, were loud in their complaints of an arrangement by which they were freed from his presence. The Spaniards complained that the liberated Frenchmen would reappear on their frontier, and it was said, with some truth, that the clause allowing private property to be removed would enable the French to carry off the spoils which formed their ill-gotten gains ; while in Britain the people had been raised to the highest pitch of enthusiasm by the decisive victories of Roliça and Vimiera, and expected nothing less than to see Marshal Junot and

[1] Or "Maciera." [2] Ensign Hector Innes. [3] Alison.

20,000 French soldiers arrive as prisoners at Spithead. So general was the condemnation that the Government consented to a Court of Inquiry. Sir Hew Dalrymple, Sir Harry Burrard, and Sir Arthur Wellesley were ordered home and appeared before the Court, which, after full investigation, arrived at the conclusion that, considering the extraordinary manner in which three general officers had been successively invested with the direction of the army, it was not surprising that the victory of Vimiera had not been more vigorously followed up ; and that unquestionable zeal and firmness had been exhibited by all three generals. But, notwithstanding this acquittal, neither Sir Hew Dalrymple nor Sir Harry Burrard were again employed in any important command, and it required all the family influence, and all the early celebrity of the hero of Assaye and Vimiera,[1] to save the future conqueror of Napoleon from being cut short on the threshold of his career, for no fault whatever of his own, by the very people upon whom he had conferred an inestimable benefit.[2]

The 1st Battalion 92nd marched from Torres Vedras to Lisbon, where they remained till the French troops were embarked, when they were moved ten miles to Campo Sancto, and encamped there, being brigaded with the 36th and 71st Regiments in the Division of Sir John Hope (colonel of the regiment).

[1] Alison's "History." Wellesley had been thanked by both Houses of Parliament for the victory of Vimiera.

[2] Napoleon was no better pleased with the Convention of Cintra. "I was about," said he, "to send Junot to a Court-martial, but happily the English got the start of me by sending their generals to one, and thus saved me from the pain of punishing an old friend."—*Alison*.

Creek of Maceira Bay, where 92nd landed.
(*From Bradford's Sketches during the Expedition.*)

MAP OF
PORTUGAL, SPAIN
and the
SOUTH of FRANCE
Scale of Miles
50
0
50
100
The places marked red indicate the various movements of the Gordon Highlanders during the Peninsular War. The lines of Torres Vedras are roughly indicated in red.
BAY OF BISCAY
BALEARIC ISLANDS
Gulf of Lions
Strait of Gibraltar
Long. West 5 of Greenwich
PORTUGAL
SPAIN
MADRID
LISBON
Gibraltar (Br.)
C.Ortegal
C.Finisterre
Corunna
16. Jan. 1809
C.Trafalgar
21 Oct. 1805
S.Vincent
14. Feb. 1797
Mountains of Asturias
Bordeaux
Toulouse
Perpignan
Barcelona
Valencia
Alicante
Cadiz
Seville
Malaga
Zaragossa
1808 1809
Salamanca
22 July 1812
Ciudad Rodrigo
Badajoz
Santander
Bilbao
S.Sebastian
Bayonne
C.Gata
C.Palos
C.S.Martin
C.Creux
C.S.Sebastian
MINORCA
Mahon
Palma
Cabrera
IVIZA
Formentera
Alboran I.(Sp.)
W. & A.K. Johnston Limited Edinburgh

CHAPTER X

On September 7th a dispatch from England arrived at Lisbon, appointing Lieut.-General Sir John Moore[1] to command the whole army which was to co-operate with the Spanish troops in the north of Spain, consisting of 30,000 infantry and 5000 cavalry, with artillery; part of them coming direct from England, and the remainder to be regiments drafted from the army then in Portugal. The difficulties of the British general were great; the roads through Portugal were very bad; the rainy season was setting in; transport was hardly to be obtained, because the Government had sent little money to pay for it; time pressed, for a great French army was crossing the Pyrenees.

Moore urged on the preparations, so that the headquarters with the main body quitted Lisbon on October 26th, and followed the road by Abrantes, Almeida, and Ciudad Rodrigo, while Sir John Hope's Division marched by Talavera and Madrid. Sir David Baird's forces had landed at Corunna on October 13th, and were to co-operate with Moore in the north.

Meanwhile the 92nd had marched along the bank of the Tagus twenty and twenty-five miles every day, through a most agreeable country, but under a very hot sun, and, crossing the river at the town of Abrantes, arrived at Portalegre on the 11th October.

Mr Innes writes from Portalegre on October 24th:—

> We passed through some of the principal towns in Portugal. They seem at present pretty miserable; war is the ruin of a country. We had a great opportunity of seeing the country and the customs of the inhabitants; they put me in mind of the people of Scotland (as I have heard) twenty years ago. We have received marked attention since we arrived in this town. It is a considerable place, almost on the confines of Spain, and in time of peace it has considerable commerce and manufactures. We are billeted on the natives, who are very civil, and we live like the sons of kings, but I am sorry to say not very economically; however, time passes agreeably. We need something to recruit our spirits; we were very low when we landed at the idea of not being engaged, but the honour now awaits us. The troops in Lisbon have been unhealthy, but "ours" has as yet been in high health and spirits. We have

[1] Sir John Moore was born at Glasgow, November 13th, 1761, and was educated at the public school and University of that city. He then travelled abroad and gained a knowledge of Continental languages, and acquired that suavity and elegance of manner for which he was remarkable; in 1776 he joined the 51st Regiment at Minorca as ensign, and afterwards distinguished himself in the West Indies under Sir Ralph Abercromby, who characterised his conduct as "the admiration of the whole army." He served in the Rebellion in Ireland, in Holland 1799, in Egypt 1801, when he was twice wounded, but declined to leave the field. He had been selected to instruct Light Infantry Regiments in an improved system of drill; he had commanded the army in Sicily and the expedition to the Baltic, from which he had just returned. He was brave, chivalrous, and high-spirited, and commanded the respect of the soldiers, to the increase of whose comforts he devoted a large part of his attention. —*Alison and others.*

not lost a man since we landed, and we are upwards of 1000 strong. I rise at seven, breakfast at nine, read or walk about the town, dine at mess at three, parade at five, and since we came we have for the most part passed our evenings at the Convent of St Bernardo, one of the richest in Portugal. Our music goes down, and we spend the night in dancing; in the convent there are several ladies for their education; I believe they are the finest women I ever saw. There seems to be much trade with Ireland, and we have eaten Irish butter mostly since we landed.

On the 26th October the battalion moved from its very pleasant quarters, most of the women of the regiment being sent back to Lisbon. The route was by Elvas to Badajos, where they entered Spain and passed the night in barracks, being received by the people with an enthusiastic welcome.

Captain Seaton of the 92nd writes from Elvas on October 31st:—

We are now under good officers, Moore and Hope, and having good heads of departments, there is no fear of the rest; the soldiers would fight the devil. Our regiment march from Lucon on the 3rd, and all the army is now in motion, moving by parallel columns towards Valladolid; ours is the right division under General Hope. We shall make a detour of sixty miles by being on the right. We go on the great road leading to Madrid, and turn off forty miles from that capital, leaving it on our right. We have underwent a great many hardships and privations already, and many more, of course, are before us, but we have the prospect of *glory*, which to a soldier is great. John Bull is a queer genius and not easy pleased; a bloody gazette he likes, and I hope our exertions in Spain will be more satisfactory to him than in Portugal. We have a fine army, well appointed and well disciplined, able and willing to perform exploits hitherto deemed impossible.

The next halt was at Truxillo, where an occurrence took place which gave the Gordons a good deal of annoyance. Being the only regiment there that wore the Highland garb, the people, struck by its novelty, wished to know to what country its wearers belonged. The 71st had lately been at Buenos Ayres in South America, where they had learned a little Spanish, and by way of a joke some of them said that the 92nd were bad characters, and that their dress was the mark of their disgrace. The Spaniards, indignant at such a set of "Ladrones" being sent to their country, refused to have any dealings with them, the Mayor of the town declining even to furnish rations. When this silly and mischievous report became known to the men through an interpreter, a deputation from each company, with the quartermaster at their head, went to the commanding officer and laid the case before him, telling him that they had been robbed of their honour, and that unless legal satisfaction was given them they would take the law into their own hands. The commanding officer went to General Hope, who was at Truxillo, as was also the colonel of the 71st. An inquiry was made, and the perpetrators of this foolish joke, which had led to more serious results than they probably intended, were punished; and so ended an affair which might have resulted in a dangerous feud between two distinguished corps, who afterwards became fast friends, and stood by each other in all the combats of the Peninsular War.[1] When the inhabitants understood the truth, they did all in their power to show their regret at the treatment the Highlanders had received, making them free of their vineyards and wine-cellars.

At Navalcarnero the men were presented by the Military Junta with red cockades to show they had embraced the Spanish cause.[2] Next day, marching through a fruitful district, they arrived at the Palace of the Escurial, which the men were allowed to visit, with its magnificent pictures and rich decorations.

While here the quartermaster was sent to Madrid, some twenty miles distant, to get a supply of shoes and necessaries, of which they stood in need; but owing to information received there (probably of the approach of the French), he had to return without accomplishing his purpose, and the men had to continue their journey in shoes worn to the welts.

Captain Seaton writes from Escurial:—

November 25th.

We entered Spain at Badajoz on the 2nd inst., and were received by the inhabitants with shouts of "Viva Gran Britannia," "Viva los Ingleses," etc. I never saw people so completely enthusiastic as they are; even the women and young children join in the general cry. We met with a great deal of hospitality from all descriptions of people; the poor were willing to give what little they had, and the rich were profuse, and heaped abundance of everything upon us. We have had long and very fagging marches, yet, strange to tell, our regiment has not left a man behind from sickness or any other cause since we left Elvas. This division will form a junction with Sir John Moore, who is at Salamanca with 15,000 men. It is six days' march from here. We have to go over an immense mountain covered with snow. There is no doubt but Bonaparte is in Spain. *So much the better*, we all say. We expect a general engagement a few days after our junction. It will be a dreadful conflict; we are all anxious for it and in the highest possible spirits.

On the 20th November Sir David Baird had with great difficulty arrived at Astorga, four days' march from Salamanca, so that the

[1] Sergeant Robertson and "Military Memoir" by a 92nd officer. The 71st had lately, on being made Light Infantry, been ordered to give up the Highland dress, with the exception of the pipers.

[2] Possibly this may be the origin of the red cockade worn by the Light Company till flank companies were done away with.

British army, only 30,000 men in all,[1] was split into three divisions —at the Escurial near Madrid, at Salamanca, and Astorga, about eighty or ninety miles distant from each other. The weather at this time was very severe and the marches long. A more robust set of men never took the field ; their discipline was admirable, and there were very few stragglers.[2]

Napoleon with 180,000 veteran troops was near Vittoria. The French had, in the meantime, defeated and dispersed the Spanish armies at the battles of Espinosa, Burgos, and Tudela, and were rapidly concentrating on Madrid. On the 2nd December Napoleon appeared before that city, and on the 4th it capitulated and was in the possession of the French troops.

While these disasters were accumulating in Spain, the three divisions of the British army had at length joined forces, but Sir John Moore was perplexed by the imperfect information he received, and by the utter incapacity of the Spanish troops to contend with the common foe. The intelligence of the defeat of all the armies of Spain appeared so alarming that orders were given to retreat. The officers and soldiers were filled with dismay, and openly declared it would be better to lose half the army than to retire without striking a blow. The gallant Moore himself recoiled from the mournful decision, and these feelings, both of the general and his troops, were wrought to the highest pitch when they heard of the advance of the French on Madrid, and of the enthusiastic preparations made for the defence of the capital by its inhabitants. These and other circumstances led him to reconsider the position; and after anxious reflection he decided to give vent to the native courage of his character by suspending the order to retire, and notwithstanding the disheartening intelligence of the fall of Retiro, and the perilous situation of Madrid, he resolved to attack the corps which kept open the communication between France and that city;

[1] Return of Sir J. Moore's Army, December 19th, 1808. Adjutant-General's morning state of that day :—

	Fit for Duty.	Hospital.	Detached.	Total.
Cavalry	2,278	182	794	3,254
Artillery	1,358	97	...	1,455
Infantry	22,222	3,756	893	26,871
	25,858	4,035	1,687	31,580
Deduct	2,275	Men composing four battalions, viz. :—	3rd Regt., left in Portugal. 76th „ 51st „ 59th „ Between Villafranca and Lugo.	
	23,583	Total number under arms.		

Note.—Of 66 guns, 42 were attached to the divisions, the remainder in reserve.—Napier, Vol. I. (Appendix).

[2] Napier's "History of the Peninsular War."

an operation which he foresaw would force Bonaparte to turn back in order to oppose the British, and would thus be a powerful diversion in favour of Madrid in case it held out, and of the southern provinces to enable them to rise and organise their forces.

He commenced his advance with 23,000 men against Soult, who commanded a division of the 200,000 men under Napoleon's orders in Spain. In case of the necessity of retreat, magazines were formed on the road to Lisbon, and at Benevente, Astorga, and Lugo.

On December 13th, the cavalry of the advanced posts surprised a body of French cavalry and infantry at Rueda, killing or taking almost the whole number. On the 14th, the news of the fall of Madrid reached the army by means of an intercepted dispatch from Napoleon.

At Toro, on the 16th, information was received that the Spanish General Romana, who had been invited to co-operate with the British, instead of doing so, was in full retreat towards the mountains of Galicia, his troops from hunger and fatigue having dwindled away to 8000 disheartened fugitives. Nevertheless, the British forces continued to advance. On the 21st December, the united divisions were established at Sahagun, at which town Lord Paget, with 400 Hussars of the 15th, had totally defeated 700 French Dragoons, killing from 15 to 20, and taking 2 lieut.-colonels, 11 other officers and 154 men prisoners, in twenty minutes.[1]

Sir John Moore's plan was to move during the night of the 23rd, so as to arrive at the Carrion by daylight on the 24th, force the bridge, and fall upon the main body of Soult's troops. But on the evening of the 23rd, trustworthy intelligence was received that the whole French armies from various parts of Spain were in movement to crush the British. Among the converging forces 60,000 men and 150 guns, having fifteen days' provisions in carts, were reviewed on the 19th by the Emperor at Madrid, and on the evening of the 22nd, 50,000 of them were at the foot of the Guadarama Mountains. Deep snow choked the pass, and after twelve hours of ineffectual toil their general reported the road impracticable; but Napoleon, rebuking him fiercely, personally urged on the columns, and the passage was effected amidst storms of hail and sleet, the cold and fatigue so intense that many soldiers and draught

[1] Dr Neales, Physician to the Forces, in his "Letters from Portugal and Spain," says that the brass helmets of the French Dragoons withstood sabre cuts, and they were seldom wounded on the head, while our men in fur caps had frightful head wounds; but our horses fairly upset those of the French, which were slighter, and many of the prisoners' horses had sore backs. Many of the prisoners' feet were frost-bitten, but those whose wounds were slight danced and amused themselves grinning through the windows at the Spaniards, who would have killed them but for the British guard. They were attended to as well as our own men. "The Town of Sahagun" is now the regimental song of the 15th Hussars, and is sung on the anniversary in memory of this exploit. Fortescue, in his "History of the British Army," gives the numbers as 520 of the 15th Hussars and 450 French Dragoons.

animals died during the two days the operation lasted. On the 26th Napoleon was at Tordesillas. "If the British pass to-day in their position they are lost," he wrote ; but Sir John Moore had become well aware that his position was untenable ; without the aid of the Spanish armies, he would soon have been surrounded by the overwhelming legions of France. He had, however, gained the political object of saving Andalusia by drawing the French power on himself.

Notwithstanding his rapid march, having scarcely rested night or day, Napoleon was twelve hours too late ; the British were across the Esla! Their heavy baggage and stores were removed to the rear, but the reserve, the Light Brigade and cavalry, remained at Sahagun, pushing their patrols up to the enemy's lines, and skirmishing to hide the retrograde march. On the 24th General Hope with two divisions fell back by Mayorga, and General Baird by Valencia de San Juan, where was a ferry-boat over the Esla, which he crossed on the 26th, and took post on the other side. Moore with the reserve and light brigades followed Hope's column to Valderas.

Sergeant Robertson's journal gives some account of the particular movements of the 92nd since we left them at the Escurial, from which quarter they resumed their route for Valladolid, crossed the Guadarama Pass, where, on the top of one of the mountains which divide the provinces of Old and New Castile, the figure of a lion is placed holding a ball in its paw. They descended into Old Castile, but diverged from the road to Valladolid, as the French were advancing by that way towards Madrid. So little information had the British, that the French advanced guard was in sight before they were aware of its approach. In the evening they reached a village, the Alcalde (or Mayor) of which came with the inhabitants to light them in with torches, and they were put into a monastery for the night. The Fathers, with the kindest intentions, lighted the charcoal stoves to warm them, and some of the soldiers, not understanding their management, were nearly suffocated, but recovered on being carried to the open air. The battalion continued its march to Alba de Tormes, where they halted ; while here they heard the report that they were surrounded by the French, and were to commence a retreat into Portugal. This report was contradicted, but as the men had no map, they did not know whether they were advancing or retiring. Continuing their movement toward Burgos, they met at Toro a party of dragoons belonging to General Baird's Division lately arrived from Corunna, who had just had a skirmish with the enemy, and had taken some waggons of stores which they were escorting.

On December 24th some of the sergeants had an exciting

adventure at Villada, where they had been sent on in advance of the battalion to draw billets. On entering the little town they found it occupied by a patrol of French cavalry. Sergeants did not carry firearms ; there were only six or seven of them armed with their pikes and claymores. Like true soldiers they never thought of retiring to wait till the troops came up, but adding Scottish caution to British bravery, they went warily. It was market-day, and the town full of people, who did not know to which side the Highlanders belonged, so all was quiet on their part. Finding that the Frenchmen, unaware that any Britishers were within miles, were regaling themselves in a wine-house—their horses being tied to the rings which were, and in parts of the Continent still are, commonly placed in the wall for that purpose—the plucky sergeants, quietly keeping the side of the causeway, reached the tavern ; and it was only when they had the bridles in their hands that the astonished dragoons, hearing some noise, looked out to find their steeds in the possession of men in feathered bonnets and kilts !

The Frenchmen had their carbines with them, but the sergeants "took the first word o' flytin'," rushed in and disarmed them ; and though a few hurried shots and many oaths were discharged at them, none of them were hurt, and they took the patrol—a n.c. officer and five privates—prisoners with their horses. When the Spaniards saw what they had done, they seemed frantic with joy, and gave the sergeants anything they liked, bringing them warm gloves, as it was very cold.

When the troops came up to Villada on the 24th, they were much fatigued on account of the state of the roads, which were deeply covered with snow. The men were ordered to clean themselves, as they were expected to halt for some time ; but all of a sudden, the same evening, they were ordered to fall in in marching order. Columns were formed outside the town, ready for the expected engagement. It was a beautiful moonlight night, but so cold that they could only keep warm by walking about till morning, but "every heart beat high with the thought that we were to measure arms with the great Napoleon. The notion entertained by the British army was, that the great victories gained by him had been over raw and undisciplined troops. Every man felt confident of his own prowess when compared with a French soldier's, and nothing was more earnestly wished than an opportunity of engaging, and an order for battle."

Such being the spirit of the British rank and file, one can imagine their surprise and mortification when a Staff officer arrived with an order to prepare to return to England. All the horrors of a retreat of several hundred miles over bad roads in winter weather appeared to their minds. "All ranks called out to stop and fight and not to run away (as we termed it), which would be a disgrace to the British army."[1] The troops were in the highest state of vigour and spirits, and an unbroken series of brilliant successes at the outposts had given rise to an unbounded confidence in their own prowess, likely to have produced glorious results, if not met by overwhelming odds.[2] But they were not aware, as their general was, that such absolutely overwhelming odds were against them.

By this timely retreat Sir John Moore reached Benevente, and the hazardous operation of crossing the swollen torrent of the Esla, over planks laid across the broken arches of the bridge of Castro, was successfully performed in the dark by General Craufurd with the rear guard, not, however, without fighting, in which the Chasseurs of the Imperial Guard rode close up to the bridge, and captured some women and baggage. Instances occurred of great bravery and devotion on the part of soldiers of the rear guard. Napier tells of two. John Walton and Richard Jackson were posted in a hollow road beyond the bridge, at a distance from their picket. If the enemy approached, one was to fire and run back to give notice if they were few or many ; the other was to maintain his ground. A party of cavalry following a hay-cart stole close up to these men and suddenly galloped in with a view to kill them and surprise the post ; Jackson fired, but was overtaken and received twelve or fourteen wounds in an instant ; he came staggering on notwithstanding his mangled state and gave the signal. Walton, with equal resolution and better fortune, defended himself with his bayonet, and wounded several of his assailants, who retreated leaving him unhurt ; but his cap and knapsack, his belts and his musket, were cut in about twenty places, his bayonet was bent double, his musket notched like a saw from the muzzle to the lock. Jackson escaped death during the retreat, and finally recovered of his wounds.

Again, at Benevente several thousand infantry slept in the long galleries of an immense convent ; the lower corridors were filled with the horses of the cavalry and artillery, so thickly packed that it was scarcely possible for a single man to pass between them, and there was but one entrance. Two officers returning from the bridge, being desirous to find shelter for their men, entered the convent, and perceived with horror that a large window shutter being on fire and the flames spreading to the rafters above, in a few minutes the straw under the horses would ignite, and 6000 men and animals would inevitably perish in the flames. One of the officers (Captain Lloyd of the 43rd), a man of great activity and presence of mind, made a sign to his companion to keep silent, and springing on to the nearest horse, ran along the backs of the

[1] Sergeant Robertson. [2] Alison.

others till he came to the flaming shutter, which he tore off its hinges and threw out of the window ; then, returning quietly, awakened some of the soldiers and cleared the passage without creating any alarm, which, in such a case, would have been as destructive as the flames.[1]

The army remained two days at Benevente, and as few of the stores collected there could be removed, after supplying the immediate wants of the troops the rest was destroyed.

Meanwhile the 92nd, with Hope's Division, had commenced their retreat on Christmas Day, and marched by Mayorga, where, arriving after dark, they found the doors and windows barricaded by the inhabitants, who would not let the men in or sell them food. Irritated by such inhospitality on a winter night, they broke open the doors, and so frightened the people that the news, spreading from village to village, had a bad effect on the feeling of the country people towards them. Spanish enthusiasm, in the north at any rate, had evaporated, says Dr Neales ; the people of Mayorga would render no assistance, and were too ignorant to know why we were in the country.

At Valderas they were put in a convent, and no one was allowed to go out. At Benevente they lay where they could on the stairs and passages of the great convent described above. Here they got a supply of shoes and blankets, and hardly had they marched, when the rear and baggage guards were skirmishing with the enemy's cavalry ; " but," says Sergeant Robertson, " when they came on, our dragoons showed that though retreating, it was not from fear."

The discipline of even the best regiments had become relaxed, and disorders equally fatal to the army and to the inhabitants had commenced. At a miserable little village where they were quartered after leaving Benevente, some of the Grenadier Company of the 92nd, either from carelessness or wanton mischief, set fire to a house, and the whole village was burnt down ; but as the French were close upon them, there was no time to investigate and bring the perpetrators to punishment. On the battalion overtaking the troops in front of them, they found them engaged in destroying a quantity of stores ; among the rest was a cask of rum. A young man named Bruce got drunk, fell into the cask and was dead before he could be extricated. Their next quarters were at Bembibre. Here in many of the houses there were Spanish soldiers dying or dead of a fever, which two of the regiment caught and died ; but the great wine-vaults of this place proved more fatal to the army than either sword or sickness ; drunkenness appeared in its most frightful colours, and when the rear guard arrived with unbroken ranks, they had to force their way through a crowd of

[1] Napier, and " Life of a Sergeant."

British and Spanish soldiers, stragglers from many regiments, who reeled out of the houses, or lay on the roadside, an easy prey to the enemy's cavalry, who thundered in close pursuit.[1] The inclemency of the weather and rapidity of the retreat had diminished the strength of the soldiers, while their spirit had been depressed by the thought of retiring before the enemy.

A 92nd pensioner used to relate that at one halt a man, overcome by wine and fatigue, went to sleep close to the fire and was fearfully burnt before his comrades noticed that he had rolled into it. However, he was put into a sick-cart and finally got safe home to Badenoch.

On the 1st January 1809, Napoleon took possession of Astorga. On that day, 70,000 French infantry, 10,000 cavalry, and 200 pieces of artillery were there united. The congregation of this mighty force, while it showed the power and energy of the French monarch, attested also the genius of the British general, who with a handful of men found means to arrest the course of the conqueror, and to draw him with the flower of his army to this remote part of the Peninsula, at the moment when Portugal and Spain were prostrate beneath the strength of his hand. "That Spain being in her extremity, Sir John Moore had succoured her, and in the hour of weakness intercepted the blow which was descending upon her, no man of candour can deny." [2]

In ten days Napoleon had brought this great army from Madrid to Astorga, 200 miles over snowy mountains, and his action was the greatest compliment that he could pay to the prowess of the British troops, and demonstrated the importance of the stroke delivered by their commander. While Napoleon was riding with the advanced posts in the pursuit, he was overtaken by a courier. The dispatch brought him intelligence of the accession of Austria to the European confederacy. He at once handed over the pursuit of the British to Soult, and returned to France.

The Gordons spent New Year's Day 1809 in anything but a cheerful manner ; marching towards Villafranca, their clothes worn to shreds, their shoes and hose worn out, officers and soldiers overrun with vermin, bearing alike the extremities of cold and hunger. At Villafranca they lay in the stables of an inn, and the straw and soil seemed to the tired soldiers so comfortable and warm, that it required all the authority of their officers to get the men to move. Sir John Moore was constantly with the rear guard, doing his best to arrest disorder and protect the retiring columns. At Calcabellos, some miles east of Villafranca, a combat took place, in which the pursuers were repulsed with the loss of their general, Colbert, and several hundred men. Whenever the British soldiers found them-

[1] Alison. [2] Napier.

selves before the foe they pulled themselves together, and the rear guard, being constantly in that position, retained their discipline. "The conduct of the soldiers generally had been good, but at Villafranca it became extremely bad, and in order to check outrages it became necessary to make an example. Three soldiers of the 7th Hussars had been detected in the act of breaking open a box and stealing wearing apparel of the inhabitants. These poor fellows drew lots, and one was shot; he had previously been a good soldier."[1]

From Villafranca to Lugo is seventeen leagues (more than fifty miles), over an immense mountain. Hundreds of stragglers who were weakened by their excesses at Bembibre and Villafranca died or were taken or shot by the enemy; want of shoes caused many to fall behind or loiter in the villages, where they were taken; want of rest, want of food, wretched roads, and heavy rain or snow filled up the sum of their miseries. The rear guard found houses filled with stragglers of various regiments, who would not or could not come on. An officer telling them the enemy would certainly shoot or take them, they said, "You may shoot us, sir, or they may shoot us, but we cannot stir." Fine fellows, with bleeding feet, totally incapable of keeping up, others whose spirit was better than their strength striving to the last to join their battalions; women knee deep in mud crying piteously for help which could not be given.[2] Sergeant Robertson says that on the mountains they found the carriage of the paymaster-general, whose wife was with him, hopelessly stuck in the mud. The Highlanders had often envied this comfortable conveyance, drawn by good English horses, as they had to open out on the march to let it pass; now it was left on the road, and the lady and her husband had to tramp on as best they could. At Nogales they passed the military chest, which the oxen could drag no further, and rather than let it fall into the enemy's hands, the casks containing £25,000 worth of dollars had the heads knocked out and were rolled down the precipice into the wooded ravine below.[3] On the roadside a soldier's wife was lying newly delivered of a fine boy—the mother was dead. A woman of the 92nd took up the helpless little one. Several men of the regiment were made prisoners this day; the next day the weather and roads were both better, and they arrived at Lugo on the 4th January.[4]

At Lugo the troops got two days' rest, and Sir John Moore gave

[1] Vivian's "Memoirs." [2] *Ibid.*

[3] Some contrived to pick up part of the money, and a woman, who had rolled a quantity in a cloth and tied it round her waist under her dress, was afterwards drowned; as she stepped on board ship at Corunna from a boat, she fell into the sea, and the weight of the gold sank her.

[4] Sergeant Robertson.

a severe and just rebuke in General Orders to the officers and soldiers for the previous want of discipline, and at the same time announced his intention to offer battle. "It has been well said that the British army may be gleaned in a retreat, but cannot be reaped; whatever may be their misery, the soldiers will always be found clean at a review and ready at a fight."[1] As if by enchantment disorder ceased, faces brightened, arms were cleaned, stragglers came up, and 19,000 men bivouacked in order of battle in a strong position in front of the town, ready and willing for the fray.[2]

The Gordons, who were lodged in a convent at Lugo, had just got a ration of flour served out, but before they had time to make scones, they turned out and formed in a field to the left of the road in front of the town, where they remained the night. "Everyone," says Duncan Robertson, "was happy that we had got the French in line, as we longed very much to fight, and abhorred the thought of running away, as we had been doing for some time past. About ten o'clock next morning, as the rain was beginning to fall, the French extended their line and beat the charge. They occupied a ploughed field, and we were posted on a heath, a small river running in the hollow between us. A farmhouse lay at the foot of the rising ground, having a few stacks of corn about it. The company to which I belonged was ordered to take possession of it, and General Hope accompanied us. While moving down, the French fired one of their cannon, the ball falling close beside us, when the general, good humouredly, took off his hat and saluted the gunner; as there was only one subaltern with the company, I got command of a section. When we came to the house we found a party of the French there before us, and a strong reinforcement coming to their assistance. However, they did not think fit to wait and receive us, and after a few rounds on both sides they fell back, and we occupied the farm, but were not allowed to remain in quiet possession; as the enemy came upon us in overpowering numbers, we fell back the breadth of a field, and posted ourselves behind a stone wall, from whence we opened fire. The French attempted to charge us, but were driven back. On our left was a cart-road between two high hedges, and while the party in front were keeping us in play, about forty went up this road, while those in front made a feint of retiring so as to decoy us between two fires. I happened to look round hoping for help, when I saw the other party of the enemy forming at the head of the field not fifty yards from us. It struck me at once to get into the lane and cut them off from their own line;

[1] Napier.

[2] 1500 men had fallen in action or dropped to the rear, but three fresh battalions, which had been left by Sir D. Baird in his advance, joined the army between Villafranca and Lugo.

my section jumped into the road, formed, and prepared to charge ; when they saw this they ran down on us, we gave them a volley, and eighteen fell killed and wounded. On seeing the fate of their comrades, twenty-two laid down their arms and were taken prisoners. Another detachment came to the rescue, and the sergeant commanding it got hold of one of our men, when I leaped back and drove my pike through his body ; the others ran off. We were now thoroughly drenched with rain, and when night came we were relieved by an equal proportion of the regiment. A number of apples were found in the farmhouse, which were greedily devoured. In the evening we got our beef served out, but having neither bread nor salt, it made rather an unsavoury supper. Although the weather was cold, we slept very comfortably beside large fires we had kindled in the open air."[1] The above experience of one company probably gives a fair idea of that of the battalion, and of the army generally.

On the morning of the 7th January, Marshal Soult had arrived at the head of some 12,000 men. His troops had suffered by the rapid marches, and he waited till his columns came up, to form them in order of battle along a strong mountainous ridge opposite the British. As he was prevented by the ground from seeing what numbers were opposed to him, he advanced some troops and guns, and opened fire on the centre, which was silenced by a reply from fifteen pieces. This satisfied him that he had more than a rear guard to deal with. He then made a feint on the right, and sent a column of infantry and five guns against the left. They were gaining the advantage over our outposts when General Moore arrived, rallied them, charged and broke the adverse column, with an estimated loss to the French of between 200 and 300 men. It being evident that the British meant battle, the Duke of Dalmatia (Soult) hastened the march of the divisions in his rear, and at daybreak on the 8th the two armies were still embattled, but Soult deferred the attack till the 9th. The British impatiently awaited the assault, and blamed their adversary for delaying the contest ; but darkness fell without a shot being fired, and with it fell the British general's hope to engage his enemy on equal terms. What was to be done ? His army had been sent to assist Spain, but the armies of that nation had been defeated before he could arrive to their assistance. There was but one day's bread for his army at Lugo, no transport for reserve ammunition, no hospitals or provisions, no second line. He was in a position to fight one battle for the purpose of defeating his enemy and being allowed to embark without further molestation ; two battles, even if victorious, would

[1] Comfort is a comparative quantity !

NOTE.—The officers and men generally got an allowance of rum at the end of a march

have been destruction.[1] For two whole days Sir John Moore had offered battle, which was sufficient to rally the troops, restore order, and preserve the reputation of the army. Marshal Ney with his column of the French army might soon turn his position ; it was impossible to remain longer at Lugo. The general determined to decamp in the night, and by stealing a march to leave Soult so far behind as to allow the British troops to reach their ships at Corunna in peace. He ordered the fires to be kept bright, and exhorted the troops to make a great exertion, which he trusted would be the last required of them. The country immediately in rear was intersected by walls and intricate lanes ; precautions were taken to mark the right roads by placing bundles of straw at certain distances, and officers were appointed to guide the columns ; but just as the army started, a terrific storm of wind and sleet blew away the marks, and the guides lost the direction. Two of the divisions lost their way, and at daybreak found themselves still near Lugo. Fatigue and depression of mind, together with the want of shoes, made stragglers numerous. The general of a leading division, thinking to relieve his men during a halt in the night, unfortunately desired them to take refuge from the weather in some houses a little way from the road. Complete disorganisation followed this imprudent act.[2] From that moment it became impossible to make the soldiers of the division keep their ranks ; plunder succeeded, the example was infectious, and what with real suffering and evil propensity, encouraged by this error, the main body of the army arrived at Betanzos on the evening of the 9th in a state very discreditable to its discipline.[3]

On the 10th the enemy's cavalry skirmished with the troops who were destroying the bridge of Betanzos. Moore now assembled the army in one solid mass. They had lost more men in the march from Lugo than in all the rest of the retreat ; still, when the French cavalry appeared, a body of stragglers, of whom a sergeant took command, defended themselves in a rather open country against

[1] Napier.

[2] The following is an extract from General Orders at Benevente, December 27th :—" The Commander of the Forces has observed with concern the extreme bad conduct of the troops at a moment when they are about to come into contact with the enemy. It is disgraceful to the officers, as it strongly marks their negligence and inattention." He refers to the Orders of 15th October and 11th November, and desires that they may be read at the head of every company in the army. " He can feel no mercy towards officers who in times like these neglect essential duties, or towards soldiers who injure the country they are sent to protect. It is impossible for the general to explain to the army his motives for the movements he directs ; when it is proper to fight a battle he will do it, and, meanwhile, he begs the officers and soldiers to attend diligently to discharge their part."

[3] The 1st Guards and 92nd Highlanders had a smaller percentage of missing, *i.e.* stragglers, than any other battalion that went through the whole retreat.—Oman, Vol. I. (Appendix).

NOTE.—No sergeant of the 92nd died or was missing.

ON THE ROAD TO CORUNNA

400 horsemen, who were following and skirmishing with them.[1] On the afternoon of the 11th the army arrived at Corunna, after an orderly march under the personal direction of the Commander-in-Chief, demonstrating that inattention and want of experience in the officers was the true cause of the disasters which had afflicted the army.

The Gordons spent a miserable night after leaving Lugo, and next day rested at noon, cold and hungry, on a heath, the men pulling turnips to eat. At evening they resumed their dreary journey, the men lame and footsore, many of them marching in excruciating pain ; others, however, had made sandals from the skins of cavalry horses that had been shot ; Sergeant Robertson's journal mentions this, and a pensioner in Badenoch, named John Cattanach, said that he and many others made " cuarans," as used in the Highlands, when their shoes were done. Another old man said he saw officers almost barefoot. On this march the wife of Sergeant Charles MacGregor, with her three boys, who were carried in creels on a donkey, fell out of the line and were never more heard of.

The battalion arrived at Betanzos in miserable plight, officers and men half asleep as they rode or walked along, the colour of their clothes hardly distinguishable through the mud with which they were encrusted. They went into good quarters in the suburbs of Corunna, where they lived in plenty and were supplied with ammunition. When the roll was called they were over 190 men short of the number who marched on Christmas Day. Of these, however, many rejoined at Corunna, others made their way into Portugal and joined the troops left there ; the rest had been killed or taken by the French, and some of them were never heard of again.[2] The army had marched about 300 miles since December 24th.

It had been at first the intention to embark at Vigo, where the transports had arrived. The general, as soon as Corunna had been decided on, ordered them round to that place, but contrary winds detained them, and thus the last exertion made by the army was rendered fruitless. When the troops reached the heights from which Corunna is visible, all eyes were directed to the bay, hoping to see the friendly fleet, but they were doomed to disappointment—nothing but a few coasters were in sight.

As the various brigades came up they were distributed in the town and at the village of El Burgo, and Moore at once caused the defences to be repaired and strengthened. The inhabitants, to

[1] Vivian's " Memoirs."

[2] Napier says with regard to stragglers from the army—" Of the whole number above 800 contrived to escape to Portugal, and being united with the sick left by the regiments in that country, formed a corps of 1876 men under the name of 'the battalion of detachments,' and did good service at Oporto and Talavera."

their honour, cheerfully joined in the work, in bright contrast to the apathy latterly shown by most of their countrymen. A large magazine was found in Corunna containing arms, ammunition, and stores, which had been sent from Great Britain to the Spaniards months before, and which were lying unused by a nation invaded by 300,000 enemies, and possessing 100,000 soldiers unclothed and unarmed,[1] a notable instance of the improvidence of the Spanish authorities. Among these stores were many thousand new muskets, for which the British troops now exchanged their old rusty and battered weapons, which, with the fresh ammunition, rendered their fire very superior to that of the enemy in the ensuing battle.

At a magazine three miles from the town, 4000 barrels of powder were deposited, and a smaller quantity was in a storehouse at some distance. To prevent these falling a prey to the enemy, they were both exploded on the 13th. The smaller one blew up with a terrible noise that shook the houses in the town, but when the train reached the larger magazine there ensued a crash like the bursting of a volcano ; the earth trembled for miles, the rocks were torn from their bases, and the agitated waters rolled the vessels as in a storm ; a vast column of smoke and dust, shooting out fiery sparks from its sides, arose perpendicularly and slowly to a great height, and then a shower of stones, bursting out of it with a roaring sound, killed several persons who remained too near the spot. A stillness only interrupted by the lashing of the waves on the shore succeeded, and the business of war went on.[2]

As the ground in front of Corunna is impracticable for cavalry, and as many of the horses left alive were foundered, these poor worn-out animals were reluctantly ordered to be shot ; their riders, whom they had carried in the constant and successful combats and skirmishes, showed great sorrow at parting with their equine friends,[3] but it was thought better than leaving them to become draught horses to the French or to wear out their existence in a Corunna cart. Of 751 horses the 7th Hussars took to Spain, they only brought 70 back to England ;[4] of the number lost, however, some were drowned by the wreck of a transport.

On the evening of the 14th the ships from Vigo entered the harbour, and the dismounted cavalry, the sick, the women and children, and all the best horses, with fifty-two pieces of artillery, were embarked during the night, eight British and four Spanish

[1] Napier. [2] *Ibid.*

[3] R. Blakeney ("A Boy in the Peninsula") says he saw troopers in tears when obliged to shoot their horses during the retreat, owing to the hoofs being worn away, and no shoes to be had for them. Blakeney, who was on rear guard, also says that excess was the cause of many men falling out, and that many stragglers were robust plunderers.

[4] Vivian.

guns being retained on shore. Meanwhile the French had appeared, and some guns had opened on the English posts at El Burgo, but had been silenced by superior fire. On the 15th, Laborde's Division arrived, the French occupied the ridge enclosing the British position, and towards evening the pickets opposite the French right were actively engaged. During the night Soult established a battery of eleven guns on the rocks which formed the left of his line of battle. Laborde's Division was posted on the right; Merle's Division was in the centre, Mermet's Division formed the left. The position was covered in front of the right by the villages of Palavia Abaxo and Portosa, and in front of the centre by a wood; the left was strongly posted on the heights where the 11-gun battery was established. The distance from the battery to the right of the British line was about 1200 yards, and midway the village of Elvina was held by the British pickets.

The late arrival of the transports, the increasing force of the enemy, and the disadvantageous nature of the ground made the difficulty of the embarkation so great, that several general officers proposed to the Commander-in-Chief that he should negotiate for leave to retire to his ships upon terms. The army had suffered, but not from defeat; its situation was dangerous, but not desperate; Moore's high spirit and clear judgment revolted at the idea, and he rejected the advice without hesitation.[1]

The encumbrances of the army having been shipped, all was prepared on the morning of the 16th to withdraw the fighting men as soon as darkness would permit them to do so unseen.

But a more glorious termination to their campaign awaited the British troops. About two o'clock in the afternoon a general movement was observed along the whole French line. The British infantry, 14,500 strong, occupied the inferior range of hills, the right was formed by Baird's Division, and approached the enemy; while the centre and left were of necessity withheld in such a manner that the French battery on the rocks raked the whole line. General Hope's Division, in which were the 92nd,[2] was on the left, and occupied strong ground abutting on the muddy bank of the River Mero; the reserve was drawn up near the village of Airis in rear of the centre, and commanded the valley which separated Baird's Division from the French cavalry. These were kept in check by a regiment from the reserve, which thus became the extreme right of the army, and a chain of skirmishers connected this regiment with the right of Baird's line. These dispositions were as good as

[1] Napier.

[2] General Hope's Division consisted of Major-General Hill's Brigade, the 2nd or Queen's, 14th and 32nd Regiments, and Colonel Crawford's Brigade, the 36th, 71st, and 92nd or Gordon Highlanders.—Stewart's "Highlanders of Scotland."

the nature of the ground would admit of, but the advantage was all on the enemy's side. His light cavalry, under Franceschi, reaching nearly to St Christopher, a mile in rear of Baird's position, obliged Sir John Moore to weaken his front by keeping back Fraser's Division till Soult's plan of attack should be completely developed. The only advantage on the side of the British lay in the new firelocks and fresh ammunition which many of them had received.

Soult's attack began by a heavy fire of artillery, covering three solid masses of troops descending to the assault. A cloud of skirmishers led the way and drove back the British pickets ; the village of Elvina was carried by the first column, which then divided, one half attacking Baird in front, the other turning his right. The second column made for the centre, the third engaged the left by the village of Palavia Abaxo.

The French guns overmatched the few British 6-pounders, and their shot swept the position to the centre. Sir John Moore ordered General Paget to carry the reserve to where the detached regiment was posted, to turn the left of the French attack, and menace the great battery. Then, directing Fraser's Division to support Paget, he threw back the 4th Regiment, which formed the right of Baird's Division, opened a heavy fire upon the flank of the troops penetrating up the valley, and with the 50th and 42nd, met those breaking through Elvina. A severe scrambling fight ensued among walls and hollow roads. A battalion of the Guards was brought up to fill the space left vacant in the line by the 50th and 42nd ; the latter, mistaking the general's intention, retired, and at that moment the enemy being reinforced, renewed the fight, and Elvina became the scene of a second struggle ; this being observed by the Commander-in-Chief, he caused the 42nd to return. "Highlanders, remember Egypt !" he said, as they gladly advanced to the attack.

The line of skirmishers, supported by Paget's Reserve, vigorously checked the advance of the enemy's troops, while the 4th Regiment galled their flank ; the centre and left of the army also became engaged. Sir David Baird was wounded, and a furious action raged along the whole line. Sir John Moore, while watching the result of the fight about Elvina, was struck on the left breast by a cannon shot, and was thrown from his horse ; he rose to a sitting posture, his countenance unchanged and his eye still fixed on the regiments engaged in his front ; when satisfied that they were gaining ground his face brightened, and he suffered himself to be carried to the rear by four soldiers of the 42nd and Guards.[1] His

[1] The descendants of Alexander M'Kay of the 92nd claim that he was one of the men who carried Sir John, stating that their informants were James M'Kay, a younger brother

shoulder was shattered to pieces, his arm hanging by the skin, the ribs over the heart broken, and the muscles of his breast torn into long strips. As the soldiers placed him in a blanket, the hilt of his sword entered the wound, and Captain Hardinge, a Staff officer, attempted to take it off, but the dying man stopped him, saying, "It is as well as it is. I had rather it should go out of the field with me," and in that manner Moore was borne from the fight.

During this time the British were rapidly gaining ground; the reserve, overthrowing everything in the valley, and obliging the French Dragoons, who had dismounted, to retire, turned the left and approached the great battery. On the left some companies of the 14th and 92nd carried Piedra Longa; the fight for Elvina ended in favour of the British, and at night the French were falling back in confusion. Sir John Hope, upon whom the command of the army had devolved, gave orders to embark, and the operation was completed without confusion or difficulty. The pickets kindled fires, and were themselves withdrawn at daybreak on the 17th.

Thus ended one of the most celebrated retreats in history, which remains an example of ability in the general and, the temporary indiscipline notwithstanding, of courage in his troops. Soult (Duke of Dalmatia), one of the ablest of Napoleon's marshals, said, "Sir John Moore took every advantage that the country afforded to oppose an active and vigorous resistance, and he finished by dying in a combat that must do credit to his memory."[1] This generous opponent erected a monument over Moore's grave at Corunna.

Napoleon declared that his talents and firmness alone had saved the British army from destruction, and Wellington was of the same opinion.

Napier in his history attributes the want of discipline during part of the retreat to want of experience on the part of regimental officers and soldiers, and a n.c. officer of the 92nd agrees with him when he says—"It is well known that a finer body of men (in appearance at least) never left the British shore. No expense was spared to complete them with every requisite necessary for the expedition. We were not inured to this sort of warfare, and did not know how to conduct ourselves. The British soldier when at home has everything provided for him, and does not know how to live without all the conveniences to which he has been accustomed

of the said Alexander, and Corporal William M'Kay, who was in the 42nd. As the 92nd were not in that part of the field, Alexander M'Kay could only have been present as an orderly, or possibly he was one of the two men who carried Sir John at Bergen-op-Zee, and would not claim the reward. Alexander M'Kay belonged to Sutherland, and died as the trusted servant of a London bank.

[1] Napier.

—and the consequence was, that when the army was reduced to live on scanty allowance, and to suffer privation to which it had not been accustomed, it went into disorder. Not one in ten of us was used to campaigning, whereas the French had lately come from Germany, where they had been inured to hardship for some years. They entered Spain as invaders, we as allies ; hence, if provisions were to be got in any way, they were not deterred from taking them."

In the battle of Corunna, the 92nd was posted on the left of the line near the sea, and protected by a steep hill in their front ; as a battalion they were not materially engaged, but Lieut.-Colonel Napier (92nd) was field officer for the day in command of the out-posts, of which a detachment of his regiment formed part.

Ensign Hector Innes thus describes what he saw when on picket with his company :—" Our sentries were close to the French. Colonel Napier sent me to General Hope, distant about two miles, to inform him of the local situation of our position, distant about three miles from our regiment and brigade. I went accoutred with my trusty claymore and pistol, with my stockings all down. I had to pass quite close to the enemy, and through a turnip field, which annoyed me a great deal more. When I returned I found they had not attacked us, so Tulloch (his cousin in another regiment) and I sat down and ate some cold beef as salt as Lot's wife, but we were soon roused from our meal by some shots which fell at our side, and my servant who had brought the dinner was mortally wounded. The business began about half four, when the enemy rushed down instantaneously in crowds in all directions, firing smartly on the pickets (ours on the left). For a while we withstood vigorously their attacks. However, being overpowered by numbers, we retired with loss, and afterwards rallied and took post behind a hedge. I do assure you we had some fun ; you would have laughed had you seen how we scampered with Jack Frenchman at our heels ; but fortune favoured us. I commanded a few but trusty men, who after three hours were successful. We charged through the village along with two companies of the 14th Regi-ment, who, I am rather piqued to find, get all the merit. We certainly gave them a complete thrashing. I had the curiosity to examine the enemy's position, and was struck with astonishment to see the awful carnage ; they were lying actually all above one another."

Sergeant Robertson says that Colonel Napier was giving orders to carry the village, when a shot struck him in the groin ; one of the men immediately bound up the wound with his shirt and carried him to the rear, but he died shortly after. A letter from Ensign Innes mentions that he delivered the colonel's sword to his mother. "Although we were twice driven from the post we returned to the attack with redoubled fury, and at last retained it in defiance of all the efforts of the enemy to dislodge us. Exasperated at being thus repulsed, the French sent down a strong reinforcement upon us to drive all before it, but General Hope, perceiving the movement, ordered two companies of the 14th to our assistance. Then came the tug of war; there was little or no firing, but fair hand-to-hand fighting; night put an end to the contest, the French withdrawing from the village, while we remained masters of it."

Captain Seton's (92nd) account is short and to the point. "The Battle of Corunna was bloody and bravely contested. The French got a devil of a drubbing, though five to one. As they were beat back, they always pressed forward with fresh troops. Night put an end to the action."

Vedette, 7th Hussars, 1808.

CHAPTER XI

BY the death of Lieut.-Colonel Napier the command of the 1st Battalion devolved upon Lieut.-Colonel Lamont.[1] Owing to the death of Moore, and Sir David Baird being wounded, Sir John Hope had assumed command of the army. Colonel Lamont had been sent for by Sir John, and Sergeant Robertson accompanied him as orderly. He says—"We received orders to keep our post till five o'clock in the morning to deceive the enemy, and then to make the best of our way to the ships. It was dusk, and Colonel Lamont and I lost our way, and, before we were aware, found ourselves amongst a French picket. They were busy killing a bullock, but we did not wait to partake of it, and made the best of our way to our own picket." At 5 a.m. on the 17th January the pickets started for the beach as fast as they could. They were observed and pursued by cavalry, but having little more than a mile to go, they reached the boats just in time. About mid-day the French established a battery on the heights of San Lucia, which fired on the shipping, causing confusion and disorder among the transports; the merchant seamen, not bargaining for this sort of thing, went below and left the soldiers to do seamen's duty as best they could. Some masters cut their cables, and four vessels went ashore. They were burned, and men-o'-war's boats removed the troops, while the *Victory* brought her guns to bear on the battery and silenced its noise. Beresford's Brigade kept possession of the citadel till the 18th, when, the wounded being all on board, his troops embarked and the fleet sailed for England.

The death of Colonel Napier was a great grief to the regiment, which he had joined as senior captain when it was raised. He had commanded the 1st Battalion since Colonel Erskine was killed in Egypt in 1801, and both battalions for a time after the 2nd was raised. Strict without nagging, just without harshness, and considerate without weakness, he was a worthy successor to such excellent commanding officers as Huntly and Erskine. His best epitaph is in the words of one of his n.c. officers, who described him as one "whom every man in the regiment adored, and to whom he was more like an affectionate father than a commanding officer."

The regiment lost, besides Lieut.-Colonel Napier, Lieutenant Archibald MacDonald, who died of his wounds shortly afterwards; two rank and file were killed, and four wounded.[2] The whole army lamented the death of the good and gallant Sir John Moore, under whom the Gordon Highlanders had served so constantly,

[1] Lamont of Lamont, chief of his clan; he entered the army as ensign, 42nd Regiment, March 1793, was promoted in the Clan Alpine Fencibles, and afterwards in the 92nd; retired as Major-General.

[2] I cannot find their names, and the Order Books and returns of this campaign seem to have been lost.

SIR JOHN MOORE, K.B.

From a portrait by Sir Thomas Lawrence in possession of the 2nd Batt. Believed to be a duplicate of the portrait in the National Portrait Gallery

both in peace and war. They had peculiar reasons to love and respect him, and the officers changed the blue line in their lace to black,[1] which is still worn in his memory. After speaking of his mother, and his "satisfaction to know that we have beaten the French," his last words were, "I hope England will be satisfied; I hope my country will do me justice."

The touching scene of the burial of Sir John Moore, as described both in prose and poetry, will ever live in the hearts of his countrymen.

"The loss of the British army at Corunna was never officially returned, but was estimated by Sir John Hope at about 800. The French loss I have no accurate account of. I have heard from French officers that it was above 3000 men." This number Napier thinks exaggerated, but that it certainly was very great. He attributes it to the arms of the British being new, and their ammunition fresh, and that "the physical strength and coolness of the men render the fire of the Englishmen at all times the most destructive known." "The nature of the ground also prevented any movement of artillery on either side, hence the French columns were exposed in their attacks to a fire of grape, which they could not return, because of the distance of their batteries."

The misfortunes of the British were, however, not over. A severe gale scattered the transports; some of them were wrecked on the shores of England, and others ran to the nearest ports; so that the inhabitants of the towns along the Channel who had seen the successive expeditions which composed Moore's army embark with all the pomp and circumstance of military display, were struck with horror and dismay when they beheld them return, reduced in numbers, with haggard faces and ragged clothing. As the news spread throughout the country, the fate of these gallant men became a subject of general commiseration, and the hopes of the nation, which had been raised by the victories at Roliça and Vimiera, and the early successes of the Spaniards at Baylen and Saragossa, were changed to despondency by the late disasters. A cry was raised that Great Britain could never contend with France except at sea; and this, though in every regular engagement her troops had defeated the veterans of Napoleon, and had only retreated because their numbers were quite inadequate to the task required of them.

But notwithstanding the misfortunes which attended the campaign, it was even more calamitous to the arms of France. One whole *corps d'armée* had capitulated, and surrendered the kingdom

[1] Mr S. M. Milne, the greatest authority on Military Heraldry, believes that this change was not made immediately after Corunna but at a later period. In this he may be right, but I know that when I joined forty-two years after Sir John Moore's death, it was impressed upon me that I wore the black line in his memory.—*Note by Author.*

of Portugal, which was still held by the British ; another had surrendered to the Spaniards at Baylen. The spell was broken, Continental nations discovered that the French armies were not invincible, and Austria prepared to renew the struggle. By drawing Napoleon to the northern extremity of Spain, Moore gave time to the southern provinces to restore their army and strengthen their fortresses, and thus to prepare the way for the campaigns of Wellington, which Great Britain supported with a determined tenacity of purpose not hitherto exhibited by the nation in her Continental expeditions.

The 1st Battalion landed at Portsmouth on the 26th of January, and arrived at Weely in three divisions on the 13th, 14th, and 15th February. The following extracts from letters and journals will give an idea of the difference between the arrival of troops from foreign service in those days, and the comforts which now await them on return to their native land.

Mr Innes, who was, with about 100 of the 92nd, along with other troops, on a ship which ran aground, but was got off again, writes from " In sight of the Isle of Wight, January 23rd, 1809.—I am in perfect health and spirits, but in rags. I lost almost everything."

Captain Seton, at Portsmouth, January 25th, after mentioning " A passage most dreadful. We are strange figures, all dirty, and the most of us almost naked. I had not a change of clothes since I was at Lugo on the 6th till last evening, when I bought a shirt and some other things," in a later letter says, " Our men are in a dreadful state " ; again, " They have got a bad typhus, which has carried off a number since our return. Our total loss in Spain and Portugal is 190 ;[1] many regiments have suffered more and few less."

Sergeant Robertson says—" When we landed at Portsmouth I had neither shoes nor stockings, but had to walk along the streets barefooted ; the condition we were in with regard to clothing and cleanliness beggars description. When we came to our billets about six miles from Portsmouth, the inhabitants would not allow us to sleep in their beds, nor sit by the fireside, on account of the vermin that infested us ; cleaning ourselves was out of the question. When we reached Weely Barracks, where our heavy baggage was lying, I got a suit of clothes I had left, and soon divested myself of my filthy raiment and reduced it to ashes, with the exception of

[1] Many of these 190 men afterwards rejoined. On 16th July 1906 Professor Oman sent the author an absolutely correct list, extracted from Pay Lists in the Record Office, of rank and file missing from the 92nd during the great retreat to Corunna between 15th December 1808 and 16th January 1809. The total is 86. Also a list of those who died in Spain, on board ship, or in hospital in England ; the total is 25, of whom 10 to 12 died in hospital in England, and 2 the day after the battle, probably of wounds.

the Highland bonnet and feathers, which I preserved."[1] "Soon after our arrival at Weely, our regiment, along with others that had been in Spain, were seized with typhus fever ; so prevalent was this fatal malady, that an order came from the War Office that we should be relieved from duty till further orders, that every article of clothing be burned, and that we should be supplied with everything new. There was also £2 given to each of us to provide necessaries. The Lords of the Treasury ordered another £2 to each man who was sickly to purchase anything the doctor thought proper."

On January 30th, Mr Innes writes from Colchester :—"We are now on our march to Chelmsford in Essex, a route of about 100 miles, which I find will be extremely expensive. We were landed a parcel of half-starved naked soldiers. You will laugh, perhaps, at the title of 'half-starved,' but believe me it is the case. Our embarkation was so rapid that we had no time for providing comforts, and previous to our embarkation we had lain six nights in the fields, feeding on spoiled ship's biscuits and salt beef." On February 16th, having arrived at Weely, he is "just about to inspect a company, some dressed in kilts, some in pantaloons. . . . We make certainly a very grotesque figure, but ere long we shall again become quite smart. . . . Our marching in England was extremely pleasant ; all people glad to see us. . . . I fell in with some strange adventures. Perhaps at a future opportunity I may relate them. . . . My mother inquires after some men. Glassan, Hay, and Patterson, as likewise Forbes, are quite well and safe in this barrack ; as to Andrew Fotheringham, I am afraid he is taken prisoner, being absent before we came to Lugo."[2]

An Order by Lieut.-General the Hon. Sir John Hope, dated H.M.S. *Audacious*, January 18th, 1809, of which the following is an extract, was published to the battalion at Weely :—

"The irreparable loss that has been sustained by the fall of the Commander of the Forces, and the severe wound that had removed Lieut.-General Sir David Baird from his station, renders it the duty of Lieut.-General Hope to congratulate the troops on the successful result of the 16th inst. On no occasion has the undaunted valour of the British troops been more magnificent. At the termination of a severe and fatiguing march, rendered necessary by the superiority of numbers which the enemy had acquired, and which had materially impaired the efficiency of the troops, many disadvantages were to be encountered. They have all been sur-

[1] The heavy baggage had been left when they embarked for Sweden in the preceding May. It is worthy of remark that the feathered bonnet, which from its appearance people are apt to think more ornamental than useful, was in this case, as it was after the Egyptian campaign of 1801, the only part of the clothing which remained serviceable.

[2] These were all men from near his home in Banffshire.

mounted by the troops themselves, and the enemy has been taught that whatever advantages of position or numbers he may employ, there is a natural bravery in the British officers and soldiers that knows not how to yield, that no circumstances can appal, and that will ensure victory when it is to be obtained by any means whatever.

"The Lieut.-General hopes the loss in point of numbers is not so considerable as might have been expected. He laments, however, the fall of the gallant soldiers and valuable officers who have suffered. He knows it is impossible, in any language he can use, to increase or diminish the regret the army feels in common with him for its late commander. His career has, unfortunately, been too limited for his country, but sufficient for his own fame. Beloved by the army, honoured by his sovereign, and respected by his country, he has terminated a life devoted to her service by a glorious death, leaving his name as a memorial and example to those who follow him in the path of honour, and it is from his country alone that his memory can receive the tribute that is his due."

G.O., Horse Guards, February 1st, 1809.—The benefit derived by an army from the example of a distinguished commander does not terminate at his death. His virtues live in the recollection of his associates, and his fame remains the strongest incentive to great and glorious actions. In this view the Commander-in-Chief, amidst the deep and universal regret which the death of Lieut.-General Sir John Moore has occasioned, recalls to the troops the military career of that illustrious officer for their instruction and imitation. Sir John Moore from his youth embraced the profession with the feelings and sentiments of a soldier. He felt that a perfect knowledge and an exact performance of the humble, but important, duties of a subaltern officer are the best foundation for subsequent military fame, and his ardent mind, while it looked forward to those brilliant achievements for which it was formed, applied itself with energy and exemplary assiduity to the duties of that station. In the school of regimental duty he obtained that correct knowledge of the profession so essential to the proper direction of the gallant spirit of the soldier, and he was able to establish a characteristic order and regularity of conduct, because the troops found in their leader a striking example of the discipline he enforced on others. Having risen to command, he signalised his name in the West Indies, in Holland, and in Egypt. The unremitting attention with which he devoted himself to the duties of every branch of his profession obtained him the confidence of Sir Ralph Abercromby, and he became the companion in arms of that illustrious officer, who fell at the head of his victorious troops in the action which maintained our national superiority over the armies of France. The life of Sir John Moore was spent amongst the troops. During the season of repose his time was devoted to the care and instruction of the officers and soldiers. In war he courted service in every quarter of the globe, regardless of personal considerations. He esteemed that to which his country called him, the post of honour, and by his undaunted spirit and unconquerable perseverance he pointed the way to victory. His country, the object of his latest solicitude, will rear a monument to his lamented memory, and the Commander-in-Chief feels he is paying the best tribute to his fame by thus holding him forth as an example to the army.

By order of His Royal Highness the Commander-in-Chief.

(Signed) HARRY CALVERT,
Adjutant-General.

On the 30th January, Sir David Baird, in a letter to the officer commanding the 1st Battalion 92nd Regiment, sent copies of letters from the Lord Chancellor and the Speaker of the House of Commons, enclosing the resolutions of both Houses of Parliament, containing the thanks of Parliament to the army lately engaged in Spain; and Sir David, in communicating this signal mark of approbation, adds his "warmest congratulations upon a distinction which you and the corps under your command on that day had a share in obtaining for His Majesty's service." The above was communicated to the officers, n.c. officers, and private soldiers of the battalion for their distinguished conduct and exemplary valour displayed in the battle of Corunna.

The following letter addressed to the Officer Commanding 92nd Regiment, from the Asst. Adjt.-Gen., Portsmouth, informs him that "The magistrates and bankers of this town having expressed a desire of aiding those women lately returned from Spain who have lost their husbands, and also the children who have lost their parents, a return of each is to be sent to the Town-major's office with the least possible delay."

The Commander-in-Chief desires that "the most vigilant care and expedition" may be used in providing every article of comfort and equipment of which they may stand in need, to the regiments from Spain, and suggests coffee as well adapted as a corrective of the maladies which are apprehended, and as the reduced price of that article places it within the soldier's reach.

The men were particularly warned against intemperance, which, after the privations they have undergone, would be seriously detrimental to their health; great care was taken as to ventilation, sick women and children not to remain in the barrack-rooms, and as cleanliness is the greatest preservative from infection, "the regiment will to-morrow be thoroughly washed with soap and water under the superintendence of an officer of a company."

On February 18th, Major-General Robertson directs that as little duty as possible may be required of the men, and that they shall be exempted from escort duty till the middle of April, and on the 22nd he ordered that the men should do no duty at all, but merely attend to their health and comfort, and by General Order they were not to be charged for necessaries lost before the enemy or ordered to be destroyed. New greatcoats were also given to all n.c. officers and men, and new knapsacks were issued, of black canvas in place of yellow, and having the *Sphinx* and number *XCII* in Roman characters on the back.

In February Captain Grant, Lieutenant Duncan M'Pherson, and Ensign and Adjutant Campbell were ordered to join the 2nd Battalion at Athlone, as were Lieutenants Archibald MacDonald and Gordon as soon as their health would permit.

In February men serving in garrison battalions under the Reserve or Additional Force Acts, on volunteering to extend their service to the line, were allowed, if for unlimited service, a bounty of ten guineas ; if for seven years, five guineas.

In March[1] Lieut.-Colonel John Cameron, from the 2nd Battalion at Athlone, joined and took command of the 1st Battalion. Colonel Cameron was a true Highland gentleman, a soldier of experience, devoted to his profession and to the regiment in which he had served since its formation.

On the 23rd of March a Court of Inquiry, consisting of one field officer, four captains, and four subalterns, assembled "to investigate and report their opinion to the commanding officer on the circumstances attending Sergeant-major Mackay being ordered for execution at Corunna so far as they can be discovered." Lieut.-Colonel Lamont presided. The opinion is not given, but the circumstances appear to have been that he was accused of plundering, but that he was released by order of Lieut.-General Hope. At any rate, he remained sergeant-major till the 25th April, when

"This showeth that I, Sergeant-major Mackay of the 1st Battalion 92nd Regiment, have resigned my situation as sergeant-major from inability of health to go on active service.

(Signed) "ARCHIBALD MACKAY.

"A true copy, Claud Alexander, Lieutenant and Adjutant 92nd Regiment."

Mackay afterwards became sergeant-major of the 2nd Battalion.

[1] In the state of the 1st Battalion on the 1st of March there appear 2 staff, 4 sergeants, left on duty in Portugal. Prisoners of war—1 sergeant, 102 rank and file. Sick in Portugal—3 sergeants, 48 privates ; in Spain—1 sergeant and 20 rank and file. On March 8th, 4 men had joined "that was thought to have been lost and not included in the return of February or the 1st inst." Sick present—2 drummers, 96 rank and file ; absent in hospital—1 captain, 2 subalterns, 1 staff, 8 sergeants, 4 drummers, 88 rank and file.

Colonel Cameron spared no pains to make the battalion again fit for service. The officers are to report when they have supplied themselves with such articles of the Highland dress as they lost in Spain. He desires the officers to explain each man's accounts to him personally ; that no man be sent to drill as a punishment without its being notified to the adjutant for the commanding officer's information. " Pay-sergeants are not to act as merchants " as to men's necessaries, or " women as hucksters " to their companies. Up to this time there had been no regular breakfast mess. " Lieut.-Colonel Cameron desires officers commanding companies to establish breakfast messes, and see that their men breakfast regularly every morning. The men in future will not dine till two o'clock ; breakfast to be at 8.30. The Lieut.-Colonel approves of the men having coffee or stirabout as they choose. Officers on duty to visit the breakfast mess and see that it is always plentiful." [1]

Battalion Orders, April 14th.—Lieut.-Colonel Cameron begs to return his best thanks to the officers for the high order in which he found their barrack-rooms and the good dinner the men had. He has never seen anything in higher order than Captain Maxwell's company's rooms, or better dinners than his men had.

The soldiers' letters were forwarded by the commanding officer, and he trusts to their honour not to put their names to a civilian's letter for the purpose of getting it franked.

He is surprised to find gambling in the barrack-rooms, and directs the n.c. officers to put a stop to it.

This year officers of Fusilier and Light Infantry Regiments, and the light companies of other regiments, were ordered to wear wings in addition to their epaulettes, and in July the following Order for ever relieved the soldier from the tail or club hitherto worn by all ranks ; and as the civil population always follow the army in the cut of their hair, the queue was afterwards seldom worn by young men of fashion.

G.O., Horse Guards, February 20th, 1809.—In consequence of the state of preparation for immediate service, His Majesty has been graciously pleased to dispense with the use of queues until further orders. The hair to be cut close to the head in the neatest and most uniform manner, and to be constantly combed, brushed, and washed. For the latter essential purpose it is His Majesty's pleasure that a small sponge be hereafter added to each man's regimental necessaries.

In order to encourage volunteering from the militia to the line, a War Office Order was issued, 29th April 1809, allowing a proportion of one sergeant and one corporal for every twenty volunteers

[1] At this time the contract price of the 4-lb. loaf was 11¼d. in Cambridge and 8 3/16d. in Essex. A breakfast mess was not general throughout the army till many years later.

to enlist, and to be supernumeraries to the establishment until vacancies occur.[1]

On May 13th Captain Archibald Campbell and Lieutenant Angus Fraser were selected for employment in the army of Portugal.

Two or three married men in a company were now allowed to work between parades, "for eight days, when others will have their turn."

By this time the fever patients had entirely recovered, and the battalion was in good condition and fit for active service; but, however attentive Colonel Cameron was to their comfort, the officers and men found him a strict commanding officer when strictness was required.

R.O., 17th May 1809.—In consequence of the great drunkenness and absence from parade among the men, the lieut.-colonel orders four parades a day instead of two, which he hopes will have the effect of making officers look better after their companies, and the men more careful of themselves.

It must not be supposed that at this period the vice of drunkenness was confined to the army. Intemperate habits were only too common in all classes; lords and lairds, statesmen and lawyers, constantly got drunk, and were not ashamed, public opinion as yet not attaching any disgrace to the proceeding. One might almost say that the army was the only profession which adopted a higher standard.

R.O., 17th May.—Sergeant Francis Murray is appointed sergeant-major in room of Sergeant Mackay, who resigned the situation. Sergeant Kennedy of the Light Company to be assistant sergeant-major, and Sergeant Sutherland, drill-sergeant, to act in the above capacities till further orders.

At this time there seem to have been few, if any, military chaplains, and the regiment attended divine service in the Established Church of England, little importance apparently being attached to particular religious tenets.

R.O.—Officers and men who profess the Catholic religion[2] will attend all church parades in future, but previous to the service commencing, they will be indulged by being allowed to fall out if they wish it, but the commanding officer cannot see the propriety of their doing so.

On the 25th of June a letter of readiness for immediate embarkation was received.

In the summer of 1809 Great Britain had hardly any allies in

[1] In April, twenty-one volunteers from the Ross-shire Militia joined the battalion, and in May, eleven from the same regiment.

[2] There were several Roman Catholic officers, and the regiment was largely recruited in Roman Catholic districts of the Highlands.

Europe, except Austria (and she was soon obliged to join the ranks of the enemy). Portugal and Spain were indeed on our side, but both were weak, and the latter country was occupied by the French. Russia was now the ally of France, Denmark was against us, Prussia was conquered by Napoleon, his brother was King of Holland, Flanders and Italy were under his sway.

But the army of France was for the moment fully occupied on the Danube and in the Spanish Peninsula, and the British Government determined to support the Austrian struggle, and to divert the French arms, by sending a powerful expedition to the Scheldt.

Napoleon had seen the great political importance of Antwerp. He had caused an immense naval arsenal and vast docks to be constructed capable of equipping and containing half the navy of France. He intended it to be the great naval arsenal of his northern dominions, from which he could at any time threaten the independence of Britain. As yet, however, the necessary repairs of its formidable fortifications were incomplete, and they were defended by a comparatively small garrison. The capture of Antwerp was, therefore, the principal object of the British. But though time was precious, the expedition was delayed till the 28th of July, when the magnificent armament put to sea, consisting of 37 sail of the line, 23 frigates, 33 sloops of war, and 82 gunboats, besides numerous transports, having on board 40,000 troops and two battering trains, all well found and equipped.[1] Four chaplains were sent to attend to their spiritual wants.

The army was commanded by General the Earl of Chatham (elder brother of the celebrated statesman, William Pitt). He was a veteran accustomed to the routine of official duty, but without experience as a leader, and without the qualities necessary to the success of an enterprise which demanded decision of character, and activity of mind and body. His appointment was not popular with the nation, or approved by the army, who would have preferred Sir John Hope.

The fleet was under the orders of Admiral Sir Richard Strachan.

The instructions to the Commander-in-Chief were:—" You are, upon the receipt of these our instructions, to repair with our said troops to the Scheldt, and carry into effect the following instructions, in conjunction with the commander of the naval forces. This conjoint expedition has for its object the capture or

[1] Twelve squadrons of cavalry, 40 battalions, besides portions of battalions of infantry, and from 80 to 100 pieces of artillery. There were 16 brigades formed into 7 divisions. The divisional commanders were Lieut.-General Sir Eyre Coote, Lieut.-General the Earl of Rosslyn, Lieut.-General Sir John Hope, Lieut.-General the Marquis of Huntly, Lieut.-General Lord Paget, Lieut.-General Mackenzie Fraser, Lieut.-General Grosvenor; Major-General M'Leod commanded the Artillery, and General Terrot the Engineers.—" Military Memoirs."

destruction of the enemy's ships either building or afloat at Antwerp or Flushing, the reduction of the island of Walcheren, and rendering the Scheldt, if possible, no longer navigable for ships of war."[1]

In order to take part in this expedition the 1st Battalion 92nd marched in two divisions on the 29th of June for Dover, where it arrived on the 6th and 7th of July. Captain Seton writes from Romford, 2nd July 1809 :—"We are so far advanced on our way to Dover, and a great difference between marching in England and Spain. We go through part of London to-morrow, and cross London Bridge to Greenwich." "We are still ignorant of our destination. Lord Chatham is certainly Commander-in-Chief. He may be a very able statesman, but I am afraid he is a d——d bad general." "Sir John Hope (our colonel) is to have a large command in it." Again, from Dover, July 11th :—"I wish we were at this moment on the banks of the Danube ; I think we would be of the greatest use there. Our destination, of course, remains a secret."

The married women who were left behind were accommodated with rooms in barracks, or, if they preferred going to their homes, they received £1 1s. and 5s. for each child for the journey.

At Dover the battalion received a draft of 220 men from the 2nd Battalion. It was placed, along with the 20th Regiment, in the 3rd Brigade, commanded by Major-General Sir William Erskine, of Lieut.-General Sir John Hope's Division or Corps of Reserve. The battalion mustered 44 sergeants, 20 drummers, 974 rank and file, and so healthy had it become, that at Dover it had not a man in hospital, and the inspecting general remarked that he had hardly ever seen a finer looking body of men.[2] There were still a number of men absent in Portugal, who had been left for duty when the battalion advanced from Lisbon, and others who had escaped from the retreat ; these were transferred to the 2nd Battalion, as well as prisoners of war in Spain and Holland, from which it would appear that some of those taken in 1799 had not yet been released.

Colonel Cameron had been exerting himself to get for his men an allowance from Government "for tear and wear of tartan in Spain," and informs them that he now hopes to succeed. He expresses his "utter astonishment" at learning that some of the companies have scarcely any hose tartan, and thanks Captains Archibald MacDonald and M'Pherson for the care "they appear to have taken of their companies in that respect." The commanding officer had previously desired officers commanding companies

[1] Alison. Lord Chatham's Instructions.—Parliamentary Debate.

[2] Sergeant Robertson's Journal.

to provide hose tartan sufficient to make gaiters of it for each man for the campaign, but the men " are never to wear their gaiters except on a march, or on escort duty."

R.O., Dover Castle, 14th July 1809.—Sergeant Ewen Kenedy of the Light Company is appointed sergeant-major till further orders.

On July 14th the 1st Battalion marched to Deal and embarked, Colonel Cameron and the headquarters being on H.M.S. *Superb*, 74 guns. An officer of the regiment gives a touching description of the heartrending scenes and affectionate partings between the soldiers and their wives and children.

The hopes of the whole country were centred on the success of this undertaking; thousands of spectators from all parts visited Deal to see the embarkation of the army. Gaily dressed ladies crowded the beach; the town was filled to an overflow with officers, sailors, and soldiers of all corps and their relations; the bands on the ships played, the crowds on shore cheered, all was enthusiasm, excitement, and expectation. This scene of revelry continued till the afternoon of July 27th, when the " Blue Peter " was hoisted, recalling the warriors from the shore.

The fleet weighed anchor on the 28th, and, whitening the ocean with its sails, arrived on the coast of Holland next day. On the afternoon of the 30th about 20,000 men, among them the 1st Battalion 92nd, were disembarked on the island of Walcheren at a place called Bree-sand. They landed in flat-bottomed boats, forty men in each boat, and each man carrying sixty rounds in his pouch, with two spare flints, and a good flint in his piece; three days' bread and cooked pork and an allowance of rum in his haversack; to land in profound silence, and in light equipment. The regimental colour to be carried in the bow of the centre boat, and boats with the flank companies to be distinguished by a red and white pendant respectively, according to order.

The boats advanced with great regularity till within a short distance of the landing-place, when a cry of " Devil take the hindmost " ran along the line, and each put on a spurt to be first. Having effected a landing, part of the troops under General Fraser were sent against Fort der Hack, about a mile off, which the enemy evacuated; the general then continued his offensive movement towards Ter-vere, a fortress which commanded a narrow entrance leading into the channel which separated South Beveland. This place was quickly taken with its garrison of 1000 men. Goes, the capital of South Beveland, opened its gates, and Sir John Hope, pushing rapidly on, appeared before the gates of Bahtz on the evening of the 2nd of August, and that important fortress was taken by the Guards, who formed the First Brigade of Sir J. Hope's

Division. Sir W. Erskine's Brigade occupied Ter Goes, St Arcut's Kerk and adjacent vilages.

At Goes the Gordons were quartered in the Frenchmen's barracks, with detachments at various places, the garrison of Goes being employed to conduct French prisoners to the place of embarkation. At Goes there were several regimental and drumhead courts-martial, invariably caused by indulgence in Hollands gin; the prisoners being sentenced to corporal punishment, of which generally only part was inflicted.

R.O., *August* 10*th.*—The 92nd to move from present quarters to Capelle and Biesling; the men are not to wear gaiters but on the march or otherwise ordered, and never to wear trousers except on fatigue or night duty.

Lieutenant-Colonel Cameron is much surprised to observe that some of the officers have presumed to alter the regimental bonnet by adding peaks to it, and desires they will not presume to make any innovation on the regimental dress till he thinks proper to order it.

R.O., *Capelle*, *August* 11*th*, 1809.—Volunteer Mr James Hope [1] is to do duty as supernumerary officer in the Grenadier Company, and to be obeyed accordingly till further orders.

R.O., *Capelle*, *August* 17*th*, 1809.—Volunteer Mr Colin M'Dougal is to do duty as supernumerary officer in the Light Company, and to be obeyed as such till further orders.

Meanwhile, the success and rapid advance of the British to within a short distance of Antwerp, had it been followed up by the Commander of the Forces, must have resulted in the capture of that place and of the French men-of-war there, which were the principal objects of the expedition; but he, instead of pressing on, unfortunately wasted precious time in reducing Flushing, a subordinate object, which, however, he prosecuted with vigour and success. The garrison of Flushing was commanded by General Monnet, who defended himself with gallantry, but the sallies of the French troops were repulsed; on the 13th August the British batteries opened on the land side with fifty-two heavy guns, while seven ships of the line and a flotilla of bomb-vessels silenced the guns bearing on the water, destroying the defences on that side. The town took fire in various quarters, and the French General was obliged to surrender on the 16th, along with 5800 prisoners and 200 pieces of cannon, besides an immense quantity of gunpowder, military stores, etc. The Gordons took no part in this action, but they and the inhabitants of the villages where they were quartered viewed the appalling spectacle from the roofs of the churches and windmills.

[1] A nephew of Sir J. Hope, colonel of the regiment. He was promoted ensign 92nd in November.

The army and navy now looked for an immediate advance to prosecute operations against Antwerp. "Language," says an officer of the Gordons, "cannot express in terms sufficiently strong the severe disappointment which all of us experienced when, instead of being ordered to act a part worthy of the land of our birth, the whole army was kept in cantonments, and in the full enjoyment of inglorious ease, till the troops became so sickly that it seemed a matter of doubt how far prudence would lend the sanction of her name to any further military operations against Antwerp."

The time lost before Flushing proved fatal to the other objects of the expedition. The French and Dutch Governments were indefatigable in their efforts to bring up troops ; 30,000 men, under the celebrated General Bernadotte, were thrown into Antwerp, the batteries were armed, the ditches filled with water, the fleet was removed to a place of safety above the town, and batteries armed with ships' guns erected along the approaches to the place. All this time the splendid British troops were kept inactive in South Beveland, almost within sight of the city which they might have taken. According to the well-known epigram—

Lord Chatham, with his sword undrawn,
Was waiting for Sir Richard Strachan;
Sir Richard, longing to be at 'em,
Was waiting for the Earl of Chatham.

About the 20th of August the troops were attacked by fever and ague, and soon it appeared in the ranks of every battalion of the army. The Gordons were quartered partly in barns, the floors of which were damp (they had arranged to give the farmers the use of the threshing floors six hours a day), and the pestilential vapours, though not hurting natives, were fatal to the health of strangers. The doctors recommended bathing, but Sir William Erskine advised a "morning glass of brandy on rising, one at breakfast, one after dinner, and another in the evening" ; a prescription more congenial to the tastes of many in his brigade, which gave them as high an opinion of his medical as of his military talent.[1]

G.O., *August 28th*.—Owing to the great increase of sickness, officers are to explain to the men that they must avoid eating unripe fruit, cucumbers, etc., or fishing or wading in the ditches, or sleeping out of doors. Smoking is strongly recommended at night and early morning and in damp weather. Battalions and detachments

[1] "Military Memoirs." In the weekly state at Goes there were :—

	On Aug. 5th,	0 Sergts.,	1 Drumr.,	9 Rank and File sick.	
At Capelle,	Aug. 8th,	0 ,,	0 ,,	39 ,,	,,
	Aug. 26th,	0 ,,	0 ,,	106 ,,	,,
After embarkation,	Sept. 3rd,	6 ,,	0 ,,	250 ,,	present.
				100 ,,	absent.

to be regularly exercised, and full allowance of spirits to be given, and every precaution taken as recommended by the medical officers.

G.O., 21*st August*.—Each regiment and company will forthwith prepare prize-lists, in which they will comprehend every one who was present on the day of the first landing effected by the troops.

The Gordons therefore participated in the prize-money.

Under the circumstances it was unanimously decided by a council of war that a further advance was impossible, and the troops were all withdrawn into the island of Walcheren, where 15,000 men were left as a garrison to retain possession of that important conquest, and also for the purpose of keeping the large French army now in Antwerp from being used otherwise. But Austria having been forced to sign a treaty of peace with France in October, and the loss to this garrison by the distemper, since known as the "Walcheren fever," being very great, there was no longer any advantage in retaining it; and after the works and naval basins of Flushing had been destroyed, the island was finally abandoned before Christmas. Meanwhile the rest of the army, including the Gordon Highlanders, was embarked in the beginning of September, their return to England being in sad contrast to the pomp which had attended their departure.

The sick who returned at various times to England from Walcheren amounted to 12,863,[1] and many of them suffered from its effects long afterwards.

[1] Alison.

CHAPTER XII

ON September 3rd, Sir John Hope's Division embarked on transports on the eastern Scheldt and dropped down to the Vere-Gat, where they anchored, and for some reason were detained till the 8th, fever and ague increasing so rapidly that before they reached England more than half the men were under medical treatment. To prevent it spreading, the worst cases were removed into boats placed under awnings on deck.

The first man who died was a tall handsome Grenadier, and, seeing that his death had a bad effect on the spirits of the other patients, the officers endeavoured to soothe their excited feelings by kindly and hopeful words. While one of them was so employed, an athletic Grenadier named Willie Milne, whose body was stronger than his mind, leaped deliriously from his boat bed, and was only prevented by the sentry from throwing himself overboard. Early next morning the officer went on deck expecting to find poor Willie in the last stage of the disease, but instead of finding him soliciting the grim King of Terrors for a respite, he found him pressing his pay-sergeant for a beef and bread breakfast, which, being against the doctor's orders, the sergeant was in duty bound to refuse. Then with an appealing look which would have softened a harder heart than that of the worthy n.c. officer, the convalescent Grenadier said, "Weel, weel, Sergeant M'Combie, an ye think the like o' that o'er strong for my puir stamack, just gang to the cook an' gar him mak' me a drap parritch, an' gude's sake, sergeant, *to mak' them gey an' thick!*" This request of a man supposed to be dying did as much good to his companions in distress by the merriment it created as the porridge no doubt did to Willie, who lived for many years on a pension which enabled him to mix his meal as thick as he liked.[1]

On the 10th, they cast anchor in the Downs, where Colonel Cameron landed, in order to get permission from the first general officer he could find to land the battalion at once. They sailed next day for Harwich, where, owing to contrary winds, they did not arrive till the evening of the 13th. Here the worst cases of fever were landed and consigned to the care of the medical staff; some were taken in boats to Ipswich, and on the 14th the battalion landed at Land Guard Fort and proceeded that day to Woodbridge Barracks. "Never did I witness," says an officer, "a spectacle more heartrending than the removal of the sick from the transports to the shore, and thence to Woodbridge. The emaciated figures and thin pale faces of the poor sufferers, the agonised groans which the jolting of the waggons drew from their death-like lips, forced tears from my eyes as I moved along."

1 "Military Memoirs."

Owing to the state of the battalion from sickness, and the men's accounts not being settled, no officer was to apply for leave of absence.

R.O.—Lieut.-Colonel Cameron strongly recommends officers of companies to provide stirabout with milk or ale for the men's breakfast, and Mr M'Intosh, acting surgeon to the battalion, is desired to be in the hospital early and late, and never to be out of barracks for more than two hours at a time, and to make constant reports to the commanding officer : "Officers commanding companies will supply each convalescent with a pint of porter daily, as Mr M'Intosh thinks it would be good for them."

For several weeks the sick list daily increased till fully 400 men were in hospital. Many died, and at first were buried with military honours ; but afterwards, as the medical officers represented the depressing effect of the "Dead March" on their patients, the music at funerals was stopped, and the mournful processions passed on in silence, broken only by the lamentations of comrades or the wailing of a widow and her orphan children. As the pathetic old song has it—"For the Lowlands o' Holland hae twined my love and me."

G.O., Headquarters, Colchester.—The general directs that the utmost attention be given to the troops lately returned from the Scheldt, and recommends the greatest attention to cleanliness and moderate exercise, but no long drills or field days.

G.O., September 19th, 1809.—In consequence of the ill-health of the 92nd Regiment, they will be excused from garrison duty for a short time, officers excepted. Battalions returned from Holland to be inspected daily by the surgeon, that the first symptoms of illness may be discovered. No band to play near the hospital.

R.O., September 18th and 19th.—All fatigue clothing to be well washed. Men not to go about the barrack square in greatcoats. Convalescents ordered exercise by the surgeon to be marched out regularly by Sergeant Duncan M'Pherson of the 4th Company.

On the 1st October there were sick in the regimental hospital—4 sergeants, 1 drummer, 4 corporals, and 87 privates ; convalescent in barrack rooms—9 sergeants, 7 drummers, 17 corporals, and 277 privates, and a large number of sick n.c. officers and privates were sent for their health to Ramsgate, Deal, Dover, etc. A memo. on the state of October 1st says:—

Of the number in hospital, there are about fifty approaching a state of convalescence ; the remainder in hospital are in the acute state of the disease. The new cases are but few, and generally slight ; the number of deaths have been twenty-one of the sick present. As the disease generally terminates in ague, the recovery is likely to be tedious with approaching season of the year. I am of opinion the greater number of the convalescents will not be fit for active service for six months at least, and even for common duty for some considerable time. There is no doubt that the disease has made more progress in consequence of the men not having been completely reinstated in their health from the effects of the fever subsequent to the campaign in Spain.

(Signed) S. D. M'INTOSH,
Surgeon 92nd.

The battalion continued sickly throughout the autumn, and Colonel Cameron, who had received leave of absence to the Highlands, writes to his parents that the sickness among his men is so distressing that they must excuse him. " I am so wedded to these poor fellows that I cannot leave them in their present state." He also mentions some of the soldiers belonging to the parts about Locheil where his father lived at Fassiefern, for the information of their families. " The little Bo-man's [1] son is alive and doing well—he is our quartermaster-sergeant—and Kenedy's son from Moy, Peter More's friend, is our sergeant-major. The Bo-man's son has begged me to forward you (enclosed) one-half of a £10 note to assist his father's family, the other half he will forward to his brother by next post. He is a very *siccar* lad." " Poor MacKenzie from Ballachulish, or rather from Ounich, was taken prisoner in the retreat from Spain, and has not since been heard of." [2] The colonel also mentions with pleasure that no one of the men who joined along with him had behaved ill or " is even a questionable character."

But though the commanding officer felt for the sick, he had no idea of allowing any want of smartness or inattention to duty among the healthy, whether officers, n.c. officers, or private soldiers.

The Jubilee of King George III was celebrated on the 25th October. One hundred men with a proportion of officers from each corps in garrison attended a thanksgiving service at church in the town of Woodbridge. The garrison paraded at one o'clock, Royal Artillery on the right, 92nd next, Volunteers on their left, the Berwick Regiment and 2nd Hussars (German) on the left of the line. The artillery fired fifty rounds, the other troops firing a *feu-de-joie*, " the men will give three ' loud huzzas.' " " The Grenadier Company of the 92nd, with the King's colour, will then take post in front a little to the right of the main guard, open ranks and present arms as the head of the column comes up to them; the line will then march past in open column of troops and companies, officers and colours saluting the King's colour of the 92nd."

The joyful occasion was marked by the release from confinement of all delinquents imprisoned for offences of a military nature.

[1] Cattleman.

[2] This man, according to his relations still at Onich, seems to have rejoined afterwards, was at Waterloo, and retired on a pension.

Dense masses of the population came to see the review, and by re-echoing the cheers of the soldiery, proclaimed the popularity of George III—the father of his people—who had completed the fiftieth year of his eventful reign ; and no doubt soldiers and farmers afterwards joined in drinking his health with equal heartiness and cordiality. The officers of the whole garrison dined together at five o'clock, and a grand ball finished the rejoicings.

The kilt was not then recognised in England as an evening costume. Sir Walter Scott had not yet thrown over it the glamour of romance, nor had George IV made it the fashion by appearing in it at Court, as he did at Holyrood in 1822.

R.O., October 24th.—Officers parade to-morrow in the Highland dress. Staff officers to be present dressed in white pantaloons and half-boots. Officers will appear in white pantaloons and half-boots at the ball on Thursday evening.

Colonel Cameron, though a very strict disciplinarian when he saw occasion for it, took all ranks into his confidence.

R.O., Woodbridge, November 2nd, 1809.—Lieut.-Colonel Cameron thinks it necessary to inform the battalion that upon the promotion of Corporal Christie, he was not aware of some part of his conduct, which reflects little to his credit as a soldier, when on the expedition to Portugal, or he would not have promoted him upon any condition.

Some months after—R.O.—Private Alexander Cameron 2nd, of the Grenadier Company, to do duty as corporal, in the 1st Company. In making Private Cameron a corporal, the commanding officer thinks it proper to acquaint the regiment that although he has had several very strong recommendations in his favour, yet from his former conduct he could not have been induced to appoint him had not Captain M'Donald, commanding officer of his company, assured him that for some time past his conduct had been most exemplary.

Again, from the long and faithful service of Private John Davie,[1] the commanding officer is induced to give him another trial as sergeant, and hopes that his future conduct will be such as to show him sensible of such indulgence ; and the Order goes on to say how much the good or bad conduct of n.c. officers may influence that of the men who witness it.

R.O., November 2nd.—Convalescents to be supplied with flannel waistcoats, as ordered by Mr M'Intosh, and Lieut.-Colonel Cameron recommends, though he does not order, officers commanding companies to supply them to all their men before the cold weather sets in.

Captains had the whole responsibility of supplying their men

[1] John Davie had been one of the original sergeants in 1794.

with all except the clothing, etc., supplied by the colonel, as appears by the following Regimental Order—"There being a quantity of tartan in store, commanding officers of companies are to see that their men do not supply themselves otherwise till this is expended; afterwards, if commanding officers of companies prefer supplying their men themselves, the commanding officer will be very pleased with it."

G.O., Horse Guards, 15th November.—His Majesty has been graciously pleased to direct that widows and children of soldiers who have died at Walcheren shall receive double the allowance which is usually given, to assist them in getting to their homes on the common embarkation of troops; and such as are natives of Scotland or Ireland shall also be provided with a free passage at the public expense.

Standing Orders of Regiment, addition to, November 19th, 1809.—When sergeantcies become vacant in the regiment, commanding officers of companies will not recommend for the succession unless called upon to do so. The commanding officer will generally consult them on that head, unless he is perfectly aware of the person most proper to be promoted.

G.O., Horse Guards, December 4th, 1809.—No officer shall be promoted to the rank of captain until he has been three years a subaltern; or major till he has been seven years in the service, of which at least two years as captain; and no major shall be appointed lieut.-colonel until he has been nine years in the service.

By Christmas the sickness had decreased so much that the daily attendance of the surgeon at parade was dispensed with, but many of the men still required care and attention; and it was positively ordered by the commanding officer that the moment the smallest symptom of the return of the ague appears, the man be at once taken to the doctor. An officer writes in praise of the unremitting exertions of the medical staff of the battalion, "particularly Assistant-Surgeon Dunn, whose humanity would never permit him to absent himself from the beds of the poor sufferers so long as his advice could be of any use; the blessings of the soldiers saluted him at every step."

Up to this time the captains and subalterns of Highland regiments had worn two epaulettes; while in all other corps only field officers did so, the company officer having an epaulette on the right shoulder and nothing on the left. This distinction was done away with by the following order:—

G.O., Horse Guards, February 19th, 1810.—The captains and subaltern officers of Highland corps are to conform to the regulations laid down for regular infantry, and are to substitute for the epaulette on the left shoulder (which is to be discontinued) a strap of the same kind as that of the epaulette, for the purpose of securing the sash.[1]

N.B.—Officers of flank companies of Guards and other infantry are to wear a wing on each shoulder with a grenade or bugle horn on the strap.

In Regimental Orders, Colonel Cameron "ventures to risk" giving officers leave to continue their two epaulettes till June 4th (the King's birthday), when all must be according to order. It appears, however, that if Highland officers gave up the left epaulette, it was for a short time only, as they appear with two epaulettes (or wings) in drawings of the Peninsula and Waterloo period.

In January the battalion took part in garrison duty, but they were excused all night duty, only going on guard at six a.m., and being relieved by a like number of the Berwick Regiment (Militia) at six p.m.; and by the surgeon's advice there was no early drill even in April, but only the usual parade for exercise.

R.O., 21*st March* 1810.[2]—This being the 21st of March, Lieut.-Colonel Cameron will not degrade the regiment by ordering a Court-martial to assemble for trying a man for so infamous a crime as desertion. A Regimental Court-martial will therefore assemble at eight o'clock to-morrow morning.

The Sphinx, which had been taken as the regimental crest on the men's bonnets, etc., when the Marquis of Huntly commanded, was now put on the bonnets of the officers.

R.O., *April 6th*, 1810.—All officers will, on the 1st of May, wear a silver Sphinx on the cockades of their bonnets in place of the regimental button now worn. The silver Sphinx to be of the same pattern as that now worn on the undressed caps. Lieut.-Colonel Cameron requests that officers, when they go into company, will appear uniformly dressed, particularly with their belts and sashes on. He calls attention to repeated orders against officers walking about barracks improperly dressed, "*particularly with round Hatts on*."

Officers are invariably, and at all times when they wear the gorget, to have their bonnets and feathers on. The commanding officer is much surprised to see men walking out improperly dressed. No man is ever to pass out of barracks without being dressed in his full uniform, with his side arms.

Reference is made to the men showing their belted plaids and kilts, evidently meaning the small belted plaid now worn separate from the kilt. The first mention of the "Highland scarf," or shoulder plaid, as worn by field officers and afterwards by other officers with trousers, is in Orders for an inspection at Woodbridge.

R.O., 26*th April* 1810.—The battalion will parade to-morrow

[1] Officers and sergeants of other corps wore the sash round the waist.

[2] The anniversary of the Battle of Alexandria.

morning at eleven o'clock ; assembly to sound at half-past ten. Every person to be present in the best possible order—sick and men on duty only excepted. The duty men to be in complete Highland dress.[1] The convalescents in grey pantaloons, with greatcoats slung. Officers in full Highland dress, with the breasts of their jackets open, and without gorgets. Field and staff officers in white pantaloons and boots, and mounted officers with the Highland scarf on. Ten rounds for Grenadier and battalion companies ; twenty rounds for Light Company. The lieut.-colonel informs the officers that he does not intend to make any alteration in the regimental purses of the officers.

A circular from the Horse Guards at this time finds fault with a fashion prevailing in many regiments of making the coats too tight to wear the waistcoat under them, and the waistcoat too small to be used as a fatigue jacket in summer, as intended. Also, that the stocks are made too high, and uncomfortable for the men.

There was no dry canteen, but married women of good character in the battalion were allowed, "for the support of their families," to sell bread, vegetables, etc., in the barracks.

As many men were still weakly, they were excused guards even by day, and the Berwickshire Regiment was ordered to find all garrison duties ; "the 92nd will, however, furnish the fatigue parties." This continued easy life seems to have been too much of a good thing for the conduct of some among the men. Idleness is the mother of mischief, and there were in those days no recreation or reading rooms to give physical or mental occupation and amusement.

R.O., 30*th April* 1810.—As it is evident to the commanding officer that whatever indulgence he might feel inclined to grant to the regiment is instantly abused, even when given with the intention of re-establishing their health, he is now determined to try the effect of moderate drill with them, and orders two drills a day except Wednesdays, Saturdays, and Sundays.

Soldiers were allowed, under peculiar circumstances, to purchase their discharge by a sum of money in place of substitutes, on the following scale :—

	For Limited Service.			For Unlimited Service.		
Soldiers finally approved,	£37	5	0	£47	15	0
First-class boys,	25	1	0	33	8	0
Second-class boys,	11	9	0	13	11	0

Calculated at twice the levy money for each recruit.

Up to this time, as far as can be gathered from the Regimental Orders, there had been no instance of mutinous conduct on the

[1] That is, with plaids and purses, neither of which were worn on ordinary occasions.

part of any soldier of the Gordon Highlanders ; unhappily, such a case now occurred ; but on the man's repentance, and at the request of Colonel Cameron, the Commander-in-Chief allowed the soldier to volunteer for service in India, and thus save himself and his regiment the discredit of a general court-martial.

R.O., *20th May* 1810.—Private George Robertson of the Grenadier Company, having conducted himself in a mutinous and disrespectful manner to Captain Dunbar in the execution of his duty, he applied to the commanding officer for a general court-martial for him. But out of regard to the regiment, the commanding officer had induced Captain Dunbar to agree to Robertson's voluntary wish, who appeared sensible of his gross breach of military subordination, to allow him to be disposed of to serve abroad, which the Commander-in-Chief was pleased to accede to.

R.O., *Woodbridge*, *4th June* 1810.—The men in future to wear their new clothing. All the old clothing is to be laid aside, with the exception of the kilt, which the men are constantly to wear, except when on duty or otherwise ordered, when they will appear in their full Highland dress. Convalescents that cannot wear the kilt are never to be seen out of barracks on any account unless regularly marched out of barracks for exercise. Men in hospital to wear old jackets and pantaloons. The commanding officers of companies to explain to their men that on no account whatever is any man to be seen outside barracks without his kilt on and otherwise regimentally dressed. On account of the late sickness in the regiment, the men will be allowed to put on their pantaloons after *nightfall*, but they are on no account to go out of barracks in that dress. Officers are to wear the Highland dress when on garrison duty, and on Sundays. The men are to be careful to preserve their plain bonnets.

These Orders to be read on three successive parades by an officer.

The colonel was evidently not an advocate for matrimony.

R.O., *June 9th*, 1810.—From the strong certificate the commanding officer has had of the young woman, and from Private John Campbell's own good character, he has granted him permission to marry ; at the same time, it is forming a connection which he strongly wishes to recommend every soldier to avoid, and his consent can never be obtained but when the most unquestionable certificate can be produced of the moral good character of the female.

It being considered that complete change of air would be of advantage to the invalids, it was determined to move the battalion to Canterbury, and on the 14th of June they were held in readiness.

Sobriety is not the virtue of northern nations, and in the early part of the century it was more, even than now, a vice common to the English, the Scots, and the Irish. Though it was more actively discouraged among soldiers, they, having ready money and spare time, were not far behind their civilian compatriots in their indulgence in intemperance; and the Gordons seem to have found the Saxons' strong ale a very sufficient, though perhaps not so rapid a means of drowning their cares as their own mountain dew. As their regimental poet sang of English quarters, "Far an dhuair sinn leann am pailteas ged bha mac-na-praisich gann oirn."[1] On the 8th of July, Colonel Cameron tells the men that should he ever observe the drunkenness he had seen that day, he will adopt severe measures, and will also immediately apply to have the battalion put on duty.

On the 11th of July the 1st Battalion marched to Land Guard Fort,[2] where it embarked, landed at Ramsgate, and arrived at Canterbury on the 20th.

R.O., Headquarters on board the "Diligent" Transport, July 19th, 1810.—Lieut.-Colonel Cameron begs leave to return his thanks to the men for their exemplary good conduct upon the march for embarkation, in consideration of which he will forgive the only two exceptions which came under his notice, though they were very glaring ones, in hopes that they will conduct themselves equally well upon disembarking. He wishes he could thank the officers for their undivided attention to the service upon the march, but he was sorry to observe a greater attention on the part of some to their private concerns than to the service. He will avoid mentioning individuals, but those that marched with loads of household furniture may be aware he alludes to them, and he can only say, next time he marches with the regiment he will have a general inspection of baggage previous to the march. He assures the officers, that he considers the regiment disgraced and affronted by the scramble about baggage, the quantity being more like a regiment of Militia than a regiment of the Line. . . . Officers are requested to be very particular at the landing at Ramsgate as to the men being well turned out. The convalescents to be regularly marched, as many of them as possible in the Highland dress.

R.O., July 22nd, Canterbury.—As the regiment is to be in-

[1] "Where we got ale in plenty, though the son of fermentation (whisky) was scarce to us."

[2] By Act passed 1810, soldiers on the march, in place of having full diet provided by the innkeepers or others, are only to have one hot meal provided, consisting of 1¼ lb. of meat weighed before being dressed, 1 lb. of bread, 1 lb. of potatoes or its equivalent in other vegetables, pepper, vinegar, and salt, and two pints of small beer. The Government is to pay 8d. for this meal. The horse soldier is to contribute 7d. and the foot soldier 5d. out of his pay and beer money towards this sum of 8d., and the soldiers are to receive the difference between this 8d. and 1s. 4d., the sum formerly paid for full diet, in order to provide themselves with other articles of subsistence.

spected by Lieut.-General Nicols to-morrow morning, the commanding officer expects every man will be in the best possible order in full Highland appointments. The convalescents that wear the pantaloons will be sent to the hospital to-morrow morning at nine o'clock, there to remain till ordered to join their companies. Those that parade in the kilt will fall in with their companies properly dressed. Pioneers with their pioneer appointments. Regimental officers in Highland dress, jackets open, without gorgets. Field officers and staff officers in white pantaloons and half-boots. Officers of companies not to pay their men till after the inspection.

24th July.—In future at the morning drills and parades the men will wear their white jackets, kilts, and new forage caps, in which dress every man to appear till after dinner, when they will resume their regimentals.[1]

The battalion was also inspected on the 28th by Major-General the Hon. Charles Hope.

On August 1st, 1810, the number of sick in hospital had been reduced to 1 lieutenant, 1 sergeant, 3 corporals, and 81 privates, and the sick absent—in England, 8 privates ; in Scotland, 3 lieutenants, 2 sergeants, 2 corporals, and 10 privates ; in Ireland, 1 lieutenant and 1 ensign.

Colonel Cameron went on leave, and Major MacDonald assumed the command on the 29th, and "is very much mortified at the very shameful excess of drunkenness to-day, and what adds to the impropriety and disgrace is its being Sunday." He calls on officers and n.c. officers "to mark in the most severe manner" this shameful conduct, and orders a regimental court-martial for the trial of the principal offenders.

R.O., August 6th.—The commanding officer grants permission to a few men, who are in debt, to work in order to clear themselves.

August 9th.—Major MacDonald is very sorry to find some of the men still persevere in irregularity, particularly some men of the Light Company. One word to the n.c. officers of the Light Company, that they had better look sharp, for this is the second instance within a few days of several men of that company being brought to punishment, which they might have prevented if they had formerly done their duty.

D.O., Canterbury, 18th August 1810.—In consequence of the very precarious state of the weather, Lieut.-General Nicols directs that every possible assistance should be afforded by commanding officers of corps throughout the Kent district for expediting the getting in of the corn harvest.

[1] The usual walking-out dress was red jacket, feather bonnet, and kilt without purse. When the purse was to be worn it was specially ordered—either "full Highland appointments" or "the men will have on their purses."

In consequence of this Order, 3 sergeants and 128 rank and file of the battalion were allowed leave of absence to work at the harvest.

D.O.—Lieut.-General Nicols finds it necessary to remind officers of one of the first things they are informed of on joining their corps, viz. that by His Majesty's Orders officers while present with their corps are to appear only in their regimentals.

At Canterbury twenty-one invalids were sent to Chelsea to go before the Pension Board for discharge, their greatcoats, purses, belted plaids and new forage caps being received from them and taken into store. They are all given characters as "good"—"a good man and a good soldier," "a very good man"—except one, "very indifferent," presumably a German musician named Ferdinand Vicarmann.

The foregoing extracts from Orders and letters give a sketch of the life of the Gordon Highlanders on home service at this period.

The 1st Battalion had, as we have seen, been employed during the last two years in Denmark, Sweden, Portugal, Spain, and Holland. It was destined soon to proceed again to the Peninsula and to take an active part in the campaigns of the next four years in Portugal, Spain, and France.

The climate of Kent restored the health of the men, and they were now fit, not only for duty at home, but for active service abroad, though the effects of the Walcheren fever and ague were afterwards felt by many when put to the test of exposure and hardship.

Since the embarkation of Sir John Moore's army at Corunna in January 1809, the French had possessed themselves of that important fortress and of a great part of Spain. The Spaniards had been defeated at all points; Saragossa and Gerona, after heroic defence, had been obliged to capitulate. The Spanish Junta took refuge in Cadiz, which, in February 1810, was invested by a French army, and 6000 British troops assisted in the defence. Marshal Soult had invaded Portugal and stormed Oporto. The small British force left by Sir John Moore, when he advanced into Spain, was concentrated under Lieut.-General Sir John Cradock for the defence of Lisbon.

The Government felt that the fate of Britain was inseparably connected with that of the Peninsula; that so long as the war was maintained there, it would be averted from the shores of Britain; and it determined to stand by Spain to the last. Large reinforcements were sent to Portugal; Portuguese troops were raised by conscription, taken into British pay, and commanded by British officers. In April 1809, Lieut.-General Sir Arthur Wellesley was appointed to command the army in the Peninsula.

By the famous passage of the Douro he compelled Soult to retreat from Oporto—a retreat which was much more disastrous to the French than was that of Sir John Moore to the British six months before. The French disgraced this retreat by the most savage cruelty,[1] and their discipline and conduct were infinitely worse than that of Sir John Moore's army. Their losses amounted to about a fourth of the whole troops that were attacked on the Douro, besides all their artillery ammunition and even a considerable part of their muskets.[2] The advance of Marshal Victor obliged Sir Arthur to desist from the pursuit of Soult and join the Spanish General Cuesta, who in March had been totally defeated by Marshal Victor at Madellin. The combined forces under Wellesley were attacked at Talavera on the 27th and 28th July 1809, and the French army, commanded by Joseph Bonaparte, was defeated.[3] For this victory Sir Arthur Wellesley was raised to the peerage by the title of Viscount Wellington.

Private John Cameron of the 92nd appears in Regimental Returns as killed on July 28th ; he was probably one of the Corps of Detachments mentioned by Napier as having done good service in this campaign. Three officers, 8 sergeants, and 71 rank and file of the 92nd formed part of the 1st Battalion of Detachments, which also contained details from the 28th, 38th, 43rd, 52nd, and 95th Regiments (Fortescue). The company was composed of those left on duty and sick in Portugal when Sir John Moore entered Spain, and also of some who had escaped from the French after being taken prisoners on the retreat to Corunna. In April 1809 it was proposed with the pay sergeant's approval to make their kilts into trousers. A deputation of soldiers went to the captain and got this order countermanded. They often met with attention on account of their strange dress. At Coimbra a number of them visited the famous library, where the professors and librarian were very civil and told them that they had books in all European languages, and that one or other of the professors could talk in each tongue. Sergeant M'Bain tried them with Gaelic, and afterwards boasted that he had defeated the most learned men of Portugal. The company formed part of the 1st Battalion of Detachments at the famous passage of the Douro and the capture of Oporto, when the battalion lost about 50 officers and men killed and wounded, but the number of casualties in the 92nd Company is not given. At Talavera, July 27th and 28th, 1809, the 92nd

[1] Alison. [2] *Ibid.*

[3] On the day after the battle General Craufurd, with the 43rd, 52nd, and 95th (Rifle) Regiments, arrived. Having been falsely informed that Wellington had been defeated on the 27th, Craufurd, withdrawing fifty of the weakest men from his ranks, pressed on with the utmost expedition, arriving on the morning of the 29th, these gallant men having in twenty-six hours covered sixty-two English miles, carrying from 50 to 60 lb. each. They left only seventeen stragglers *en route*, and immediately took charge of the outposts. The above time and distance are from Napier. Oman, however, gives "43 miles in 22 hours," which is probably correct.

Company, then about 60 rank and file, lost 6 killed and 24 wounded, of whom Nicol was one. The British having retired after their victory, their wounded were left to the care of the French, who treated them well. They lay in the same rooms with the French wounded in the town of Talavera. When the French retreated and the Spaniards arrived, the wounded Frenchmen feared they would be murdered. One of their convalescents, whose duty it was to fetch water from a well in the street, always borrowed a Highlander's kilt and hose, saying he was safe now as a "Grenadier sans-culottes." The 92nd officers of the company being on the sick list, it was commanded at Talavera by Captain M'Pherson of the 35th.

After the battle of Wagram (July 1809), Napoleon had reached the height of his greatness, and the peace with Austria enabled him to throw into the conflict in the Peninsula the veteran troops which had been employed on the Danube. In April 1810, he had espoused Maria Louisa, daughter of the Austrian Emperor. He now proposed to subjugate Portugal, as well as Spain, and the army of Portugal, 90,000 strong, under Massena,[1] threatened to carry out the proposal.

Ciudad Rodrigo and Almeida were captured by the French ; and Viscount Wellington was obliged to fall back on the strong mountainous position of Busaco, where, on the 27th September 1810, with his British and Portuguese troops, he gallantly repulsed the French under Massena and Reynier, with a loss of 1800 killed and 3000 wounded—among them the Generals Foy and Merle. It was the first time the Portuguese troops had stood victoriously beside the British against the French, and the moral effect on their conduct was most important. Wellington, with the foresight and caution which are as necessary to success in a general as gallantry and activity, had, during the previous twelve months, employed the British engineers in forming the *lines of Torres Vedras* ; he now retired his army to this stronghold, which the advanced guard reached on the 8th of October, and by the 15th the whole army was collected within the lines, now completed and armed with 600 guns.

This was the state of affairs when the 1st Battalion 92nd, commanded by Major Archibald MacDonald, landed at Lisbon on the 8th October 1810. It had been held ready for embarkation by District Orders, Canterbury, September 9th, and embarked at Deal, 21st September, on board the ships *Audacious*, *Apollo*, and *Vestal*.[2]

[1] Massena, a Marshal of France, Prince of Essling, and Duke of Rivoli.

[2] Embarkation Return of 1st Battalion 92nd Regiment of Foot at Deal, 21st September 1810 :—Colonel, 0 ; lieut.-colonel, 0 ; majors, 2 ; captains, 7 ; lieutenants, 14 ; ensigns, 5 ; paymaster, 0 ; adjutant, 1 ; quartermaster, 1 ; surgeon, 1 ; assistant-surgeons, 2 ; sergeants, 48 ; drummers, 17 ; rank and file, 821 ; women, 16 ; children, 4.

The officers who landed at Lisbon on October 8th, when the Gordon Highlanders entered on this memorable and romantic campaign, were:—

Major Archibald MacDonald.
,, Peter Grant.
Captain Donald M'Donald.
,, John MacPherson.
,, Samuel Maxwell.
,, Robert N. Dunbar.
,, James Lee.
,, Geo. W. Holmes.
,, Ronald MacDonald.
Lieutenant J. L. Hill.
,, John Hill.
,, Samuel Bevan.
,, John Warren.
,, John Ross.
,, William Fyfe.
,, John Cattanach.
,, J. A. Durie.
Lieutenant Allan M'Nab.
,, J. J. Chisholm.
,, Robert Winchester.
,, Thos. M'Intosh.
,, Donald MacDonald.
,, Andrew Will.
Ensign Patrick Leitham.
,, George Mackie.
,, A. M'Pherson.
,, Ewen Ross.
,, T. R. Meade.
,, George Gordon.
Adjutant Claud Alexander.
Qmr. Duncan M'Farlane.
Surgeon Duncan MacIntosh.
Assist.-Surgeon Henry Dunn.
,, J. P. M'Robert.

Officers Absent

Colonel the Hon. Sir John Hope, Lieut.-General.
Lieut.-Colonel Cameron, on the way to join.[1]
Captain J. Seton, duty at Canterbury.
,, Ewen MacPherson, duty in Portugal.
,, Dugald Campbell, not joined from 2nd Battalion.
Lieutenant Angus Fraser, duty in Portugal.
,, H. H. Rose, ,, ,,
,, W. Baillie, sick at Inverness.
,, Dougald M'Pherson, sick at Laggan, Kingussie.
,, Ronald M'Donell, ,, Fort-William.
,, Thomas Hobbs, ,, King's County.
,, George Marshall, duty at Canterbury.
,, Alexr. Gordon, ,, ,,
Ensign John Mackie, not joined from 2nd Battalion.
,, Luke Higgins, absent without leave.
Paymaster James Gordon, on the way to join.

[1] Lieut.-Colonel Cameron had gone on his long-deferred leave, and was grouse-shooting at Cluny in Badenoch when he heard of the battalion being ordered for service, and at once started to join. The following extract from a letter from him on arrival in London shows the difficulty and expense of travelling in 1810:—"I arrived here at five o'clock this morning. I had an outside and inside place on the mail, and by changing with Ewen (Ewen M'Millan, his soldier servant) stage by stage, we made it out without stopping at York. Guess my travelling expenses, without including one other iota, from Cluny to London, Ewen and I, £36. Ewen is in bed all day; but I have not been in bed since we left Edinburgh, having been at all the offices during the day." (In 1810 the time occupied by the mail between Inverness and London was six days.—*Hay's Post Office Recollections.*) After all, he was too late for the battalion, but joined with "Ewen" on the 20th October at Torres Vedras.

The celebrated lines of Torres Vedras, having the sea on their left, the great river Tagus on the right, and the city and harbour of Lisbon in the rear, consisted of three ranges of defence. The first, extending from Alhandra on the Tagus to the sea at the mouth of the Zizandre, was twenty-nine miles long ; the second stretched from Quintella on the Tagus to the sea at the mouth of the St Lorenza, at a distance varying from six to eight miles in rear of the first ; the third, intended to cover a forced embarkation, extended from Passo d'Arcos on the Tagus to the Tower of Junquera on the coast. Here an outer line, constructed on an opening of 3000 yards, enclosed an entrenched camp, designed to cover the embarka-

Girl of Guarda, Armed Peasant, Officer of Infantry, Peasant of Torres Vedras
(*From "Sketches in Portugal," by the Rev. Mr Bradford, Chaplain to the Forces,* 1809)

tion with fewer troops, should the operation be delayed by weather ; and within this second camp, Fort St Julian's, whose high ramparts and deep ditches defied an escalade, was armed to enable a rear guard to protect the army. The nearest part of the second line was twenty-four miles from Passo d'Arcos, and some parts of the first line were two long marches distant; the principal routes led through Lisbon. The second line was the strongest, and it was there that Wellington had originally intended to make his stand ; the first being meant rather to retard the enemy's advance, and to enable the army to take up its ground on the second.

But Massena had delayed so long in reducing Ciudad Rodrigo and Almeida, that the British engineers had time to render the first line so capable of defence, that Wellington resolved to abide his opponent there. It consisted of thirty redoubts placed on a ridge of heights, on which were mounted 140 guns ; the great fort on Monte Agraca in the centre was perched on an eminence that overlooked the town of Sobral and the exterior lines, and from it signal posts communicated over their whole extent. A road ran along the front of the position from Torres Vedras by Runa, Sobral, and Aruda to Alhandra, and the five highways leading through the barrier were all palisadoed ; the left, from the sea to the town of Torres Vedras, was defended by the river Zizandre, which was unfordable ; the redoubts were armed with *chevaux de frise* ; while the intervening spaces which were not fortified (except where a scarp was executed along the brow of the ridge) were formed into encampments for the troops, under shelter of the guns of one or other of the works. On the whole fifty miles of fortifications, no less than 600 pieces of artillery were mounted on 150 forts. "Neither the Romans in ancient, nor Napoleon in modern times have left such a proof of their power and perseverance ; and they will remain in indestructible majesty to the end of the world, an enduring monument of the grandeur of conception in the chief who could design, and the nation which could execute, such a stupendous undertaking."[1]

[1] Alison. Yet at this time, Members of Parliament, backed by a considerable part of the British public, entirely without experience of war, and ignorant of the special difficulties he had to contend with, were clamouring against Wellington because he had not followed up his success at Talavera.

CHAPTER XIII

ON landing, the battalion found Lisbon in the greatest confusion, and thronged with country people fleeing with their goods from the horrors of war. Here it had ammunition served out and camp equipment, such as blankets, camp kettles, bill-hooks, spades, pick-axes and felling axes, medicine panniers, etc., with fifteen mules for the battalion with a Portuguese muleteer. The officers had also thirteen private mules. They marched from Lisbon on the 10th October, halting on the 12th at Sobral, where they met that part of the army under the immediate command of Lord Wellington, who arrived at this time, and fixed his headquarters at Peronegro, a short distance in rear of Sobral, a telegraph being erected on a high rock, so that he could communicate with every part of the lines.

Several skirmishes took place, especially near Sobral on the 14th, when, on attempting to dislodge the 71st[1] from a field-work, the French troops were repulsed and pursued to their entrenchments. In these affairs the Allies lost about 150 killed and wounded, among them General Harvey; but it does not appear that any of them belonged to the 92nd. The loss of the enemy was greater.

The weather was wet and they had no tents, and several men suffered from a return of the Walcheren ague; for the first few days, also, they had cause to curse the commissariat, but afterwards the rations were good and regular, and quantities of luscious grapes in the deserted vineyards were to be had for the picking. On the 15th October they were quartered in some dilapidated houses at Crozendera, and were brigaded with the 50th and 71st Regiments and a company of the 5th Battalion 60th Rifles, under Major-General Howard, whose brigade was attached to the First Division of the army, which, with the Fourth and Sixth Divisions, was composed of troops just arrived from England and Cadiz. These were posted in a position seven miles long from Zibreira to Torres Vedras, under the immediate command of Lord Wellington himself.[2] There were about 30,000 British troops in the front line, besides 25,000 Portuguese and 5000 Spaniards. In the rear was a reserve, consisting of a superb body of British Marines, also Portuguese artillery and Militia, while the navy manned the gunboats on the river Tagus. Such were the resources provided by the British Government, that not only did 130,000 fighting men receive rations within the lines, but the multitude who had taken refuge from the surrounding country occupied by the French, amounting, with the population of Lisbon, to at least 400,000 more, were

[1] The 71st had left Canterbury at the same time as the 92nd.

[2] The First Division was under Lieut.-General Sir Brent Spencer.

provided with subsistence,[1] and the troops of every description were never so healthy nor in such high spirits.[2]

For a month the battalion was employed with the other troops in outpost duty and in strengthening the fortifications. Meanwhile Massena, whose advance had been delayed by the battle of Busaco, arrived in the middle of October in sight of the formidable barriers, the existence of which he had only learned shortly before. Astonished at the strength and extent of the works, he devoted some days to reconnoitring the lines ; when, finding them to be impracticable, he disposed his army along their front, when the above-mentioned fighting took place at Sobral, in which town the whole of his Eighth Army Corps eventually established its position. General Foy was sent to report the circumstances to the Emperor, and ask for instructions.

The French at once set to work to raise entrenchments and redoubts, and the war was reduced to a species of blockade. Massena hoped that the British Government and the Portuguese Regency would be intimidated and abandon the contest,[3] or that he could feed his army on the country, by spreading movable columns to the rear to seek provisions, and forming magazines at Santarem till reinforcements arrived to his assistance. Lord Wellington, on the other hand, hoped to starve him out before his succour could arrive. Between the two armies there were a number of deserted farms and houses unoccupied by either party ; and in this neutral ground the British and French soldiers constantly met when looking for potatoes and other vegetables, without molesting each other—those who knew a little Portuguese or Spanish using these languages in their intercourse ; and they were to be seen shaking hands and drinking from each other's canteens—" Drink with you to-day ; fight with you to-morrow," they would say laughing ; for the French, though brutal and cruel in their treatment of the country people, respected the red-coats, and when not actually fighting with them, were generally friendly and civil.[4]

The increasing strength of the works, reported by British deserters (unhappily very numerous at this period),[5] added to his

[1] Corn was bought in Ireland, America, Egypt, and Algiers, and purchased at any price.

[2] Alison.

[3] Massena's hope was not without foundation. The fears of the British Government had been plainly disclosed ; and the factious opposition of some members of the Portuguese Regency greatly increased Wellington's difficulties, and this was known in France through the British newspapers. It was on Wellington's calm confidence and firmness alone that the fate of the Peninsula and of Europe depended.

[4] J. Fergusson and other old Gordon Highlanders.

[5] Napier.

own observation, convinced Massena that it was impossible to force the lines with the troops at his command. His army suffered from sickness and from the vengeance of the Portuguese, excited by the wanton excesses of his foraging parties, who had reduced the country far behind him to a desert ; and yielding to necessity, he at length resolved to fall back upon the strong position of Santarem, where a fertile country would afford supplies to his army.

His dispositions were made with ability. The morning of the 15th of November was foggy, and it was only some hours after daybreak that the British outposts discovered that the French army was retiring. All was joy and excitement in the British lines. Wellington immediately directed the Second and Light Divisions to follow the enemy, whose intention was not clearly developed ; it might have been to attack the lines at Torres Vedras ; so he kept the principal part of his army stationary at first, but on the 16th, it being clear that no attack was intended, the First Division was brought on to Alemquer, being followed by the Fourth, Fifth, and Sixth, while Hill with the Second Division was ordered to cross the Tagus to head the French from passing that river.

The 92nd had marched on the evening of the 15th into the town of Sobral, which the French had occupied till the previous night. Close to the town they saw a soldier hanging on a tree and another apparently flogged to death—signs of the revenge which the French had brought on themselves by their treatment of the inhabitants. They also noticed the cleverness of these old campaigners, who had made the enormous wine casks of the country into sleeping places for their outposts. Continuing the pursuit, they marched three days, following the Light Division by Alemquer to Cartaxo, where they came up with the French rear guard. During this advance many prisoners were made, principally stragglers or marauders, and a remarkable exploit is recorded by Napier, performed by Sergeant Baxter of the 16th Light Dragoons, who, with only five troopers, came suddenly upon a picket of fifty men who were cooking. The Frenchmen ran to their arms and killed one of the dragoons, but the rest broke in amongst them so strongly that Baxter, with the assistance of some countrymen, made forty-two captives.

On the 19th, Wellington, supposing Massena to be in full retreat, made dispositions for assaulting Santarem with a small force, thinking he had only to do with a rear guard. The Light Division advanced between the Rio Mayor and the Tagus, Pack's Portuguese Brigade and the cavalry were to turn the French right, while the First Division (in which was the 92nd) was to attack the causeway which ran through a marsh from the Rio Mayor to the heights on which the enemy had placed artillery. The columns were formed for the attack, the skirmishers were exchanging shots with the enemy, when it was found that the guns had not arrived ; and Lord Wellington, not quite satisfied with the appearance of his adversary's force, cautiously manœuvred till evening, when he ordered the troops to retire to their former ground. During the day our men could see the French occupying every advantageous spot, their advanced sentries returning the fire of our skirmishers ; while large bodies of troops could be seen, some under arms, some cooking, while others were felling trees and forming abattis and entrenchments on the opposite hillsides. On the 20th the demonstrations were renewed, but as the enemy's strength and intention to fight on his admirably chosen position was now evident, they soon ceased ; and Massena having advanced his Second Corps under Clausel towards Rio Mayor, Wellington was obliged instantly to withdraw the First Division to Cartaxo to check it, the 92nd being quartered near that place at Almostal, where they remained some days. On this march they found more evidence of the retaliation by the peasantry on the invaders, in the mangled bodies of French soldiers captured as they had fallen out of their ranks from fatigue or for plunder.

Massena, though he had retired about forty miles, had no intention of retreating further, so long as he could find food for his troops. He posted a strong rear guard at Santarem, a walled city situated on a high hill near the Tagus, one of the strongest positions in Portugal, approachable from the west only by the narrow causeway through the marsh formed by the Rio Mayor, and flooded in winter ; the main body of his army was cantoned behind in the rich valley of the Zezere, so disposed as to menace a variety of points, and at the same time to command two distinct lines of retreat. There he determined to await additional troops from other parts of Spain. He calculated also on the effect his maintaining his position at Santarem would have on the suffering inhabitants of Lisbon, and on the British Parliament, by rendering the final success of the British so doubtful in appearance.

Wellington on his side had many difficulties, both military and political, to consider. A successful battle was desirable ; it would relieve the horrible sufferings of the people of Portugal, and would silence opposition both in London and Lisbon ; but the attack must be made in a difficult country, the rivers and even the roads rendered impassable by winter rains. But the loss of a serious engagement might cause the opposition to triumph, and the troops to be withdrawn from Portugal. If Massena lost even a third of his force, there were troops at hand to replace it ; if Wellington failed, the Lines were gone, and with them the whole Peninsula. He would not risk a battle except on advantageous terms, and these were at present not to be had. He determined to remain on the defensive, and to strengthen his position, watching the roads and mining the bridges, so as to render impossible any sudden incursion of the enemy. Torres Vedras was still occupied in force lest Massena might make an attack on that side. Hill was with two divisions on the opposite bank of the Tagus, and by his activity prevented the French from adding to their resources by foraging on that side ; the rest of the British army was cantoned along the front opposite the French.

In these dispositions the headquarters of the First Division were at Cartaxo, the 1st Battalion 92nd being stationed at the neat little village of Alcantrinho, where they arrived from Almostal on 28th November. The place had been deserted by most of its inhabitants, who had fled from the French and taken refuge in Lisbon; but the Gordons made themselves very comfortable in these winter quarters, and parades, drill, etc., went on in the usual routine, varied, however, by picket duty ; while their interest was kept up by the constant rumours of the movement of the French armies in Spain, and the engagements which took place between them and the Spanish commanders.

Lord Wellington, though he was himself fully occupied in improving the discipline and organisation of the Portuguese troops and militia, as well as in quieting the political troubles occasioned by certain disloyal Portuguese, thoroughly understood that "all work and no play makes Jack a dull boy," and that soldiers, like other people, are the better for a little amusement in their leisure hours. He encouraged games and entertainments among the troops, and, recognising that the chase is the best training for war, he kept a pack of foxhounds, which hunted the country in rear of the army, but the followers were desired on no account to cross the line between the two armies. The huntsman was a private in the Coldstream Guards, who had been in a hunting establishment before he enlisted. One day, after a long run, the fox crossed the line of demarcation ; the officers pulled up and called to the soldier huntsman to do the same. "Where my hounds go I go," said Crane as he galloped on. Having killed his fox, he was about to return to the British outposts, when he was pounced upon by a picket of French dragoons and carried off to headquarters. Massena, however, sent him and his pack back with a courteous note to Lord Wellington.[1]

Thus the time passed merrily enough ; even outpost duty was not without its interest. The pickets could see at Almeyrim on the left bank of the Tagus the splendid residence of the Marquis

[1] The Duke of Wellington afterwards bought his discharge, and "Tom Crane" became a well-known authority on hunting, and died as huntsman of the Fife Foxhounds.

de Alorna, once a general in the army of Portugal, who had espoused the cause of the French and now returned as their guide and counsellor. This miserable man was one of those who fomented Wellington's political troubles. He was resident at Santarem, where, once respected by all, he was now treated with looks of contempt and indignation as a traitor to his country. Some of the outposts were so near those of the enemy, divided only by a small river, that they could see the dragoons exercising, men cleaning their arms, and on a calm day hear what was said by the vigilant officers as they patrolled the opposite bank, with whom our officers often courteously conversed. "One day some of them saluted us from the opposite bank—'Bon jour, Messieurs'—asked after Lord Wellington, said he had done wonders with the Portuguese, and praised his conduct of the campaign. Then they asked if our King was not dead,[1] and on our replying, 'No,' one said, pointing to another, 'Le général dit que tout le monde aime votre roi George, qu'il a été bon père de famille et bon père de son peuple' (the general says that every one loves your King George, that he has been a good father of his family and a good father of his people).

"We quizzed each other; they asked us how we liked *bacallâo*[2] and *azete* for dinner instead of English roast beef; and we, what they did at Santarem without the cafés and Salles de Spectacle of Paris. They replied, laughing, that they had a theatre, and asked us to come and see the play of that evening, 'L'entrée des Français dans Lisbon' (the entry of the French into Lisbon). One of our party quickly answered that he recommended to them, 'La répétition d'une nouvelle pièce—la fuite des Français!' (the rehearsal of a new play—the flight of the French). They burst into a loud, long and general laugh, the joke was too good—too home. Their general then pulled off his hat, and wishing us good day with perfect good humour, they went up the hill."[3] On another occasion our men were astonished by a French soldier calling over the stream, "How are the jolly old Buffs getting on?"—evidently a deserter.

One day some French troops were about to kill a bullock, when it broke loose and galloped towards the 92nd, one of whom shot it, and they proceeded to cut up their prize in view of their hungry and disappointed foes. Two French soldiers, waving a white handkerchief by way of flag of truce, came over with a message from their officer that he was sure the Scottish soldiers were too generous to deprive his men of their only provisions;

[1] King George III was at this time suffering from the mental malady from which he never recovered. Shortly after, the Prince of Wales (afterwards George IV) was made Regent.

[2] Salt fish, etc.

[2] "Recollections of the Peninsula," by Sherer.

on which half the beef, with some bread and a bottle of rum, was sent back.

Although the French were ready to chaff and laugh with the British, their conduct towards the natives of the country was very different. Foraging parties of great strength scoured the districts within their reach, driving off herds of cattle, flocks of sheep, and spoil of every description. These excursions gave rise to horrible cruelty and excess, which broke down the discipline of the French army, nor were they always executed with impunity. Often they were interrupted by the British cavalry, who, concealing their movements, would watch for a favourable opportunity, and when they saw a party of the enemy busy with the plunder of a village, would swoop down on them, redeeming the cattle, and taking many officers and soldiers. Our dragoons often made large sums of money by these adventures, as all horses captured from the enemy were sold for the benefit of the captors. The French also felt the effects of their system in the relentless vengeance of the exasperated people, who lost no opportunity of killing, often with torture, any stragglers or small detachments they could master.[1]

Organised plunder, even in an enemy's country, has never been the system of the British army. It is prevented by stringent rules, and though these may occasionally have been broken under circumstances of privation and temptation, there is a general confidence in our honesty, resulting in more regular supplies than can be obtained by desultory plunder. Still there are black sheep in all flocks, and even in the Gordon Highlanders there were men who, like Donald Caird, sometimes " Found orra things where Allan Gregor fand the tings."[2] While at Alcantrinho, four of the 92nd went into the country to see what they could pick up in the houses which they believed to be deserted, when to their surprise they found the inhabitants had returned to their homes since the British had occupied the district. Having gone thus far on the downward path, and loath to return empty handed, they flung aside restraint and proceeded to plunder. One of the inhabitants immediately ran off and informed the officer commanding. The roll was called, the men were found absent and confined on their return ; were tried by court-martial, and sentenced to be shot. Colonel Cameron, however, sorry for the untimely fate which he feared awaited these young men, and believing from his knowledge of them that they might still become trustworthy and good soldiers, before the sentence was promulgated made a strong representation, with any

[1] The marauders of our army, at this time very numerous, committed a thousand excesses. The cruelties committed against us seem to the Spaniards legitimate vengeance. Their hatred was profound, ardent, irreconcilable.—*Fezensac, general in Napoleon's army.*

[2] Sir Walter Scott.

extenuating circumstances, in their favour. I find among his correspondence the following letter :—

CARTAXO, *December 25th*, 1810.

DEAR SIR,—Sir Brent Spencer has directed me to inform you that he has spoken to Lord Wellington respecting the four men of the 92nd Regiment, and that he has every reason to believe that should the result of the Court-martial be serious, Lord Wellington will give it his lenient consideration. The Lieut.-General has thus taken the earliest opportunity of communicating to you the result of his application, which he trusts will prove satisfactory.

(Signed) T. DRAKE,
Captain and A.D.C.

To Lieut.-Colonel Cameron,
92nd Regiment.

The result was that the men, after being sentenced to death, were pardoned.[1]

While his troops were enjoying comparative rest throughout the winter, their general was far from being idle. He had, indeed, besides the care and discipline of the allied army under his immediate control, many and great responsibilities on his shoulders. On the 1st of January, Napoleon had caused an army of 70,000, including the young guard, to enter the north of Spain. The army of the centre, under King Joseph Bonaparte, numbered 27,000 ; Marshal Soult, Duke of Dalmatia, was in the direction of Cadiz with another large body of all arms, with which he was ordered by the Emperor to march upon Badajos, and, if possible, to assist Massena ; but such were the precautions of Wellington and the activity of the Spanish guerillas, that no communication could be effected between the two Marshals, and each remained in ignorance of the situation of the other. Lord Wellington corresponded with and advised the Spanish generals as to their operations ; they neglected his advice and were defeated. Soult besieged Badajos, and Wellington sent two Spanish divisions to help their countrymen who composed the garrison, but the incapacity of their commander brought destruction on his own force, and finally resulted in the fall of the fortress.

Meanwhile the British Government, roused by the wise and statesmanlike representations of their indomitable general to the far-reaching importance of the war, acceded to his request for more troops, without which it was impossible to carry the struggle to a

[1] In a Monthly Return at Alcantrinho the following note occurs :—" Alteration in number of sergeants. The drum-major struck off and placed upon the strength of the drummers, agreeable to General Orders of October 20th, 1810." The " Memoir of Colonel Cameron " mentions that at this time the bandsmen were put in the ranks ; the music of the battalion alternating between " the ear-piercing fife and spirit-stirring drum," and the swinging rhythm of the pipe, which was not broken by the accompaniment of a drum till fifty years later.

successful issue. General Hill with his two divisions was still in the Alemtejo, on the left bank of the Tagus, where not only did he prevent Massena from crossing to forage in that district, but was ready to render assistance to the Spaniards, who were hard pressed by Soult in the neighbouring province of Estremadura. Wellington's army was still cantoned between Santarem and Torres Vedras, watching Massena.

That general had during the last three months entirely exhausted the resources of the country he occupied, and when he heard that British reinforcements had landed at Lisbon on the 2nd March he at once resolved to retreat, and he carried out his resolution in a manner worthy of a great commander. Of the various lines open to him he chose that which led to the river Mondego and Almeida. First he destroyed the munitions of war and all the guns which could not be horsed; he sent on his sick and baggage, keeping only his fighting men (reduced by sickness to 40,000) in front. When his *impedimenta* had gained two days' march to the rear, he commenced his retrograde movement with the main body of his troops on the night of the 5th March, but he had previously caused an army corps, and his cavalry under Marshal Ney, to assemble near Leiria, threatening Torres Vedras, thus preventing Lord Wellington from taking a decided step lest he should open the lines to his adversary.

The British general was aware that a retreat was imminent, but it was only at daylight on the 6th that the empty camps at Santarem showed that it had begun. Wellington immediately followed the enemy with his own army, and the Gordon Highlanders marched from their comfortable quarters at Alcantrinho to the appropriate air of "Gabhaidh sinn an rathad mòr, olc air mhaith le càch e,"[1] and with the First Division moved by Golegao to Assentisse, near Thomar. The British advanced guard soon came up with the rear guard of the enemy under Ney, and were constantly in contact with them.

Napier, who was with the Light Division, and from whose history the account of these operations is principally taken, gives a horrible instance of the calamities which war may bring upon a country. Near Thomar they discovered a large house in an obscure part of the country filled with starving people. Above thirty women and children had sunk, and sitting by the bodies were fifteen or sixteen survivors, of whom only one was a man, but all so enfeebled as to be unable to eat the little food the British soldiers had to offer them. The youngest had fallen first—all the children were dead; the man seemed most eager for life, the women appeared

[1] "We will take the high road, whether (it leads to) bad or good." This tune is believed to have been composed on the march of the Royalist Clans to the battle of Inverlochy, though afterwards adapted to English words on the battle of Sheriffmuir. For many years it was used by the Gordon Highlanders in marching past to the pipes.

patient and resigned, and, even in this distress, had arranged the bodies of those that died first with decency and care.

That part of the army to which the 92nd belonged continued the close pursuit of the French, in moderate weather, through a country where the hollow roads were confined by wooded mountains on either hand, with villages here and there offering strong positions to the enemy in which to make a stand. He did so on the morning of the 9th, on a table-land in front of Pombal, where skirmishing took place at the advanced posts, and the cavalry of the King's German Legion took some prisoners. Lord Wellington had detached part of his army to relieve Badajos, and they were already on the march, when intelligence being received that morning that Badajos was capable of holding out longer, he recalled them. That night the Gordons lay at Peyalvo. Their division (the First) with the Third, Fourth, Fifth, Sixth, and Light Divisions, and the Portuguese which were attached to each British division, were assembling in front of the enemy on the 10th, when Massena suddenly retired through the town of Pombal. He was so closely followed by the British that, the streets being still encumbered, Ney drew up a rear guard on a height behind the town and threw a detachment into the old castle. He had waited too long. The French army was moving in some confusion in a narrow defile between the mountains and the river Soure, which was fordable, and the British divisions were in rapid motion along the left bank with the intention of crossing lower down and cutting off Massena's retreat ; but night came on, and the operation terminated with a sharp skirmish, in which, after some changes of fortune, the French were driven from the castle and town with such vigour that they had not time to destroy the bridge, though it was mined. About forty of the Allies were wounded, and the French loss was somewhat greater.

Daybreak on the 12th saw both armies in movement, and eight miles of marching and skirmishing brought the head of the British into a hollow way leading to a high table-land on which Ney had disposed 5000 infantry, a few squadrons of cavalry, and some light guns ; behind him was the village of Redinha, situated in a hollow; it covered a bridge over the Soure, and a long and dangerous defile. Beyond the stream some rugged heights commanding a view of the position in front of the village were occupied by a division of infantry, a regiment of cavalry, and a battery of heavy guns, all so disposed as to give the appearance of still greater force. An open plain extended between the French and the position of the British.

Lord Wellington, after examining the enemy's position, directed the Light Division to attack the wooded slopes covering Ney's right. They cleared the woods, and their skirmishers advanced into the plain ; but the French opened a heavy fire, and their 3rd Hussars charged and took fourteen prisoners. The British had meanwhile seized the wooded heights protecting the French left, but Ney, though he saw that Lord Wellington was bringing the masses of his troops into line, continued to hold his ground with astonishing confidence, and even charged the skirmishers, though there were already cavalry and guns on the plain enough to overwhelm him. In this position both sides remained for an hour, when three shots were fired from the British centre as a signal to advance, and a most splendid spectacle of war was exhibited. The woods seemed alive with troops, and in a few moments 30,000 men, forming three gorgeous lines of battle, were stretched across the plain; while horsemen and guns springing forward simultaneously from the centre and from the left wing, charged under a general volley from the French battalions; these were instantly hidden by the smoke, and when that had cleared away, no enemy was to be seen.

Ney had opposed Picton's skirmishers on his left, and at the same moment withdrew the rest of his troops so rapidly that he gained the village ere the British cavalry could reach him; he personally superintended the carrying off of a dismounted howitzer, which he effected with great danger and considerable loss. The British Horse Artillery thundering on his rear, and their infantry "chasing like heated bloodhounds, passed the river almost at the same time as the French,"[1] and they fell back on the main body at Condeixa. The British had 12 officers and 200 men killed and wounded in this combat, and the enemy lost as many, and might have been destroyed had Lord Wellington not been deceived as to his strength by his skilful arrangements, and so acted with more caution than afterwards appeared to have been necessary. Yet the extraordinary facility and precision with which the British general handled so large a force was a warning to the French commander, and produced a palpable effect upon the after operations.[2]

On the 13th the pursuit was renewed. The French had obstructed the road by felling trees and constructing palisades, which impeded the advance of the Allies. They burned the town of Condeixa, and lighted a number of fires, which covered the retreating troops with their smoke; but the British skirmishers and cavalry closed with their rear, and Massena himself only escaped by taking the feathers out of his hat and riding through the skirmishers. At dark the pickets were posted close to the enemy. On the morning of the 14th the mist was so thick that nothing could be seen, but from the sounds on the hill in front, it was evident that the French were there in force, and General Erskine rashly sent the 52nd Regiment forward; they unconsciously passed the

[1] Napier. [2] *Ibid.*

enemy's outposts and nearly captured Marshal Ney, whose bivouac was close to his pickets. The rattle of musketry and the booming of round shot were heard, and the mist slowly rising, discovered the 52nd engaged without support in the midst of the enemy's army.

At this moment Wellington arrived, and sent the Light Division to repair Erskine's error and aid the isolated 52nd. The First Division, including the 92nd, the rest of the infantry, the heavy cavalry and artillery came up, and Ney retreated, covering his rear with guns and light troops, retiring from ridge to ridge at first with admirable precision, till as the British got within range, the retreat became less orderly; but he gained the strong pass of Miranda de Corvo, which had been secured by the main body of Massena's army. The British loss was 11 officers and 150 men killed and wounded; the French loss was greater, and 100 prisoners were taken by the British. The French army was now compressed between the higher mountains and the Mondego River, and Massena destroyed great quantities of ammunition and baggage, while Marshal Ney still covered the retreat.

The "morning state" of the 92nd on the 15th is dated "Camp, near Condeixa." Massena had burned the town of Miranda and crossed the Ceira the preceding night, but the fog was so dense on the morning of the 15th, that the pursuers could not reach that river till late in the afternoon, when they lit their fires, thinking nothing more would be done; but Lord Wellington, seeing his opportunity, attacked Marshal Ney, who occupied a strong position on the hither side of the river. The Horse Artillery galloped to a rising ground and opened with great effect; the Third Division charged and overthrew the left of Ney's troops, while his right was engaged with the Light Division and other troops. Darkness coming on, the French in their panic began to fire on each other, and finally fled in confusion to the river, in which many were drowned, some were crushed to death on the bridge, and in all their loss was about 500 men. An eagle was afterwards found in the bed of the river. Four officers and sixty men fell on the side of the British.

Sergeant Robertson of the 92nd mentions that at the place where Massena had destroyed his baggage, he had ordered a number of asses to be destroyed, and the person charged with the execution had, instead of killing them outright, cruelly ham-stringed 500 and left them to starve in agony. Napier also mentions this circumstance. "The mute but deep expression of pain and grief visible in these poor creatures' looks wonderfully roused the fury of the soldiers." British soldiers are notoriously kind to dumb animals, and at that moment no quarter would have been given to a Frenchman. Indeed, Massena's retreat was marked by unlimited vengeance on the part of the Portuguese peasantry. Towns, villages, and corn stacks were burned ; wine which could not be consumed was left running in the gutters ; the people were murdered if they remained, and their property destroyed if they fled from their homes. They formed themselves into bands sworn to vengeance, who hung on the rear of the French, revenging themselves on any stragglers from the ranks. The Gordons saw not only the dead, but the wounded, stripped naked by them, and Napier mentions a peasant whom he saw hounding on his dog to devour the dead and dying.

On the 16th the Allies halted ; they had in some respects suffered greater privations than the enemy, who had cleared the country of all supplies, and had also carried fifteen days' bread with them, while the Allies depended on a commissariat which broke down under the difficulties. The men of the 92nd had no bread, only beef killed as required, and eaten without salt ; and one day a little rice was served out as a great treat. The soldiers grumbled to each other as they stumbled along in the darkness of a foggy morning. "The Parliament and people at home hear all about the grand Lord and the movements of the army, but they don't know anything about us individuals ; they don't know, for instance, that you are d——d tired and that I have no *pao*"[1] (Portuguese for bread), and they blamed the Commissary-General, but he was not really in fault. The Portuguese Government, notwithstanding the representations of Wellington, would neither feed the Portuguese troops regularly even at Santarem, nor collect the means of transport for the march ; after passing Pombal, many of the Portuguese soldiers were actually without food for four days ;[2] many died, and to save the whole from destruction, the British supplies were shared with them ; thus the Commissary-General's means were overtaxed, the whole army suffered, and their general was obliged to call a halt.

On the 17th of March the 92nd were in "camp near Lasisoa." The French had again taken up a strong position behind the river Alva and on the Sierra de Moita, having destroyed the bridges behind them ; and Massena sent out foraging parties, intending to halt for several days ; but on the 18th the First, Third, and Fifth Divisions were ordered to advance over the mountains to menace the French left, and they carried out these instructions with wonderful perseverance and strength, while the other divisions cannonaded the enemy on the Lower Alva. The Gordons had hard work during this movement, and the women of the regiment suffered a good deal of hardship. The Staff Corps made a raft

[1] "Recollections of the Peninsula," and Sergeant Robertson.

[2] Napier.

bridge over the Alva, and the pursuit continued. Massena recommenced his retreat with great rapidity; again destroyed baggage and ammunition, and abandoned his more distant foraging parties, of whom 800 were taken. Wellington assembled the whole army at Moita on the 19th; and on the 20th the pursuit was continued by the cavalry and Light Division, while the 92nd, with the greater part of the troops, halted at Moita till provisions, which had been sent by sea from Lisbon to the Mondego, could come up to them. The French reached Celerico on the 21st, where they were reinforced by a division 9000 strong; and Massena proposed to send his sick to the fortress of Almeida, then to pass the river Estrella at Guarda, make a countermarch through Sabugal to the Elga, and establish communications across the Tagus with Marshal Soult and with King Joseph Bonaparte, with a view to mutual co-operation. But he could no longer command the obedience of his generals, who were at variance with each other and with him. To such a height did the discord rise between Massena and Ney, that the latter was superseded in his command by General Loison.

By the 28th the troops (including the 92nd) had come up from Moita, and with them the reinforcements, which were formed into a Seventh Division. At Guarda a battle was expected on the 29th, and Wellington disposed his army for the attack of that nearly impregnable position—the 92nd, in the First Division, being placed in the centre. The absence of Marshal Ney, who had so ably commanded the rear guard of the enemy, was at once felt by both armies. On the appearance of the allied troops, the position was abandoned without firing a shot; and the French retired in confusion to the Coa, a considerable river running northwards through ravines and rugged banks to join the Douro.

On the 1st of April the allied army descended the mountains and reached the Coa—the 92nd starting from Alviria, about twenty miles distant. The enemy's troops were disposed on the right bank of the river, and the position occupied by General Reynier's Army Corps at Sabugal suggested to Wellington the possibility of cutting it off and compelling it to surrender. The enterprise failed of complete success, owing to the attack being prematurely delivered by the British advanced troops before the other columns, who were confused by thick mist and rain, had reached their posts. A fierce encounter took place, in which the 43rd and two battalions of the 52nd distinguished themselves by the presence of mind of the officers and the furious bravery of the soldiers. They sustained and repulsed repeated attacks of cavalry and infantry, and by the steadiness of their musketry fire silenced two French guns, which opened with grape at 100 yards distance. Reynier had put 6000 infantry, with cavalry and artillery, in motion to storm the contested height, when at the critical moment the five British divisions passed the bridge of Sabugal, the British cavalry appeared on the hills beyond the enemy's left, and the Third Division, issuing from the woods, opened fire on Reynier's right, which decided the fate of the day. The French general retreated on Rendo, where he met the Sixth Corps coming to his assistance, but they fell back together, pursued by the British cavalry.

The First and Seventh Divisions being in reserve, the 92nd did not take part in this bloody encounter, which, though it lasted less than an hour, cost the Allies nearly 200 killed and wounded; 300 dead Frenchmen lay heaped together on the hill, and more than 1200 were wounded, so true and constant was the British fire.[1]

Massena sought rest for his army behind the cannon of Ciudad Rodrigo, and retreated across the frontier with the bulk of his army to that fortress and to Salamanca, at the same time throwing a garrison into Almeida. One more affair, very creditable to the French, occurred as a French brigade was marching from Almeida to Ciudad Rodrigo. Suddenly two cannon shots were heard, and six squadrons of British cavalry, with Bull's troop of Horse Artillery, came sweeping over the plain. The French immediately formed squares and retreated, their cavalry on the flanks; and though the cannon shots ploughed through them, they retained their military order and coolness; and, gaining rough ground, escaped over the river Agueda, but with the loss of 300 men in killed, wounded, and prisoners. Massena entered Portugal with 70,000 men; 10,000 joined him at Santarem, and 9000 on the retreat. He brought only 45,000 of all arms out of the country. He lost, therefore, the enormous number of 45,000 men during the invasion and retreat by want, sickness, and the sword. The British were not weakened to the extent of a fourth of that number;[2] and Wellington stood victorious on the confines of Portugal, having executed what, to others, appeared incredibly rash and vain to attempt.[3]

During these operations the 92nd had neither officer nor soldier killed in action.

The army was now cantoned in villages on both sides of the Coa, the headquarters of the Gordon Highlanders being at Aldea Ponte, on the frontier of Portugal, about ten miles from Sabugal on the road to Ciudad Rodrigo. On the 9th of April the battalion entered Spain and was quartered at Albergaria, where they were joined by a draft of one lieutenant and forty-four rank and file from the 2nd Battalion.[4]

[1] Napier. [2] Alison. [3] Napier.

[4] At this period there was a small depôt of the 1st Battalion still at Canterbury; the 2nd Battalion being in Ireland.

In these places there were good markets, and they enjoyed rest and abundance of good things. Lord Wellington's headquarters were at Villa Formosa on the frontier not far from Almeida, which fortress he closely invested. The remarkable escape at this time of one of Lord Wellington's staff is recorded by Napier. Colonel Waters had been taken prisoner during the retreat; he had refused to give his parole, and was consequently kept in the custody of *gens d'armes*. When the French army was near Salamanca, he waited till the chief *gens d'armes*, who rode the only good horse in the party, had alighted, when, giving the spur to his own mount, he galloped off, an act of incredible resolution and hardihood, for he was on a large plain covered for miles around him with the French columns. His hat fell off, and, thus distinguished, he rode along the flank of the troops, some encouraging him, others firing at him, and the *gens d'armes*, sword in hand, close at his heels; suddenly breaking at full speed between two columns he gained a wooded hollow, and, having thus baffled his pursuers, reached headquarters on the third day, where Lord Wellington had caused his baggage to be brought, observing that he would not be long absent.

Guerillas under Don Julian Sanchez, and Spanish Grenadier

(*From Booth's "Military Costumes of Europe,"* 1812.)

CHAPTER XIV

WHILE the Gordons with the army under Wellington's immediate command were driving the French before them from Portugal, General Graham had, on the 6th of March, gained a glorious victory at Barrosa, near Cadiz, which, though it did not relieve that city, obliged Marshal Soult, who, thanks to the pusillanimity of the commander of the Spanish garrison, had taken Badajos, to return southwards to Andalusia. Wellington now felt safe in detaching a considerable part of his army to co-operate with Beresford in besieging Badajos and Campo Mayor, and went himself to conduct the operations. But Napoleon had no intention of allowing the British general to gain possession of the frontier fortresses, and he ordered Massena to retrace his steps and advance against him. Accordingly the French commander, having received large reinforcements, marched from Salamanca in the end of April, and on the 2nd of May crossed the Agueda with an army of 50,000 men,[1] including 5000 magnificent cavalry, of which part belonged to the Imperial Guard, and thirty pieces of artillery, for the purpose of relieving Almeida. Wellington, who had hastened back from the neighbourhood of Badajos, determined to dispute his passage and to put the question to the ordeal of battle.

On the 1st of May the 92nd turned out from Albergaria and lay under arms till evening, when they marched, and on the 2nd took post with the First Division a short distance behind the village of Fuentes d'Onor, where Wellington placed the right of his army, consisting of 32,000 infantry, 1200 cavalry, and 42 guns.[2] His line extended to Alameda in the centre and Fort Concepcion on the left, a distance of five miles along a table-land lying between the river Turones in his rear and the river Dos Casas in his front, this position interposing between the enemy and Almeida.

Fuentes d'Onor is situated at the bottom of a ravine, having an old chapel and other buildings on a rocky eminence overhanging one end of the village. The 92nd bivouacked on the night of the 2nd, and on the morning of the 3rd it was one of five battalions of chosen troops[3] detached from the First and Third Divisions to occupy the village. The French came up in three columns ; the Eighth and Second Corps against the centre and left ; the cavalry, the Sixth Corps, and Drouet's Division against Fuentes d'Onor ; the lower town was vigorously defended, but the violence of the attack and the tremendous cannonade compelled the British to fall back to the high ground at the chapel, where they gallantly, though with difficulty, maintained their ground. The officer commanding fell, and the fight was becoming critical, when Wellington sent

[1] Napier and Alison. [2] *Ibid.* [3] *Ibid.*

the 24th, 71st, and 79th to their assistance. By a splendid charge the French were forced back, and after a severe contest were finally driven over the Dos Casas, which flows through the lower part of the village. During the night the 92nd and other detachments rejoined their divisions, the 24th, 71st, and 79th being left in the village. On this occasion the Light Company of the 92nd distinguished itself, and Lieutenant James Hill and nine rank and file (of whom six belonged to the Light Company) were wounded.

It was here, I believe, that, as Cameron the pipe-major was giving forth his most warlike notes, a bullet pierced the bag of his pipe, causing it to emit a piteous and unwarlike skirl. Filled with wrath at the insult to his music, and with the desire to avenge the wound of his beloved instrument, he first tied it round his neck, and then, exclaiming "Bheir sinn ceol dannsaidh eile dhaibh!" (We will give them a different kind of dance music), seized the musket of a wounded man, and discharging it at the offending foe, drew his sword and rushed into the thick of the fight amid the laughter and cheers of his comrades.

The 4th of May was partly spent by the troops in marching and countermarching to new positions, during which manœuvres occasional shots were exchanged with the enemy. The left and centre of the British position was covered by the ravine through which the Dos Casas runs, but the right was comparatively open and exposed to the attack of all arms; therefore Wellington occupied ground to the right of Fuentes d'Onor by Poço Velho, near the Dos Casas, where the ground is comparatively flat, and having a swampy wood near the stream. At Poço Velho the Seventh Division, under General Houston, was posted; to their right rear 3000 Spanish guerillas, under Julian Sanchez, were placed on the height of Nava d'Aver on the extreme right; the British cavalry were on a plateau between Nava d'Aver and the village of Poço Velho, and immediately to the left of the cavalry was the First Division, including the 92nd, which was employed to cover a brigade of artillery, the Light Company and a sub-division of each of the others in its front being warmly engaged throughout the day. The whole line of battle was now about seven miles long.

Meanwhile Massena reconnoitred Wellington's position, his design being to hold the British left in check with his Second Corps, and to turn their right with the remainder of his army. He intended to make his dispositions at night so as to commence the attack at daybreak on the 5th, but a delay of two hours occurring, the whole of his movements were plainly seen.

The Gordons were without provisions when they stood to their arms, and, but for the hospitality of the Coldstream and 3rd (now Scots) Guards, who generously shared with them the contents of their haversacks,[1] they would have been ill prepared to act their part in the drama to which the affair of the 3rd had been the prelude.

About eight o'clock they could see masses of the enemy's infantry and all his cavalry marching on Poço Velho, and they were wheeled into line ready to receive them ; the left wing of the Seventh Division, consisting of British and Portuguese, was driven from the village with loss, and the French were gaining ground, when the First and Third and part of the Light Divisions moved in support. Then the French cavalry passed Poço Velho, forming in order of battle towards Nava d'Aver, from which the guerillas[2] retired across the Turones, when Montbrun, the French cavalry general, turned the right of the Seventh Division, and charged the British and Portuguese cavalry, about 1000 strong, with about 4000 Horse. The combat was unequal, and after one shock, in which the enemy were partially checked, and the French Colonel Lamothe was taken fighting hand to hand by General Charles Stewart, our cavalry was driven behind the Light Division. Montbrun then swept with his terrible Dragoons round the infantry, now exposed to his attack, but they, rapidly forming squares, treated the horsemen with the confident contempt with which steady infantry in that formation may always regard the onset of cavalry. Such, however, was the swiftness of these magnificent horsemen, that they fell upon part of the Seventh Division before these could form square, but with admirable steadiness, though some were cut down, they took advantage of a loose stone wall, received the attack in line, and repelled it by the excellence of their fire-discipline. A stirring scene was now witnessed by those among the Highlanders who had leisure to admire it. In the *mêlée*, Captain Ramsay's troop of Horse Artillery was surrounded, and the spectators gave them up for lost, when presently a great commotion was observed among the glancing throng of cavalry, "officers and men closing in on a point where a thick dust was rising, and where loud cries, the sparkling of blades and flashing of pistols, indicated some extraordinary occurrence. Suddenly the multitude was violently agitated, a British shout arose, the mass was rent asunder, and Norman Ramsay burst forth at the head of his battery, his horses breathing fire, and stretching like greyhounds along the plain, his guns bounding like things of no weight, and the mounted gunners in close and compact order protecting the rear."[3]

[1] McKinnon's "Origin and Services of the Coldstream Guards," and Sergeant Robertson.

[2] Guerillas were irregular troops organised in bands under chiefs who were not officers of the army. Some joined for plunder, some for patriotism, or to revenge the wrongs suffered from the French. Scherer mentions one who told him the French had burned his house and killed his father and mother, and that he had sworn not to plough a field or dress a vine till the murderers were expelled from Spain.

[3] Napier. Captain Norman Ramsay's tomb is in the kirkyard of Inveresk, Musselburgh.

While this brilliant feat of chivalry was being enacted, the enemy was making progress ; the British divisions were separated, and the right had been turned, so that if the lost ground was not regained, or a new and strong position taken up, the battle would be lost. Wellington under these circumstances determined to change his front to the right ; his left remaining firm at Fuentes d'Onor, the other divisions wheeling backwards on that pivot ; the Seventh Division was directed to cross the Turones to Frenada, the Light Division to retire over the plain, the cavalry to cover the rear ; the First and Third Divisions and Portuguese were drawn back and placed in line on the rising ground which runs at right angles to the ravine of Fuentes d'Onor, extending towards Frenada, where the Seventh Division formed the right of the new position, which thus faced the right of the original line and still barred the way to Almeida. Such an operation could only have been attempted by a general who had absolute confidence in the discipline and valour of his troops, or carried out by troops whose matchless steadiness was worthy of his confidence. The outer extremity of the line had to retire four miles over a level plateau, which was covered with a confused multitude of camp followers and baggage. Slowly, but in perfect order, the retiring squares of the First, Seventh, and Light Divisions made their way, often hidden from view by the press of the formidable French cavalry, who outflanked them, charging whenever opportunity offered, while their infantry were in order of battle behind them ; but undismayed and unbroken the red-coated squares emerged from every difficulty, and soon their line was formed, with guns in the intervals, upon the appointed ground. It was indeed a hazardous game to play, but it succeeded, and with comparatively little loss—" Yet," says Napier, " in all this war there was not a more dangerous hour for Britain." [1]

The Light Division formed a reserve to the right of the First, sending riflemen among the rocks to connect it with the Seventh at Frenada, which was joined by Sanchez' Guerillas.

At the sight of this new front the French stopped short and commenced a heavy cannonade.

In describing the action as he experienced it, Sergeant Duncan Robertson of the 92nd says that the detachment of the battalion covering the guns near Poço Velho was ordered to lie down in order to escape the enemy's shot as much as possible. A body of cavalry charged, attempting to take the guns, " but we started up and gave them a few rounds, which made them wheel to the right about. They repeated this different times, but with as little

[1] In a private letter Wellington wrote, " If Bony had been there, we should have been beaten."

success as the first. The French did us considerable damage with their shells, which were beginning to fall fast and thick about us. One of them burst among the company to which I belonged, when we were in the act of lying down, and killed and wounded four of our men." He also tells how a shell fell among another company and killed or wounded an officer and eleven men. "While this was going on, another attempt was made by the enemy to carry off the guns, but all their efforts were unavailing; for so firmly did we maintain our ground, and so well directed was our fire, that they were compelled to retire with considerable loss."

Colonel Cameron, in a letter dated "Plains near Almeida, and over Fuentes del Honore, 8th May 1811," writes of the action of the 5th as being particularly hotly contested about our right and centre. "The 92nd were most unlucky, as they have been all this campaign; they were placed in a situation where it was utterly impossible for them to get their revenge, but where they were exposed for upwards of four hours to the hottest fires of grape and shell I ever witnessed. Our poor fellows comported themselves well, scarcely even murmuring, and only muttering that they wished they could have satisfaction. Had we not been in line, we must have been annihilated." At length the enemy's fire (which was replied to by the British artillery) abated, their cavalry drew out of range, and their infantry was repulsed in an attempt on the right centre by the Rifles and Light Companies of the Guards.

Meanwhile a fierce fight had been going on at Fuentes d'Onor. Two hours after Montbrun's cavalry had turned the British right, Drouet, with his whole division, had attacked the three regiments at that post. They made a desperate resistance, but, outnumbered, they were forced back to the upper part of the village; two companies of the 79th were taken and their colonel, Cameron,[1] mortally wounded.

Wellington, having concentrated his force (for his second line was very much shorter than the first), was now able to send considerable support to the regiments at Fuentes, while the French also reinforced their troops, and the contest continued, sometimes in the lower town and on the river banks, sometimes on the rocks round the chapel, the enemy even penetrated beyond; but the British never entirely abandoned the village, and in a charge of the 71st, 79th, and 88th, led by Colonel MacKinnon, against a column which had gained the chapel, they drove them down the street.

[1] The late Mr Lindsay Carnegie, of Boysack, then in the artillery, and acting as A.D.C. was sent with instructions to Colonel Cameron early in the day. Having delivered his message and entered for a few minutes into conversation, the colonel asked if he was the Lindsay who was known as a great billiard player, and finding he was, a match was arranged to come off when next they met within reach of a billiard table; but Cameron had played his last game, he died of the wound he received shortly after.

The 71st had recruited a good deal in Glasgow, and their commanding officer incited them by calling out, " Glasgow lads, clear the Gallowgate ! " which was received with a cheer, and the Imperial Guard was driven back. The ensign who carried one of the colours of the 79th was killed ; his covering sergeant called out, " An officer for the colours ! " One came forward, but was immediately struck down. " An officer for the colours ! " again shouted the sergeant, and a third and fourth officer were wounded, till at last the adjutant carried the regimental colours safely till the fight ended with the day.

Such was the battle of Fuentes d'Onor, in which neither side could claim a decided advantage, but Wellington gained his object of preventing the relief of Almeida ; and the great convoy of provisions which was waiting at Gallegos in rear of the French, till they could open the way for it, was unable to enter the fortress, which soon afterwards was abandoned. The loss of the Allies was 1500 men and officers killed, wounded, and prisoners ; that of the enemy was estimated at nearly 5000, but Napier considers this to be an exaggeration.

The loss of the 1st Battalion 92nd on the 3rd and 5th of May was 7 rank and file killed, and 3 officers, 2 sergeants, 1 drummer, and 41 rank and file wounded, of whom Sergeant Alexander Cameron, 1 drummer, and 15 rank and file belonged to the Light Company ; 1 officer and 8 rank and file died of their wounds. No less than 13 of the wounded and 1 killed were married men. One private was taken prisoner.

Officers Wounded

Major Peter Grant.
Lieutenant Allan M'Nab, died of his wounds.
" James Hill.

Killed in Action

Corporal John Whiteford.
Private John Lyon.
" Joseph Leith.
" David Lownie.
Private John M'William.
" Alex. Robertson.
" Thos. Southwell.

Died of their Wounds

Private Robert Colquhoun.
" George MacLeod.
" James Fleming.
" Donald Urquhart.
Private Donald Macdonald.
" William Turrill.
" Thomas Murray.
" James Dingwall.

In Colonel Cameron's letter alluded to above he says—" Almost all the wounded are lost to the service for ever, as most of them have lost legs or arms. Major Grant has lost his left leg, Lieutenant

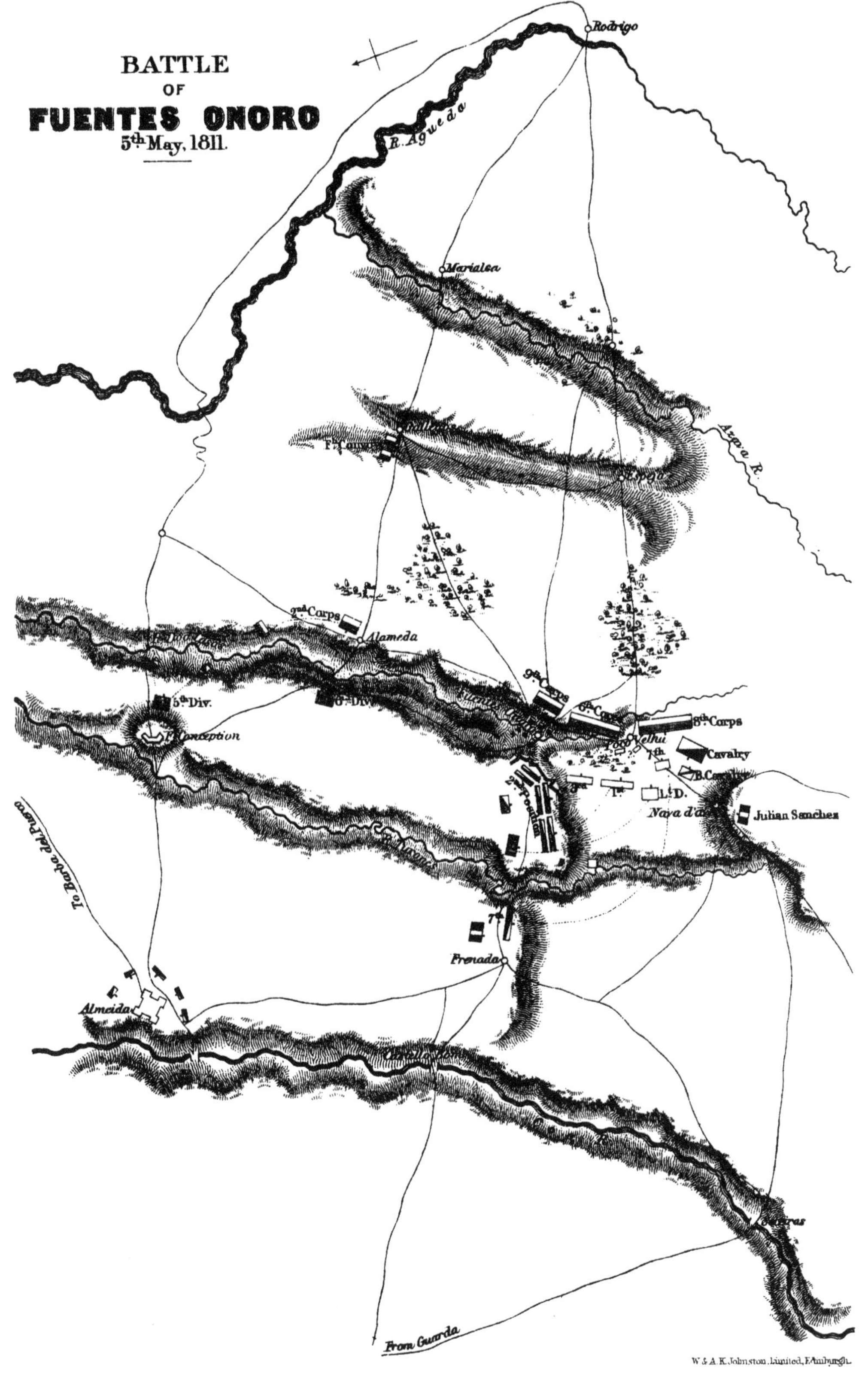
BATTLE
OF
FUENTES ONORO
5th May, 1811.
Rodrigo
R. Agueda
Marialva
Azava R.
2nd Corps
Alameda
5th Div.
6th Div.
Concepcion
6th Corps
8th Corps
Cavalry
B. Cavalry
3rd
1st
Lt. D.
Julian Sanchez
To Barba del Puerco
Frenada
7th
Almeida
From Guarda
W. & A. K. Johnston, Limited, Edinburgh.

M'Nab his right arm from the socket, and the left splintered the whole length. Lieutenant Hill severely wounded in the hip, and ball not extracted. Allan M'Nab, the laird's son, has proved himself of the genuine old Highland stamp by an exhibition of fortitude barely human, and which has attracted the admiration of the whole army here. Write the laird [1] that he is a credit to his country, and that I hope in God he will survive. They all promise well. The French, you will see by the papers, were baffled in all their efforts, and have not meddled with us since, and seem to-day (the 8th) as if retiring. The 71st and 79th Highland Regiments had an opportunity of distinguishing themselves most nobly. Charles Cameron, Errachd's son, behaved most admirably; he was dangerously wounded. I am much exhausted, as we have been for seven days without bag or baggage under the canopy of Heaven without covering of any kind. I with difficulty borrowed this paper. None of the lads from the country killed or wounded, except Claives,[2] wounded, but not dangerously."

Major Archibald M'Donell, who commanded the detachment in front which greatly distinguished themselves, was promoted to the brevet rank of lieut.-colonel, and the 92nd afterwards [3] received the Royal authority to bear the words *Fuentes d'Onor* on the regimental colours and appointments in commemoration of the gallantry of the 1st Battalion in that battle.

The Gordons remained bivouacked close to Fuentes d'Onor, and were occupied on the 6th and 7th along with the Light Division in strengthening the position, in case the enemy, who remained in their position, should renew the attack. Massena, however, withdrew part of his forces on the 8th, though still maintaining posts at Alameda and Fuentes.

The morning state of the battalion, 8th May, Plains of Almeida, was—

	Sergeants.	Drummers.	Rank and File.
Present and fit for duty	34	12	607
Wounded and sick	8	2	176
On command	7	4	49
	49	18	832

On the 10th Massena retired his army across the Agueda towards Salamanca. He was himself recalled to France, and Marmont assumed the command of the army of Portugal. Almeida, the only place in the country still in possession of the French, was abandoned to its fate; and its brave commander, Breunier,

[1] M'Nab of M'Nab.

[2] A Cameron, son of a tenant of Claives.

[3] In 1829.

after destroying part of the defences, evacuated the place during the night of the 10th, and with great bravery and skill, aided by the darkness, broke through the blockade, losing, however, 400 killed or made prisoners by the 4th and 36th Regiments, who, throwing off their knapsacks, overtook the rear of the column as it was descending the deep and difficult pass of Barba del Puerco. On the other side, the Second French Corps was waiting in order of battle to receive the gallant garrison, of whom 1100 got clear off.

Next day, Wellington took possession of Almeida with its artillery.

Thus ended the French invasion of Portugal. Although the inefficiency of the Government at Lisbon had paralysed the efforts of the country, the British were supported with fidelity and cordiality by the rural population, and the peasantry had agreed admirably with our soldiers. Their sufferings had excited warm sympathy in Great Britain, and Parliament unanimously granted £100,000 for the relief of sufferers by the French invasion, while private subscriptions trebled the amount.[1]

On the 14th of May the battalion returned from Fuentes to its former cantonments at Albergaria, where orders were received increasing its establishment to 1200 rank and file.

Meanwhile, Marshal Beresford, who was blockading Badajos, having received information that Marshal Soult was advancing from Seville to its relief, raised the siege on the 15th of May, and marched to meet the enemy. On the 16th he gained the hard-fought victory of Albuera, and Soult, leaving Badajos to its fate, retreated towards Seville. Wellington now gave orders for Badajos to be again closely invested ; and to assist in the operations, the brigade of which the 92nd formed part was appointed the first of the Second Division of the army, which was commanded by Lieut.-General Rowland Hill, and was at this time in Spanish Estremadura covering the siege of Badajos.

The battalion, under Lieut.-Colonel Cameron, marched at five o'clock on the evening of the 25th of May, and continued for two days and nights almost without intermission ; the heat during the day was excessive, and the men were a good deal fatigued ; but Soult having withdrawn from Badajos, they were allowed a day's rest, and crossed the Tagus at Villa Velha, arriving at Niza on the evening of the 31st. The country about Niza is fertile and beautiful : gardens, vineyards, and cornfields for miles round the city, which is situated on a lofty eminence commanding splendid views towards Badajos and Elvas. Though the days were hot, the nights were delightful, and sleeping in the open no hardship. They bivouacked generally on the edge of a wood for shade, and near a

[1] Alison.

stream. The troops halted in open columns ; arms were piled, pickets and guards paraded and posted, and in a few moments all appear at home. Some fetch stones to form fireplaces ; others, water for the camp kettles ; while the wood resounds with the blows of axes and bill-hooks. Under the trees officers are seen dressing, or arranging green boughs to form shelters. Milk, water-melons, bread, and vegetables are brought from the villages, and soon all are discussing dinner ; the favourite dish being Irish stew of ration beef with onions, pumpkins, and tomatoes. "Nothing could be worse than the clothing of some of the regiments : as red could not be procured, the jackets were patched with grey, white, or brown ; but the perfect order and cleanliness of their arms and appointments, their steadiness on parade, their erect carriage, and their firm and free marching, were all admirable."[1]

From Niza they proceeded by easy stages ; and crossing the Guadiana above Badajos, near Talavera la Real, on the 8th June, joined the Second Division in camp near Albuera on the 10th, about ten days after the commencement of the second siege of Badajos. While here they witnessed a fearful fire ; a cornfield was accidentally set alight by some Portuguese soldiers, and the flames spread over the burning plains for a distance of about thirty miles.[2]

Soon after, Marmont, having reorganised the army of Portugal, effected a junction with Marshal Soult. Wellington was prepared to fight Soult alone, and had entrenched the position of Albuera ; but the arrival of Marmont gave the enemy such an overwhelming superiority in artillery, cavalry, and infantry, as made it necessary to raise the siege, and withdraw the Allied army. Accordingly, the Second Division broke up from its bivouac on the 16th ; marched by Valverde, recrossed the Guadiana, and arrived at Torre de Mouro on the 20th of June. The united French armies had entered Badajos on the 19th, just in time to save the garrison, who would have been driven by want of provisions to abandon the place they had twice so bravely defended. Wellington, prepared to accept battle, drew up his army in front of Elvas—his right being below that fortification at the bridge over the Caya, and his left extending along the bank of that river to beyond the fortress of Campo Mayor, a distance of about ten miles. On the 21st the enemy's cavalry, having crossed the Guadiana, drove in the Allied outposts ; and by a feigned retreat drew the British cavalry, who had at first been successful, into an ambuscade, and took a squadron of the 11th Light Dragoons.

A decisive battle was now expected. The enemy had a superi-

[1] Recollections of the Peninsula.

[2] Alison and Sergeant Robertson.

ority of one-fourth in artillery and infantry, and the strength of his cavalry was double that of the British,[1] whose strength was greatly reduced by the number of their sick and wounded;[2] but the men who lined the banks of the Caya were the same stern soldiers who had fought at Albuera and Fuentes d'Onor; and though his own troops were the flower of the French army, including a large number of the Imperial Guard, they were conscious of late defeats; the British had the weight of moral superiority, and Soult felt it unsafe to tackle them. He withdrew without fighting. Wellington, aware that till he had possession of all the frontier fortresses, he could not advance into Spain, resolved to move with the bulk of his army, which had been reinforced from England, in the direction of Ciudad Rodrigo; leaving Hill with 10,000 infantry, 1500 cavalry, and four brigades of artillery, in the rich district between Villa Viciosa and Portalegre, which was free from the Guadiana fever, and where the troops could rapidly concentrate for offence or defence. Therefore, the Second Division broke up from the bivouac at Torre de Mouro on the 21st July, marched to Elvas, and next day went into quarters about Villa Viciosa; the 92nd, along with the rest of the First Brigade and a brigade of Portuguese, occupied the neat little town of Borba, where was a plentiful supply of wine, vegetables, and fruit to supplement the commissariat rations—a market being held every day during their stay. The 1st of September saw them again on the move, and on the 3rd they marched into the city of Portalegre, where the whole division was quartered, and where the Gordons no doubt renewed their agreeable relations with the inhabitants, whom they found so friendly when they were quartered here under Sir John Moore. General Hill kept hounds, the officers had partridge shooting and fishing, and there were entertainments and amusements for all ranks.

Meanwhile Wellington, with the main body of the army, had invested Ciudad Rodrigo. Towards the end of September, Marmont, having received large reinforcements, especially of the Imperial Guard, advanced to relieve the fortress with 60,000 men, of whom 6000 were cavalry, and he had more than a hundred pieces of artillery.[3] Wellington had 45,000 men of all arms,[4] and was weak in artillery.

On September 25th the heroic combat of El Boden furnished another instance of how impossible it is for even the most daring cavalry to overcome the resistance of steadfast infantry. Another combat took place at Aldea de Ponte under the immediate direction

[1] Napier.
[2] A fever is prevalent at that season in the valley of Guadiana.
[3] Napier.
[4] Alison.

of Lord Wellington, who, on the morning of the 28th, occupied a strong position in an angle formed by the Coa River, which could only be attacked on a narrow front, and Marmont, who had only brought a few days' provisions, retired, having victualled and placed a new garrison in Ciudad Rodrigo. Wellington cantoned his army on both sides of the Coa, and resumed the blockade of the fortress. In the above-mentioned actions the Allies lost about 300 men and the French about 600. Alison relates, as an instance of the generous spirit which animated the brave men of both armies, that a French officer was in the act of striking at Captain Felton Harvey of the 14th Light Dragoons, when, seeing that he had but one arm, he at once lowered his point to the salute and passed on. Major Gordon, who had been sent with a flag of truce by Wellington to Marmont, was hospitably received, dined with the French Marshal, rode with him round the outposts, and they freely discussed the prospects of the campaign and the qualities of the troops on both sides ; while General Regniaud, Governor of Ciudad Rodrigo, who had been led into an ambuscade by Sanchez' Guerillas and taken prisoner, was a frequent guest at Wellington's table, and entertained the company by his anecdotes of the French generals and armies. A story was told me many years ago by a brother of a Peninsular Gordon Highlander, showing that these chivalrous feelings were by no means confined to the higher ranks. I forget on what occasion it happened, but after a battle a French colonel was lying wounded on the field, when one of the 92nd approached him. The Frenchman could speak English, and, supposing the soldier was coming to plunder him, offered his purse and gold watch, asking his assistance, but the man indignantly answered that he was a soldier and not a robber, that he would gladly help him without reward, and carried him to the surgeon. The colonel showed his gratitude by reporting the circumstance very favourably to the British general, and the man was promoted corporal. At Elvas, where the hospitals were, the crowded wards were occupied by British and French side by side, who were to be seen performing little kind offices for each other, using Spanish to express their wants or their thanks.

While Wellington remained to oppose Marmont and Dorsenne, and was occupied in the still more difficult task of pacifying party disputes in the Portuguese Government, and inducing them to feed their own troops, who were paid by Britain but ought to have been rationed by their own country, Hill was still at Portalegre on the frontiers of Estremadura, watching the French under Drouet.

On the 7th of October a draft joined from the 2nd Battalion, consisting of 1 major, 2 captains, 4 subalterns, 1 assistant surgeon, and 199 n.c. officers and soldiers. They had marched from Athlone for embarkation at Cork on the 17th August with 248 men.

An officer who was with them mentions that at Fermoy an order reached them for forty-eight of the detachment to return to Athlone, being considered too young to undergo the fatigues of a campaign, and relates how some of the poor lads "absolutely shed tears" in their sorrow and disappointment at losing the chance of active service. He also describes the ruined state of the villages which had been occupied by the French, through which they passed in their march from Lisbon, and the warm welcome they received from the 1st Battalion on their arrival.

When Drouet had retired to Zafra, Hill received orders from Lord Wellington to drive General Gérard's Division of the Fifth Corps away from Caceres, and to re-establish the Spanish troops under Morillo in that district, from which Gérard had forced them to retire; accordingly, the Second Division broke up from their comfortable cantonments at Portalegre. The 92nd left the women and baggage in the quarters of the corps, with a guard of one subaltern, one sergeant, and twelve rank and file of the most weakly soldiers. The men were ordered "at all times to march with the wrong side of their kilts outermost unless when ordered to the contrary." Passing Allegrete, the last Portuguese town, they bivouacked that night at Codeceira, the first Spanish village. An officer of the regiment notices the effect of want of knowledge of camp life on the part of the draft lately joined, and remarks that all soldiers should be taught practically field cookery and other incidents of the bivouac before being sent on active service. The arms on this occasion were no sooner piled than the old stagers were off in search of comforts, but the "Johnny Raws," as all young soldiers were called, sat shivering on their knapsacks in expectation that dinner would be provided for them as at home, without personal trouble, and the old campaigners were enjoying their meal almost before the new hands had got their kettles on the fire. At night a pouring rain fell on them, as officers and soldiers sat or lay round smoky fires with their heads resting on each other's legs—some warming wine in the lid of a camp kettle, others crouching for shelter behind a wall, with their forage caps pulled down over their ears, cursing the commissariat, the rain, and the French; but the cheering rays of the morning sun brought warmth and good humour, and the battalion advanced to the ancient city of Albuquerque, where they learned that the French had retired from Alesada.

On the 24th October they reached Cantellana, and the following forenoon found the Third Brigade at Alesada, where they also met some Spanish irregular cavalry in quaint old-fashioned costumes, under Colonel Downie. The road here was so bad that the guns had to be removed from their carriages and carried up a steep place by the men of the First Brigade. At Alesada they were ordered to cook as quickly as possible, an order which is generally ominous of a long march or a fight. "The bullocks," says an officer, "on which we were to dine were running and jumping about us, but in less than an hour they were amusing us with more interesting leaps in our camp kettles. The soup, just off the fire, having been placed before us at the same time that the bugle called us to arms, we were compelled to dispatch it in a state little colder than boiling lead." They started at 4 p.m. in hopes of surprising part of Gérard's Division at Malpartida. The afternoon was fine, but as the sun went down the rain descended in torrents. Marching at night is not pleasant in the finest weather, but in storm and darkness, over roads deep in mud and water, with deep ruts and stones over which the weary soldier trips, to fall prostrate in a puddle, it is the greatest trial of endurance; so worn out were some of the men at dawn on the 26th, that had they not expected to meet the enemy they would have broken down. The French, however, had given them the slip, and on arriving at Malpartida they found that the enemy, warned of their danger by three Spanish deserters, had retired from that town, where General Hill ordered his exhausted troops under cover for twenty-four hours of much-needed rest and refreshment.

CHAPTER XV

EARLY on the 27th October, Hill, hoping to intercept the enemy's march, though without being certain of its exact direction, pursued by a cross road through Aldea del Cano and Casa del Antonio, when he received intelligence that Gérard, ignorant of the movements of the British general, whose pursuit he expected on the main road by Caceres, had posted a rear guard at Átbala, and halted at Arroyo dos Molinos in fancied security. Hill at once made a forced march to Alcuesca, three miles from Arroyo dos Molinos, which he reached in the evening. The two armies, in fact, had been marching on parallel lines not far apart, but each concealed by hills and woods from the sight of the other.

On crossing an extensive plain near Aldea del Campo, the mounted officers started a hare. Several greyhounds were kept by the regiment, and after a successful course puss was killed. Finding that the whole battalion enjoyed the fun, and hares being plentiful, the officers continued the sport, which so wonderfully beguiled the time, that many of the men thought they had not marched above half the distance they had really come.

The 71st occupied the village of Alcuesca, and placed pickets all round to prevent any intelligence being carried to the French ; but the people felt the British to be their friends, and though every Spaniard in the neighbourhood knew they were at hand, not one betrayed the fact. The 92nd with the other troops bivouacked in rear of the village—rain and wind continued through the night, and no fires were allowed to be lighted, but the troops without a murmur, though wet to the skin, made the best of their discomfort, keeping their muskets as dry as they could, and bore their plight like men and soldiers.

At two o'clock on the morning of the 28th, the orderly sergeants went round their companies, and in whispers bade them prepare for action, the utmost silence being necessary to ensure success. Though the distance was short, the broken state of the roads, the inclemency of the weather, and the darkness of the morning made progress slow and caused great loss of time ; but at last the sight of the enemy's fires delighted the troops by showing that this time the birds had not flown. The whole moved in one column till they arrived within half a mile of Arroyo, where, under cover of a rising ground, they were formed into columns of attack unseen by the enemy, Gérard, with culpable negligence, having no pickets out on that side.

The left column, commanded by Lieut.-Colonel Stewart of the 50th Regiment, was composed of Howard's Brigade, viz. the 1st Battalions of the 50th, 71st, and 92nd, Captain Blassier's company 5th Battalion 60th Rifles, and three field pieces, supported by

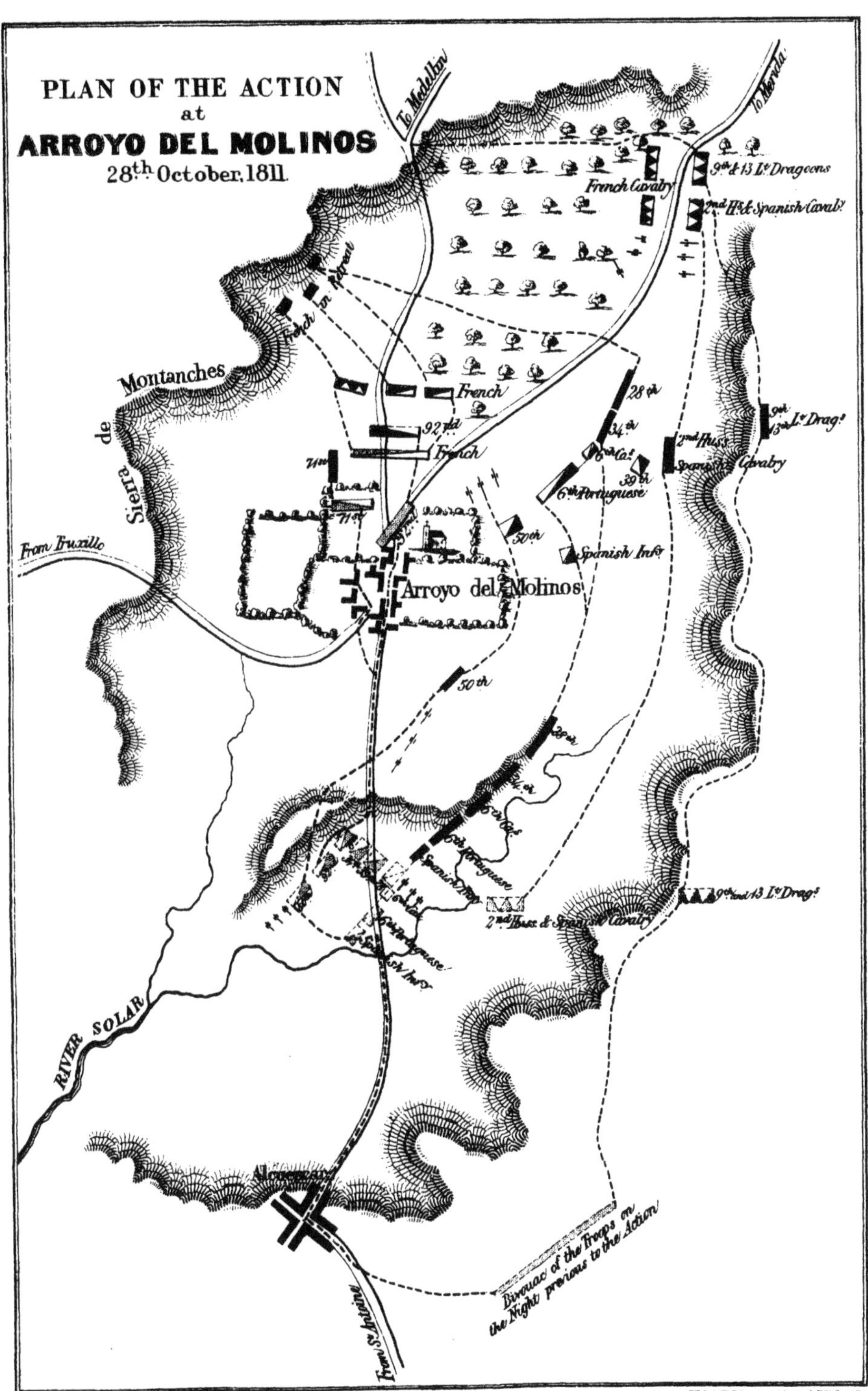
PLAN OF THE ACTION
at
ARROYO DEL MOLINOS
28th October, 1811
To Medellin
To Merida
French Cavalry
9th & 13 Lt. Dragoons
2nd H. & Spanish Caval.
French in Retreat
Montanches
Sierra de
French
92nd
French
71st
28th
34th
6th Caç.
39th
6th Portuguese
2nd Huss.
Spanish Cavalry
9th 13th Lt. Drag.
From Truxillo
50th
Spanish Infy
Arroyo del Molinos
50th
9th and 13 Lt. Drag.
2nd Huss. & Spanish Cavalry
6th Portuguese
Spanish Infy
RIVER SOLAR
Bivouac of the Troops on the Night previous to the Action
From St. Antoine
W. & A. K. Johnston, Limited, Edinburgh

Morillo's Spanish infantry. The right column consisted of Colonel Wilson's, the Third Brigade, viz. the 1st Battalion 28th, the 2nd Battalions 34th and 39th Regiments, one company 60th, and the 6th Portuguese Caçadores, with two field pieces and a howitzer. The cavalry, 9th and 13th Light Dragoons, the 2nd Hussars of the King's German Legion, and the Conde-de-Penne Villamur's Spanish Horse, formed a third column under Sir William Erskine, and was in the centre.[1] Major-General Howard commanded the whole infantry. The formation being quickly and quietly completed, the 71st and 92nd and the company of the 60th moved in profound silence directly on the village at quarter distance, the 50th in close column and the artillery being a little in rear as a reserve. The right column, having the 39th as a reserve, moved across the plain to the right of the town, so as to cut off the retreat of the enemy by the roads leading to Truxillo, Merida or Medellin. The cavalry moved between the other two, ready to act wherever its services might be required.

At this moment the storm was succeeded by a calm, the dense fog cleared away about half-past six, just as the left column was entering the town. The 71st went to the left through some olive groves, where they surprised and took a body of cavalry in the act of bridling their horses. The 92nd silently entered the main street leading to the market square, where the French were forming for the march, while some battalions were filing out of the village, one brigade having already marched; then the pipers, entering into the humour of the situation, struck up the appropriate air of "Hey, Johnnie Cope, are ye waukin' yet?" and the music of the bagpipes and the cheers of the charging Highlanders were the first intimations of their presence. Colonel Cameron, in a letter to General Hope, says—"They (the French) were quietly getting under arms for the purpose of proceeding to Merida, and had the attack taken place two hours earlier as intended, I think every man of them would have been taken in bed."

The French troops, cavalry and infantry, who were formed threw themselves across the street, disputing the passage, and firing on the leading sections of the Highlanders, who dashed forward, and amidst the greatest uproar and confusion drove out of the village those they did not bayonet or capture. The 50th in rear were not idle, and secured the prisoners.

Gérard, who was a brave officer, made the best disposition in his power; he formed the infantry in two large squares just outside the village, where the roads branch off to Merida and Medellin, with the cavalry on the left flank of one of them. The 71st extended, lining the walls of the village gardens and vineyards, and

[1] The Second British Brigade had been ordered to make a detour and had not come up.

peppered their antagonists in good style, having skirmishers opposed to them, who returned their fire most briskly. On getting to the end of the village, Colonel Cameron found a considerable opening between the vineyards, which admitted about one-half of the battalion in column. The fire of the French riflemen annoyed the head of the column considerably, nor could they advance without being exposed to the fire of the 71st. The men were ordered not to fire, which, says an officer, "was extremely galling to the soldiers, who saw their officers and comrades falling around them, but knowing that the success of an enterprise often depends on the manner in which such orders are attended to, the Highlanders, with a patience not very peculiar to their countrymen, waited the arrival of the decisive moment."

Quoting again from Colonel Cameron's letter to General Hope—"However, we got the 71st to cease firing, and to allow us to advance. As each company had cleared the opening, I caused them to deploy to the right into line with the utmost rapidity, as we were then in the presence of the French square of five battalions, which I proposed charging the moment my line was formed. Three companies were scarcely in line when the French square moved off, throwing out a cloud of skirmishers to impede our advance. By this time the other British brigade, composed of the 28th, 34th, and 39th, the British cavalry and Portuguese artillery having got round the town, were coming up upon their left flank, and the whole followed in pursuit. Your regiment" (General Hope was colonel of the 92nd) "conducted themselves much to my satisfaction during the whole time, and under the trying circumstances of not one firelock in ten giving fire when attempted to be used." After mentioning that he had been wounded in the hand while preparing to charge, and had his finger amputated "after the business was over"—"poor Captain Donald MacDonald is most dangerously wounded, having his left leg broke, and wounded under the right knee. Captain John Macpherson severely through the left wrist, but in no danger; Major Dunbar is also wounded; we had three men killed and seven wounded. The Light Company was not in action, being in occupation of a village some little distance in rear under Lieut.-Colonel M'Donnell. I am not aware what people in Britain will think of this business, but this corps of the army fancy that having obtained such results with so small a loss renders it unique of its kind, particularly on comparison with the fatal and miserable battle of Albuera."

Another officer describes how, as the Highlanders were forming for the charge, the French troops showed something like a wavering in their squares, and, declining the honour intended for them, wheeled to the right about and retired towards the steep hill in their rear. He mentions that there were several wonderful escapes, and that Colonel Cameron's life was saved by the hilt of his sword.[1] The captain of the Grenadier Company having been wounded, the senior lieutenant, on assuming command of it, made a false movement in deploying, which the Colonel perceiving, he repeated his order in a voice of thunder, and, as was his custom when displeased, he struck his left breast with his right hand, which grasped his sword. The last word of his orders had just been spoken, and his hand had hardly touched his breast, when a rifle bullet shattered the middle finger, passed through the handle of the sword, and struck the breast so violently that, thinking he was shot through the body, he relinquished the command to Major Mitchell ; but, recovering directly from the faintness, again took his place with his hand streaming with blood and the finger dangling by the skin, and remained at the head of his Highlanders to the close of the action.

The guns, having been brought up, made havoc in the French ranks as the 92nd and 71st re-formed and advanced ; our dragoons and hussars charged and dispersed their cavalry with great loss, but Gérard, wounded though he was, kept his infantry together and continued his retreat along the Truxillo road. The right column of the British was, however, already there, the artillery and cavalry were close on the French flank, the victorious Highlanders were coming fast behind him, his men were falling by fifties. Gérard's position was desperate, yet he would not surrender ; but, giving the word to disperse, endeavoured to escape by scaling the almost inaccessible rocks which overhung the road. The 39th and Portuguese turned the mountain by the Truxillo road, while the 28th and 34th Regiments, led by General Howard, followed the enemy, scrambling up the rocks and taking prisoners at every step. An officer of the right column says—" Our share of the business among the rocks was a scene of laughter and diversion rather than of bloodshed and peril, for though some of the enemy's Grenadiers discharged their muskets at us before they broke them, the danger was too inconsiderable to be thought of."

The 34th had here an amusing experience of French character. They were leading up the mountain, and got mixed up with the French regiment of the same number, whose officers, as they tendered their swords on being made prisoners, embraced the officers of the British 34th, saying—" Ah, messieurs, nous sommes des frères, nous sommes du trente-quatrième régiment tous les deux." " Vous êtes des braves." " Les Anglais se battent toujours avec

[1] Colonel Cameron sent the sword home and it is now at Callart, in the possession of his great-niece, Mrs Cameron Lucy. It had been a present from his friend the paymaster, son of Mr Gordon, of Croughly, in Strathaven, a family who have given several officers to the Gordon Highlanders.

loyauté et traitent bien leurs prisonniers." "Ah, messieurs, la fortune de la guerre est bien capricieuse."[1]

The 92nd and other British regiments desisted from the pursuit only because they, with knapsacks and arms, could not overtake the 600 Frenchmen who alone escaped with Gérard by throwing away their arms and accoutrements.

Gérard's troops were said to be the finest then in Spain, and selected for his expedition, and, as Napier remarks, their resolution not to surrender in such an appalling situation was no mean proof of their excellence.

No enterprise during the war was better planned or executed, and Hill's Division rejoiced in the triumph of their general, which, to one of his humane disposition, was the more grateful that such solid success was gained with little loss to his troops. General Gérard's Corps was almost totally destroyed or dispersed, and the trophies of the action were 1300 prisoners, including the cavalry General Le Brun, Colonel the Prince d'Aremberg,[2] who was a connection of Napoleon, Lieut.-Colonel Voirol and another lieut.-colonel, Gérard's aide-de-camp, one commissary and thirty captains and subalterns. All the artillery, waggons, baggage, ammunition, and numbers of horses and mules were also taken ; in the baggage was the military chest with a large sum of money, and another chest with all the paraphernalia of the Order of Masonry.

The loss of the 92nd on this occasion was four officers and seven rank and file wounded, and three rank and file killed, viz. Privates John Denoon, William MacDonald, and — Campbell, a large proportion of the total loss of the British and Portuguese troops, which amounted to only seven men killed, and sixty-four officers and men wounded, and Sir W. Erskine's aide-de-camp taken prisoner.

The account of the action given above is exactly in accordance with General Hill's dispatch dated at Merida, October 30th, 1811.

[1] "Recollections of the Peninsula."—"Ah, gentlemen, we are brothers ; we are both of the 34th Regiment." "You are brave men." "The British always fight loyally and treat their prisoners well." "Ah, gentlemen, the fortune of war is very capricious."

[2] "The Prince d'Aremberg is a great card, being a member of the Confederation of the Rhine and a Prince of the Imperial family."—*Letter from Wellington to Hill.*

NOTE.—Aroused by the sound of the pipes, the Prince came out half-dressed, when a sergeant of the 92nd seized him. He resisted, but the sergeant, applying the point of his sword, compelled him to move forward as his prisoner.—*Sergeant Robertson.* The Colonel of the French 34th had been preceded by his regiment when the First Brigade entered Arroyo. Rushing from his quarters he mounted his charger and galloped along one of the streets to join it. Colonel Cadogan of the 71st happened to meet him and tried to stop him by a blow of his sword, the Frenchman guarded and hit Cadogan on the head, but the head-dress saved him ; the gallant French Colonel was, however, made a prisoner. He had on various occasions shown great kindness to British prisoners, and this conduct was repaid by the marked attention shown to him when he was himself a prisoner, and soon after his arrival in England he was allowed to go home to France.—*Memoir of a 92nd officer.*

It is therefore remarkable that the 34th Regiment (which was in the right column, and whose active duties, admirably performed as they were, chanced to be required only in making prisoners of the troops already defeated by the left column) is the only one which bears *Arroyo dos Molinos* on its colours, an honour which was granted to them after the death of General Hill about 1840, in lieu of a distinction in dress of which they had been deprived, but which had no relation to this battle. The 92nd, without in any way questioning the good service done by the 34th, felt very strongly that the facts as reported in dispatches show that the 34th had not the same opportunity of distinguishing themselves on this occasion as fell to the lot of the 92nd (with the cavalry and 71st), and that to single out a regiment for so conspicuous an honour has the effect of making it appear to the army at large that the conduct of other corps engaged had been less distinguished, and this impression is naturally strengthened as time goes on and the circumstances are forgotten. It seems, therefore, very desirable that the subject should be again brought forward by the colonels of corps concerned, when no doubt the slight inconsiderately cast on other troops who were at least equally engaged will be removed by granting them the same honour.[1]

The French seem, under all circumstances, to find some method of softening their fate, and are very ready in framing excuses for any disaster which may befall them. The officer of the Third Brigade, above alluded to, mentions that being on duty over some of the prisoners a few days after the battle, at the close of a day's march, a chapel was allotted to them for the night ; and to see them taking possession of it, he would have thought they were still marching free and in arms, singing, " Grenadiers ici, Grenadiers ici ; Voltigeurs là là, Voltigeurs là là " ; the grenadiers running to the altar, the voltigeurs to the gallery. In ten minutes they were all at home—some playing cards, some singing, some dancing ; now a man performing Punch behind a greatcoat with infinite drollery. Quieter men were occupied in repairing their clothes and shoes ; while in another part of the church, a self-elected orator was addressing a group on their late capture. " Gentlemen," said he (in French, of course), " you are not dishonoured. We have been betrayed. That spy—that Spaniard—sold us." " How ? " said a rough voice. " Who told you that ? " " Sir," replied the orator, " you will permit me to know. I am a Parisian, and I understand war." This speech was highly applauded. " Yes ; he is right. We have been sold by this villain of a spy. We should have beaten the British in a stand-up fight." " Why, certainly," said the little Parisian. Just then

[1] See Appendix VII.

the rations appeared, and all rushed to the door singing a chorus about "Bonne soupe, bonne soupe"; but some of the sergeants and older soldiers, decorated for service, were very sulky, and vented their anger in a sort of muttering, smothered swearing. They were all very fine soldierlike men.

The Gordons, when they discontinued the pursuit and the prisoners were collected, bivouacked, with the First Brigade and cavalry, in a wood at St Pedro, about six miles from Arroyo, where double allowance of soft bread and spirits was served out to them with their rations. They had not been long in camp when they became spectators of a most interesting scene. A party of twenty-three French dragoons was observed moving fast across the plain towards Medellin; immediately, a picket of seventeen British dragoons dashed down to intercept them; both parties were in view all the time the *mêlée* lasted, which was but a few minutes, when the enemy agreed to accompany our dragoons into camp, where they were received with three hearty cheers.

At three o'clock on the morning of the 29th they quitted St Pedro, and after a long march entered Merida, wet, weary, and hungry. The 30th was a rest-day; and all the horses, mules, and asses captured on the 28th were sold by auction in the market-place, and the produce, together with the money found in Gérard's military chest, was afterwards divided among the troops actually engaged that day.

Report by Lieut.-Colonel Stewart, 50th Foot, commanding left column of attack at Arroyo dos Molinos, addressed to Lieut.-Colonel Brooke, assistant-adjutant-general, dated Merida, 29th October 1811:—

SIR,—In conformity to the orders I received, the village of Arroya de Moulina was attacked at half-past six o'clock yesterday morning by the 71st and 92nd Highlanders—the 50th Regiment being held in reserve. We were opposed by French cavalry and infantry; but our troops, behaving in their usual style of bravery and discipline, and using the bayonet only, soon cleared the town of the enemy. I distinctly saw the real gallantry, and in my humble opinion, the good military conduct of Lieut.-Colonels the Hon. H. Cadogan and J. Cameron, at the head of their respective corps. The latter officer was wounded early, but did not for a moment quit the command of his regiment until the affair was entirely concluded. The 71st have taken, and are in possession of, a flag which seems to be a colour of the 40th French Regiment of the Line.

(Signed) CHARLES STEWART,
Lieut.-Colonel 50th Regiment.

On the 31st they commenced their return journey, halting at Montejo; on November 1st they had a tramp of twenty-one miles to Campo Mayor. On the way they crossed a stream swollen by the late rains, which took the men up to the waist; then halted at midday to refresh, dancing themselves dry before rousing fires, which, with a bright sun and a stiff glass of grog, banished all traces of their ducking. In high spirits they started again, but only to find, within half a mile, another river, broader, deeper, and more rapid than the first, where they got a worse wetting than before. It may be imagined that the choice of such an unfortunate halting-place brought on the quartermaster-general's department a torrent of objurgations as deep and broad as the river that caused it. At Campo Mayor they halted a day; and on the 3rd moved to the old fortress of Arroncho, and next day, about noon, they re-entered their old quarters at Portalegre after twelve days' absence, and terminated their memorable trip into Spain amid the acclamations of the multitude. The bells were rung, the city was illuminated at night in their honour, and the inhabitants welcomed the soldiers with every demonstration of gratitude, admiration, and hospitality.

On the 7th of November, Lieut.-General Hill issued the following General Order at Portalegre:—

"Lieut.-General Hill has great satisfaction in congratulating the troops on the success which has attended their recent operations in Estremadura, and in doing so, he cannot but endeavour to do justice to the merits of those through whose exertions it has been obtained. A patient and willing endurance of forced and night marches during the worst of weather and over bad roads, of bivouacs in bad weather, oftentime without cover and without fire, and a strict observance of discipline, are qualities, however common in British soldiers, which the lieut.-general cannot pass unnoticed. Having on this occasion witnessed the exertion of them in no ordinary degree, he feels that nothing but the most zealous attention of commanding officers, the goodwill and zealous spirit of the n.c. officers and soldiers, could produce such an effect; and he requests that they will generally and individually accept his warmest thanks, particularly those corps which were engaged in the action of *Arroyo del Molinos*, whose silent attention to orders, when preparing to attack and when manœuvring before the enemy, could not but excite his notice and give them an additional claim on him."

Letters from the Secretary of State, dated December 2nd, and from H.R.H. the Commander-in-Chief, dated December 6th, were promulgated, expressive of H.R.H. the Prince Regent's approbation and thanks to Lieut.-General Hill and the troops under his command.

In November the officers and n.c. officers belonging to the 2nd Battalion, who had come out with the last draft and had been doing duty with the 1st, were sent home to rejoin their battalion, except the medical officer. The 1st Battalion remained quietly at

Portalegre, where they were on quite friendly terms with the people, till Christmas. That festive day was spent as usual, and the officers at mess had just picked the bones of the Christmas goose, the glasses were set, and a Merry Christmas to those at home was being drunk, when an orderly entered. The battalion was to be at the alarm post next morning before daybreak ![1] and having bid adieu to their fair friends of Portalegre, on the evening of the 26th the First Brigade occupied Codeceira. Next day they were at Albuquerque, and on the 28th quitted that city *en route* for Merida, which had been taken possession of by the French under General Dombrousky. The troops bivouacked near the village of La Rocha under a hill, so that their fires might not be seen, the object being, if possible, to repeat at Merida the surprise of Arroyo.

At daybreak they started under cover of a dense fog. The cavalry, under General Long, preceded the infantry at a considerable distance, and about half-way to Merida they came bump upon one of the enemy's vedettes, belonging to a foraging party of 300 infantry and 150 hussars. Before he was made prisoner, the vedette warned his invisible friends by discharging his carbine, and on the fog rising the French were seen in square ; they received the charge of our cavalry with admirable firmness, and beat them off with loss ; then retiring in square, intrepidly sustaining charge after charge and taking every advantage of the inequality of the ground ; and, notwithstanding the fire of two pieces of artillery, by which he lost forty men killed and wounded, Captain Neveux and his gallant little band traversed several miles, till, on their being reinforced from Merida, our dragoons gave up the pursuit.

General Hill in his dispatch expressed his admiration of the courage and prudence of the commander and of the coolness and bravery of the Frenchmen he commanded. It furnished another example of how little a body of well-disciplined infantry, under an officer of courage and experience, has to fear from cavalry, even if very superior in numbers.

In the evening the whole corps closed up and bivouacked in front of La Nava, and next morning advanced to Merida, to find that the French had abandoned their unfinished works during the night, leaving behind them a quantity of bread and corn. Here Hill rested his troops on the 31st. Intending to fight Drouet, he marched from Merida on the 1st of January 1812, crossed the Guadiana and moved towards Almendralejo, where he expected to

[1] To conceal his design of taking Ciudad Rodrigo, Wellington ordered General Hill to assume the offensive in Estremadura, which had the effect of making Soult, supposing Badajos to be threatened, concentrate his forces in that direction.—*Alison.*

During the period his army was resting in cantonments, Wellington had been incessantly at work improving every department of the British and Portuguese armies, especially the transport and commissariat service and the military hospitals.

meet the enemy. Their way led over a barren heath, then over a rich cultivated plain, till, having ascended about five miles, the Gordons halted and enjoyed a delightful view of the country, with the long columns of horse, foot and artillery winding below; then jogged along again as cheerily as the fatiguing nature of the march would allow, till, near Almendralejo, it was time to prepare for action.

At one o'clock a thick fog enveloped the surrounding objects, and prevented the general from seeing what the enemy were about, or what were their numbers. The First Brigade (in which was the 92nd) was ordered to move towards the town in order of battle. They had not proceeded far when the cavalry came into contact with the French pickets, which were instantly attacked and driven in; the rear brigades now moved up quickly on the left. It appears that when the first collision took place, the French in the town were busy cooking their New Year's Day dinners and thinking more of beef than of bullets, but part of them instantly moved in support of their pickets, the remainder retreating to a rising ground in rear of the town, where, after a few rounds from the British artillery, the others joined them, and the whole retired to a height at a considerable distance, skirmishing all the way with Hill's Light Companies. When the main column came up, expecting to be hotly received, and the fog lifted, they found the French general had withdrawn the main body of his troops to Zafra, leaving only a strong rear guard to cover his retreat.

During this little affair the Highlanders were much amused by the conduct of the Marquis of Almeida, an elderly Spanish nobleman, who, having suffered severely from French rapacity, was their bitter enemy. Being a warm admirer of the British, he had attached himself to Hill's suite, and in compliment to the general he laid aside his Spanish habit and donned a long scarlet coat and a cocked hat, with very long red and white feathers dangling to his shoulders; but his singular appearance soon ceased to attract notice, and he became rather a favourite with all ranks. On this occasion he had no idea he was so close to his foes till bang, bang went the guns close in front of him. Neither he nor his horse had bargained for taking so active a part in the redress of his grievances. He stared, his horse reared, and on a third gun being fired, the old marquis, calling on "Jesu, Maria, Jose," and casting a glance towards those around him, which plainly said, "I 'm off!" put spurs to his willing nag and was lost to view in the fog.

When the mist cleared away the rain began falling in bucketfuls; night was approaching, and, there being no chance of bringing the enemy to close quarters, General Hill gave orders for strong pickets to be posted on all the roads round Almendralejo, and the rest of the troops to march into the town. On taking possession of their quarters, many of them found the savoury stews still on the fires which the French had been preparing, an acceptable addition to their rations, which their Spanish hosts were delighted to accept their invitation to join in discussing, adding their country wine to the soldiers' rum, and thus their New Year's Day ended more agreeably than it began.

On the 3rd two detachments were sent to dislodge the enemy from Villafranca and Fuente del Maestro. The one which moved against Villafranca consisted of the 9th and 13th Light Dragoons, two pieces of artillery, the 50th, 71st, and 92nd Regiments, and Captain Blassier's Company of the 60th, under Major-General Howard. The other, under the Hon. Lieut.-Colonel Abercromby, was composed of the 2nd Hussars K.G. Legion, 4th and 10th Portuguese Cavalry, and the 28th Regiment (British). Both detachments started at noon. On reaching a height about halfway to Villafranca, General Howard's Brigade saw the enemy drawn up on a commanding position near the town ; Colonel Cameron, with the 92nd, the Rifle Company, and one gun, advanced against the French left flank, while the remainder, under General Howard, moved direct upon the enemy's centre and right flank. Everything looked like a tough tussle. The enemy were prepared, and stood their ground until the British were getting near enough to prepare to charge, when, to their great disappointment, the French faced about and walked off towards Los Santos, pursued by the cavalry. General Howard had no intention of letting the enemy pay him back in the coin of Arroyo dos Molinos ; he placed strong pickets on every road leading to Villafranca, the rest being thrown into the town, with orders to remain accoutred so as to turn out at a moment's warning, and to be on the alarm post two hours before day. It was an awful night—rain, hail, and a hurricane of wind—so that the poor fellows on picket joined their battalions in the morning in a deplorable condition.[1] Those who had been in quarters proceeded to the alarm post, where they were drenched in ten minutes, and after being exposed to the fury of the elements for four hours, the whole returned to Almendralejo. Colonel Abercromby's detachment had had a skirmish with a regiment of cavalry which crossed their path, of whom they killed and wounded twenty-one and took thirty-two prisoners.

On the 5th, General Hill retraced his steps to Merida, where the men arrived looking as if they had been six months in the field

[1] Colonel Cameron, in visiting the picket of his regiment, found that the young officer commanding it had placed his men in a hollow sheltered from the storm, but without keeping a proper lookout, for which he received a severe reprimand.

BRITISH LIGHT DRAGOONS AND FRENCH DRAGOONS, 1812

instead of ten days. The marches had been often through adhesive clay, into which the soldiers sank ; many tore their gaiters to pieces, and some actually left their shoes behind them and trudged along in their hose, which, with knapsacks and accoutrements to carry, made it pretty hard work.

Apropos of this very severe march, a 92nd officer tells the following story of an infantry private who, having on January 5th partaken too freely of the strong wine of Almendralejo, found himself very hard put to it to keep up with his column on the subsequent march. He contrived to fall out unnoticed at a halt and remained behind when the march was resumed ; then, composing himself to sleep, he enjoyed so long a nap that it was already night when he awoke. He was quite recovered from his intoxication, but his fevered brain at first failed to comprehend how he came to be alone in the dark and in a seemingly deserted place ; but as he gradually realised the situation, the thought flashed across him that he would be supposed to have deserted to the enemy. He saw in fancy his name sent in to the adjutant-general—the court-martial assembled to try him—the sentence of death passed upon him ! During the hours of darkness no satisfactory plan presented itself for getting back to the colours unnoticed, and he started at dawn to follow the troops, full of dread for the future. Afraid to join by daylight, he made his way to a village three miles from headquarters, and on the opposite bank of the Guadiana. Here he heard that two French stragglers were concealed, and a brilliant idea struck him—he would capture the pair and present them as a propitiatory offering to offended discipline. No sooner said than done ; the villagers helped him to secure the Frenchmen, and he marched them off in triumph, their hands tied behind their backs. The officer who relates the incident met him at the bridge of Merida, and, seeing him thus escorting two prisoners from a district where no British soldiers were quartered, naturally asked for explanations, whereupon the soldier burst into tears, unable to utter a syllable, but at length poured forth the foregoing story. It is added that, as he had hoped, it was considered a sufficient atonement for his escapade.

Sir Rowland Hill (General Hill had been made a Knight of the Bath), having received an order early on the morning of the 12th of January to retrace his steps to Portalegre, the 92nd marched from Merida the same day at 9 a.m. and bivouacked behind La Nava. Next night they reposed on the banks of a stream under the castle of Zagala ; on the 14th they halted at Albuquerque, Alegrete on the 16th, and arrived at Portalegre on the 17th. Next day the men replenished their kits, and on the 19th moved to Alpalhao on the way to assist in Lord Wellington's operations. At Niza, on the 21st, they heard of the fall of Ciudad Rodrigo, but as

Marmont still showed a disposition to fight, the battalion advanced to the Tagus on the 25th, crossed that river by a bridge of boats at Villa Velha, and occupied several villages. Next day they entered Castello Branco, where they had the pleasure of meeting the French garrison of Ciudad Rodrigo on their way to British transports. They remained at Castello Branco till February 1st, when, as Marmont had withdrawn his army to Salamanca, they retraced their steps, arriving at Portalegre on the 4th.

Here they remained, occasionally finding officers' escorts for prisoners of war, till the 3rd of March, when Sir Rowland Hill's Division moved to cover the siege of Badajos, which Lord Wellington had now determined to undertake. The battalion halted for some days in Albuquerque, and on the 16th, with the rest of Hill's Corps, moved on Merida ; they spent the night in bivouac. On the 17th, finding that some French cavalry, supported by a battalion of infantry, occupied Merida, Sir Rowland sent his cavalry across the Guadiana by a ford below the bridge, to intercept the enemy and enable the infantry to get up. On the first alarm the French cavalry fled, some by the bridge and some by a ford above it, giving notice of the danger to the infantry by discharging their carbines.

No time was to be lost, and the First Brigade, 50th, 71st, and 92nd moved to the town at the double, and, knowing the place well, they made straight for the bridge as fast as they could run. Crossing it, they halted a minute to gain breath and then continued the pursuit, but failed to bring the enemy to action, and, fast as the British ran, the French ran faster, and at last, when our men were completely blown, they gave up the pursuit, having, however, taken a certain number of prisoners who could not keep up with their comrades. The 92nd bivouacked on the far side of the river.

Sir Thomas Graham had crossed the Guadiana on the 16th at the head of the right wing of the covering army, consisting of most of the cavalry and the First, Sixth, and Seventh Divisions, and was posted on the heights of Llerena.

Lord Wellington invested Badajos the same day, and broke ground before it on St Patrick's Day in the morning. The left wing of the covering army was to be on the Guadiana at Don Benito. At daybreak on the 18th, Hill's Corps crossed the Guadiana, and after marching and countermarching to Almendralejo, etc. etc., with the view to prevent the relief of Badajos, on the 28th the cavalry, a brigade of artillery, and the First Brigade of infantry advanced to Medellin and Don Benito. *En route* the detachment was formed into two columns, of which the left, under Colonel Cameron, consisting of the 92nd and two field-pieces of artillery, moved against Medellin ; the right, under Major-General Howard, composed of the cavalry, the 50th and 71st, the company of the 60th, and the remaining guns, against Don Benito.

Medellin was occupied without opposition, and General Howard was informed that the enemy had also retired from Don Benito; he, however, sent Captain Blassier with his riflemen to make sure. The captain was a gallant soldier, but fond of good living, and was making his way through the streets thinking more of the good dinner he expected to find there than of an enemy, when, on turning a corner, to their mutual astonishment, he found himself close to a French cavalry patrol, who had entered the town unconscious that the British were so near. First they gazed at each other, and then proceeded to settle the right of way by sabre cuts and rifle bullets; the latter had the best of it, and the dragoons wheeled to the right-about, leaving some wounded men in the hands of the German riflemen.

The 92nd were posted in an old castle on a hill near the town, from which they could see the memorable plain and hills about Arroyo dos Molinos, and hear the thunder of the guns at Badajos. On the 31st the whole retreated to Quarena, and on the 2nd April they were back at Merida; on the 6th they had again crossed the Guadiana and bivouacked on an eminence at the village of Lobon, near Badajos, all these manœuvres being for the purpose of preventing Soult from relieving that fortress.

In the camp at Lobon, it was known that the assault was to take place that night, and with intense anxiety the troops there listened for the sounds of the terrific conflict; they feared their friends could never surmount the obstacles placed by the garrison to prevent their passage; but such was the bravery and determination of the stormers, that though beaten back, again and again they mounted the breaches, maintaining the murderous conflict, till the ditches were literally filled with dead and wounded piled above one another, and at daybreak on the 7th, the garrison surrendered, having inflicted on their conquerors a loss in killed and wounded of 3800 British and over 1000 Portuguese.[1]

Marshal Soult had arrived in the vicinity confident that Badajos could hold out, and he was advancing to give battle to Hill's covering force, which was close in front of him, when on the 7th he heard of Wellington's successful assault of the fortress, and after giving way to a violent fit of passion, at once retreated towards Seville, for he was not prepared to fight the united armies of Wellington and Hill. He was pursued by 2000 British cavalry, who, after a brilliant encounter at Villa Garcia, defeated an equal force of French horse with a loss of 300 killed, wounded, and prisoners. As an instance of the feeling of revenge which actuated

[1] Lieutenant J. Cattanach, 92nd, was employed as an engineer at Badajos.

the Spanish peasantry, a 92nd officer mentions that on the 6th, while walking near Lobon, he spoke to a man who was resting in the shade of a tree, and found that he had come from a village twenty miles distant, for the sole purpose of killing as many Frenchmen as he could *after* the battle, which was expected to take place at any moment between Hill's force and Soult's, and he showed the long knife with which he had assassinated eleven Frenchmen on the morning after Albuera. The officers left him, after expressing their disgust at his cold-blooded project.

On the 10th April, Wellington, finding that Soult had finally retired into Andalusia, broke up his encampments and moved north to keep Marmont in order, while the 92nd, with part of Hill's Corps, which now consisted of two divisions of infantry and three brigades of cavalry, marched to Almendralejo to look after their old opponent Drouet, the Count d'Erlon. Here they were joined by a draft of thirty-nine men from the 2nd Battalion, described as "of good appearance," and here they remained from the 13th April till the 11th May, utilising the time in repairing damage to clothing and appointments, in drill, and in making themselves agreeable to the fair residents.

Lieut.-General Chowne, who had lately assumed the command of the Second Division, inspected the battalion in heavy marching order—"The men will have on good hose with their rosettes"—on May 10th; and next morning at three o'clock they marched along with the First Brigade, forming part of a force of 6000 men under Sir Rowland Hill, including 400 cavalry, two brigades of field artillery, a battering train and pontoon equipage, for the purpose of breaking the chain of communications between Soult's army of the south and that of Marmont in the north. The bridges over the Tagus had been destroyed by one or other of the belligerents, and the only means of crossing possessed by the French was a bridge of boats laid by Marmont at Almarez, where, to secure it, he had constructed three strong forts—Fort Napoleon on the left bank of the Tagus, Fort Ragusa on the right bank, which contained many stores and provisions, and Fort Mirabete, about three miles from the river, commanding the only road from Truxillo practicable for artillery. There was also a fortified bridge-head or *tête-de-pont*.

It was the surprise of Almaraz that Hill was now ordered to attempt. On the 15th they reached Truxillo, in which town the women of the battalion were left with the baggage, only one mule per company being taken for the officers and one for the camp kettles. As no fires would be allowed on the following night, the men cooked provisions for two days, and they were also to carry three days' bread. At midnight the warning pipe resounded through the streets, and at 1 a.m. on the 16th April they moved off towards Almaraz. By sunrise they halted in the shade of a forest which concealed them from view, where they rested during the day, except that those who were to be stormers were prepared for their part by running many times up the scaling ladders placed against an old stone bridge. A young officer complaining of stiffness in his legs after this unwonted exercise, another answered, "Be thankful, old fellow, if your legs are not stiffer to-morrow!"

There being three points of attack, the troops were divided into as many columns. The left column, composed of the 28th and 34th Regiments and the 6th Portuguese Caçadores, under Lieut.-General Chowne; the centre, under General Long, consisted of the 13th Light Dragoons, the 6th and 18th Portuguese Infantry; while General Howard led the right, viz. the 50th, 71st, and 92nd, and a company of the 60th. Each column was provided with scaling ladders. Howard's Brigade was destined to attack Fort Napoleon and the *tête-de-pont*.

The whole moved from their woodland bivouac about eight o'clock on the evening of the 16th, but marching at night over mountainous tracks, with cross paths and difficult passes intersecting them, made such operations extremely uncertain. The right column lost its way, and at the hour intended for the assault it was still five miles from its destination. Sir Rowland, who was with this right column, deemed it prudent, therefore, to halt on the ridge of Lina, and to order the other columns to withdraw from Mirabete. It being found impossible to get the artillery forward except by the guarded pass of Mirabete, the 17th and part of the 18th having been spent in fruitless attempts to discover an opening through the hills practicable for the guns, Hill determined to leave them on the mountains with the centre column, to make a false attack upon the tower of Mirabete with General Chowne's troops, while he himself with the right column secretly penetrated by a scarcely practicable sheep path, where in some places only one man could move along at a time; resolving with the infantry alone to storm works defended by eighteen guns and a powerful garrison!

About 10 p.m. on the 18th, Howard's Column, reinforced by a second company of the 60th, the 6th Portuguese and some artillerymen, and led by Sir Rowland Hill, descended the mountains, but though the distance was only about six miles, the difficulties were great. Eight men at a time carried each scaling ladder, and when one lost his footing, down went the whole party on the steep hillside along which the path lay, causing constant delays, so that the sun was up before the rear had closed up and formed for the attack. This second disappointment caused great depression of spirits among the troops, but on Sir Rowland Hill making it known that he would there and then lead them to the assault, the men resumed their usual gaiety. The foregoing circumstances having altered the situation, Sir Rowland altered his original plan and divided the detachment. The 50th and a wing of the 71st were to attack Fort Napoleon, under General Howard; the 92nd and the other wing of the 71st were ordered to be ready to help them, or to attack the *tête-de-pont* and Fort Ragusa ; while the 6th Portuguese and two 60th companies were held in reserve.

Formed behind a little height a hundred and fifty yards from the fort, the 50th, on a given signal, moved from their concealment between six and seven o'clock on the morning of the 19th April, and advanced to the attack aided by a wing of the 71st. The French were not altogether unprepared, for they had heard that British troops were near, and Fort Napoleon had been reinforced during the night ; but just then the garrison heard the guns of Mirabete answering General Chowne's attack, and were crowding the ramparts gazing in that direction, when a British cheer made them quickly turn their eyes to see the gallant "Dirty half-hundred" bounding over the nearest rising ground. At once they opened a heavy fire of artillery and musketry on the assailants in front, while the guns of Fort Ragusa took them in flank from the opposite side of the river ; but a rising ground close to the ramparts soon covered the British, and General Howard, leading the foremost men into the ditch, commenced the escalade. The scaling ladders had been cut in halves to get them round the short narrow turns in the precipitous descent of the mountains, but General Howard's presence of mind, and the pluck of the soldiers, overcame the difficulty. They first ascended to the berm, which was broad ; then drawing up the ladders, planted them there, and thus by a second escalade, though many fell back into the ditch, they forced their way over the rampart, and after hard fighting, the 50th and their friends of the 71st established themselves in little more than ten minutes in Fort Napoleon.

Meanwhile the 92nd and the other wing of the 71st moved forward to attack the *tête-de-pont*, taking advantage of every little knoll which sheltered them from the fire of Fort Ragusa, till they got to a point opposite the left-flank face of Fort Napoleon, when they dashed forward to the *tête-de-pont*, entering it with the garrison of Fort Napoleon, who, their retreat being cut off, rushed to the bridge, but the guns from the forts (for the British artillerymen had turned the guns of Fort Napoleon on Ragusa) were now sharply cannonading each other, and stray shots had cut away some of the boats, making a gap in the bridge. The fugitives had either to jump into the river and swim for it, by which many were drowned, or to surrender to the Highlanders, who were pressing them up to the edge of the chasm.

The passage of the river was now the difficulty. It was solved by some Gordon Grenadiers, who immediately leaped into the deep swirling stream, swam to the other side under fire, and brought back boats, with which the bridge was secured ; the enemy, seized with panic, then abandoned Fort Ragusa, and the success of the enterprise was complete.

The individual gallantry of Privates James Gauld and Walter Somerville, the two men of the Grenadier Company who first leaped into the river, was brought to the notice of Sir Rowland Hill, as having materially forwarded his object upon Fort Ragusa, and he ordered two doubloons (equal to £8) to be given to each of these soldiers on the field. As a seven years' man, Gauld was not entitled to a pension, but the then Duke of Richmond, having heard the above story, wrote to the 92nd Depôt in 1853 for confirmation of it. His Grace brought it to the notice of the War Office, and Gauld received a pension for the rest of his life. He was a man of exceptional physique, and had distinguished himself on other occasions. His family held the farm of Edinglassie, and were a branch of the famous fighting family known as the old Gaulds of Glass, in which parish James' descendants still live. What became of Walter Somerville is not recorded.

As soon as the enemy had fairly taken flight, permission was given to the men to help themselves to the good things stored in the fort and officers' mess-room, and soon hams, bacon, pickled beef, and other delicacies decorated the bayonets or were stuffed in the haversacks of hundreds of soldiers, and none were without a bottle of wine or brandy to wash down the good cheer, in the bivouac to which they were ordered in the neighbourhood. The satisfaction of the Gordons in talking over the stirring events of the morning round their camp fires was not on this occasion marred by regrets for the death of any of their comrades; they had only two wounded, though the loss of the Allies amounted to 2 officers and 31 n.c. officers and soldiers killed, and 13 officers and 131 n.c. officers and soldiers wounded. The total loss of the enemy was about 450, 259 of whom were taken prisoners, including 16 officers and Major Aubat, the Commandant of Fort Napoleon, who, on finding further resistance useless, had surrendered and been allowed to keep his sword. An officer of the storming party rushed in, and, being ignorant of the terms on which the commandant retained his sword, and also of the French language in which he tried to explain, made a lunge at him. A mortal wound was the result of this rash and excited act, which caused great commiseration for the unfortunate Frenchman, to whom every attention was paid, but he died in ten days, and was buried in the great church of Merida with military honours, the whole of the British officers attending the ceremony.

After removing the wounded, the forts, bridge, and the cannon (eighteen guns were taken) and stores which could not be carried off were destroyed. The position was of the greatest importance to the French, and Wellington, in a letter dated Fuente Guinaldo, 28th May 1812, says, "I doubt their having the means of replacing it (the bridge), or that they will again form such an establishment at that point, however important it is to their objects."[1]

Their general expressed in Orders his admiration of the discipline and valour of the troops, and in July a letter from the Secretary of State was published expressing the Prince Regent's approbation of their conduct. Sir Rowland Hill gained great credit for the decision with which this daring attack was planned and executed, and in memory of the part taken in it by the 1st Battalion, the Gordon Highlanders afterwards received the Royal authority to bear *Almaraz* on their colours.

Next morning they began their return journey, and on the 21st they re-entered Truxillo, where the inhabitants entertained them at a grand bull-fight prepared in their honour. Here they halted for two days, which the men employed in washing and cleaning their clothing and accoutrements, the officers being desired to see this thoroughly done, for it has always been well understood in the 92nd how much a clean and smart appearance increases that feeling of self-respect in the individual soldier which compels the respect of others for his regiment. On the 24th they marched for Merida, and three days later re-entered that pleasant city amid the cheers of the populace and the congratulations of the army left to protect that part of the country.

After drinking the King's health on his birthday, the 4th of June, they took leave of Merida the next morning and moved to Almendralejo. Sir Rowland having received intelligence that Marshal Soult was likely to pay him a visit, concentrated his troops at a more advanced point, the whole of the infantry moving forward to Zafra, Los Santos, and Sancho Perez, the 92nd being cantoned in the latter. On the 13th of June the interest of the troops was kept alive by a severe and gallant action near Llera, between the French cavalry and the British heavy dragoons, in which the latter had a loss of 150 men, part of which they repaid in a skirmish two days afterwards.

[1] Gurwood.

CHAPTER XVI

NAPOLEON had withdrawn from Spain a considerable body of troops, including 10,000 of the Imperial Guard, but his generals had still in that country four times as many men as Wellington's allied forces amounted to.[1] Although the report that Soult had left Andalusia was erroneous, he had reinforced Drouet's army, and that general was advancing against Sir Rowland with 21,000 men, of whom 3000 were first-rate cavalry. Hill's army was slightly stronger in numbers, though partly composed of Spanish and Portuguese troops, and he was weaker in horse by 1000 sabres.

The enemy having moved from the south, seemed inclined to bring on an action in this advanced position, and thus to divert Wellington's operations towards Salamanca against Marmont, with whom Soult had no direct means of communication since the loss of the bridge at Almaraz. Hill determined to fall back on the position of Albuera, and there to give battle. The cavalry retired as the French advanced, and with them the 50th and the 92nd marched to Los Santos on the 16th, and the following evening the whole corps retired and halted behind Santa Martha at sunrise on the 18th June. On the 19th they marched and took post in a wood in front of the position of Albuera on the right bank of the stream of that name. The village was defended by barricades in the street and loopholes in the houses, and the ridge by breastworks, every precaution being adopted by the prudence and military skill of Sir Rowland to strengthen his position. Cavalry pickets were posted in front of the wood where the infantry were encamped, and the First Brigade, 50th, 71st, and 92nd were held in readiness to retard the enemy in front of the wood, and give time to the other brigades to take up their ground ; but the French general was too strongly impressed with the recollection of the dreadful battle of last year at Albuera to risk another attack on the same ground ; while Hill, who had shown himself so daring at Arroyo and Almaraz, refrained from the temptation of a battle which promised him unbounded fame, "simply because he was uncertain whether Lord Wellington's operations in Castile, then in full progress, would warrant one." "His forbearance must be taken as a proof of the purest patriotism."[2]

Thus things remained pretty quiet, though, with opposing armies in such close proximity, scenes of interest are sure often to occur. One day some of the 92nd being on outpost duty with a troop of Light Dragoons, the horse of one of the troopers becoming restive (probably provoked by the intention of the rider to desert), he asked permission to take it to the front to quiet it. On leaving

[1] Alison.

[2] Napier.

the ranks Paddy said to his comrades, " Now by J——s, boys, I 'll show you something you never saw before " ; then he began to thrash and spur his horse, always moving further to the front, his comrades chaffing him and his officer ordering him to desist ; but he kept on spurring and swearing, till, having got about one hundred yards from the troop, he turned his horse's head, galloped off towards the enemy, and gained the French lines before his enraged comrades could catch him.

The British Government had neglected to supply money for the needs of the army, and Wellington wrote in July, " I have never been in such distress as at present, and some serious misfortune must happen if the Government do not attend seriously to the subject and supply us regularly with money." " As it is, if we don't find means to pay our bills for butcher's meat, there will be an end to the war at once." The pay of the muleteers was a year, and of the officers and soldiers six months, in arrear. While at Albuera, however, the 92nd received a supply of necessaries, and pay up to the 24th of the preceding March.

Education was not neglected during their periods of rest, and here Corporal MacEwen, of the Grenadier Company, was appointed schoolmaster-sergeant. No one was allowed to be absent from the camp, and, by Sir Rowland's order, the roll was called by officers of companies four times by day and night at uncertain hours. The men were also ordered to make huts with branches for protection from the sun.

For ten days the enemy gave them little trouble ; but on the 30th June, as the First Brigade were busy with their breakfasts, the alarm was given, the kettles were emptied, and away they went to dispute the passage of a body of cavalry approaching the advance post ; but after two hours the enemy retired, and the brigade returned to their camp and set about cooking dinner ; but just as it was ready, a second alarm forced them to stand to their arms and double to the front, where, this time, a skirmish ensued, in which some of the cavalry lost their lives, but the Gordons only lost their dinners ! That night officers and men remained accoutred.

Hill having no longer the same reasons for remaining on the defensive, assumed the offensive on the 1st of July. Moving in one column, his army encamped at Santa Martha, and next day advanced in two columns ; the left, in which was the 92nd, marched against the enemy at Villa Alba, where a combat of cavalry took place, the enemy retiring to a height, where skirmishing went on after the arrival of the First Brigade, who, when the enemy finally retired, waded the Guadacira and moved some miles along its bank. In the evening they were proceeding to bivouac, when the enemy's artillery cannonaded them from an opposite height ; but after an exchange of shots and some manœuvring, the French withdrew, having wounded some of the Portuguese infantry and British artillerymen. A 92nd officer relates how at Villa Alba the Highlanders admired the conduct, and regretted the fate, of a hussar of the King's German Legion, who, being attacked by a powerful French dragoon, after a deal of dexterous sword play, killed him just as a second arrived to his assistance. To it they went, cut and thrust, till a third dragoon ran his sword through the gallant German, at the same moment that the point of the latter pierced his second antagonist.

From the banks of the Guadacira the battalion marched for two days through an enclosed country by Los Santos, and reached the town of Bienvenita at sunrise on the 5th. An officer and a n.c. officer with a spy-glass were posted in turn from each battalion of the brigade to look out from the steeple and report the approach of any body of troops. As they entered Llerena on the 7th they were welcomed by the "Vivas" of the fair inmates of a suburban convent, who waved their handkerchiefs till the bonnets were lost to sight. At Llerena the spy-glass on the steeple was again in use. On the morning of the 8th, the troops were at the different alarm posts half an hour before dawn, and remained till it was completely daylight, when they were dismissed to cook; after which, Sir Rowland Hill, at the head of his whole corps, marched from Llerena to dislodge the enemy from the position he had taken up at Berlenga. The infantry moved direct upon that place, preceded by the Spanish Horse, while the British cavalry went round to the left in hopes of turning the French right. Six miles from Llerena the Spaniards drove in the enemy's advanced pickets, the French retiring in extended order, and skirmishing very prettily till they joined their main body. The 92nd, with the other infantry, continued to advance so as to arrive at a given point at the hour fixed for the British cavalry to attack. As they approached the height on which the French left was drawn up, the latter retired over a deep ravine and took post at a bridge, while their right wing, seeing the left retiring, evacuated Berlenga and conformed to their movement. Meanwhile the enemy's artillery saluted each battalion as it took up its ground, which compliment was returned by the British; but the hopes of an engagement were disappointed by the non-arrival of the cavalry till the enemy had withdrawn beyond pursuit. While the 92nd were forming, the parish priest came beside them to see the expected battle, but paid for his curiosity with his life, being killed by a stray shot.

At the close of this affair, which in the morning had promised more brilliant results, the troops entered Berlenga, and in the evening bivouacked in its vicinity. Next morning they retraced their steps to Llerena. It was piping hot, for at that season the sun in Spain has almost tropical power, and not a drop of water was to be had except when a half-dried pond enabled the men to scoop out a mouthful of mud and water with the wooden *quaichs* which most of them carried; some who could not get even that refreshment were completely overpowered by the burning heat. It may interest some to know how the quartering of a division in such a town as Llerena was performed. The quartermaster-general preceded the troops, accompanied by the brigade-majors, the quartermasters of battalions, and a n.c. officer and intelligent soldier from each company. After marking off several houses for his general and staff, he split the rest of the town between the majors of brigades; they, in their turn, provided for their generals and staff, and then made a wholesale division of streets among the quartermasters of regiments, who, after providing for their commanding officers and staff, retailed the remaining houses in equal proportion among the companies, so that when the regiment arrived the quartermaster had simply to tell each captain, "Here 's a certain number of houses for you." Each battalion provided a subaltern to take charge of its own baggage on the march, each brigade a captain who took charge of the baggage of the brigade, and a field officer looked after the whole baggage of the division.[1] That of the lieutenant-general and his staff led the way, then that of the brigadier of the leading brigade and his staff, followed by that of the colonel of the leading battalion of the brigade and regimental staff, then that of the leading company of the battalion, and so on in regular succession. At the close of the day's march, each baggage officer reported to his superior up to the lieutenant-general.

Officers and men soon made themselves at home in their quarters, their comfort depending a good deal on the cordiality of their hosts, some of whom were surly, not without cause, suffering as they often did from the misconduct of their self-invited guests, both French and British. But the soldiers, in addition to their beef and bread, had generally plenty of vegetables, fruit, and wine, and in the smaller towns and villages, where the battalion remained for a time, acquaintances were made, and sometimes romantic attachments were formed with the Spanish girls, whose natural ease and graceful manners made them equally at home both with officers and soldiers. Little dances were arranged in the houses or in the open air, where the Highlanders learned the bolero and the waltz, while they initiated their partners in the mysteries of a foursome reel, or showed themselves in the sword dance to be as light-footed as the Spaniards in the fandango. Parade was regularly formed at seven in the morning and at 6.30 in the evening. Inspections in

[1] Captain J. Kincaid, and an officer of the 92nd.

marching order by general officers frequently took place, the men having their blankets and greatcoats neatly rolled and folded, their dress and accoutrements "in the best possible order," "best hose and rosettes" being always particularly mentioned; and on great occasions their purses on; but these they never wore when marching or fighting, being left with the heavy baggage, or were strapped on the back of the knapsacks. The batmen with the ten company mules, the mule for the entrenching tools, and the sergeants' pannier mule, formed part of the parade. Every Sunday, when stationary, the whole division invariably attended divine service.

The battalion furnished its share of the pickets which guarded every approach to Llerena, and remained undisturbed (except by an alarm which kept them under arms the night of the 18th July) till the 20th, when they retired to Bienvenita in the morning, and, resuming their march the same evening, arrived at Zafra at sunrise on the 21st. Here they remained till the 28th, when General Hill marched at midnight with the First and Second Brigades towards Villafranca, where his cavalry were stationed, and which was threatened by the French, who had been largely reinforced. The battalion bivouacked at daylight in an olive grove near Villafranca, about three miles from the enemy, and expected an immediate trial of generalship between Hill and Drouet. In the forenoon a brigade of French artillery astonished them by drawing up on a height opposite their bivouac and firing a royal salute, which the officers at the outposts ascertained was in honour of a victory they supposed to have been gained by Marmont over Wellington at Salamanca. But Sir Rowland Hill immediately afterwards received an express with more authentic intelligence, and at once published the following Order:—

Villafranca, July 29th, 1812.—Captain Maxwell's Brigade of Artillery will fire a salute of twenty-one guns at twelve o'clock, being in honour of the glorious and important victory gained over the enemy at Salamanca by the army under the immediate command of the Earl of Wellington on the 22nd instant. An extra ration of wine or spirits is to be issued to the British and Portuguese troops to drink the Earl of Wellington's health.

(Signed) J. C. Rook,
Asst.-Ad.-General.

The announcement was received with cheers loud enough to satisfy the neighbouring French outposts of their mistake. They did not await further proof, but retired before Hill, when he advanced and took possession of Fuentes-del-Maestre, where the 92nd bivouacked.

Next morning, however, the enemy attacked our cavalry in front of Villafranca, and the First Brigade was dispatched in haste to their assistance. The day was spent in manœuvring under a broiling sun, marching and countermarching according to the movements of the enemy, but without bringing him to close quarters ; in the evening the battalion went into cantonments at Villafranca.

During their marches the troops were accompanied by the wives and children of the n.c. officers and soldiers, who received rations ; and though they sometimes caused anxiety, both to their husbands and the commanding officer, were of great use in nursing the sick, washing the linen of the officers and men, etc., while their presence gave something of a homelike appearance to the camp or cantonments. They generally had donkeys, which they rode, or which carried panniers with their children and possessions ; they were capital foragers, were as full of *esprit de corps* as the men, and bore the fatigues of a campaign with the patient fortitude of their sex. I knew well an old lady who used to tell with pride how, when a sudden order to march came while the linen of the men she washed for was in the tub, she took advantage of the fact that she was billeted on a wood merchant to make a roaring fire, and succeeded in giving every man his dry shirt as he stood on parade, emerging, like Wellington at Fuentes d'Onor, undefeated by the difficulties of the situation. She gave brandy to the wounded in the ensuing engagement, made her husband's breakfast before the fight of the next day, and ended her eventful life as the respected hostess of a hotel in Argyllshire. Apropos of domestic life in the regiment is the story of a soldier who returned from picket duty just before they marched from Bienvenita, to find himself the father of twin boys ; gazing at the helpless babes, and overwhelmed at the doubly increased responsibility, he exclaimed in accents of pathetic despair, " Gude preserve me, Betty Watt, what can I do wi' them ! "

R.O., July 23rd.—In future, when any officer has occasion to stop the allowance of spirits to any man (as a punishment), the circumstance to be always reported to the commanding officer for his concurrence.

The above Order was probably occasioned by a story of Captain Dugald Campbell (the handsome sergeant-major of 1799), a gallant officer who had been in every action with the regiment, but who was as " drouthy " as he was brave, and liked at least double allowance ; at some bivouacs, however, no liquor was to be bought. The stopped grog was supposed to be taken to the surgeon for the use of the patients, but Dugald, thinking himself as good a judge as a doctor of its disposal, applied it to assuage his own thirst. On one occasion his company was in such good order that he had difficulty in finding any fault, but he pitched on a quiet young fellow, and stopped his spirits for having his belts dirty. On the parade being dismissed the quiet young fellow marched straight up to Colonel Cameron, who, surprised at being approached in such an unusual manner, asked what he wanted. "With your leave, sir, I want you to inspect me." The Colonel asked what he meant, and the lad told him his grievance, saying that he did not so much mind losing his liquor as his character as a clean soldier. The Colonel investigated the matter, and it was understood that "Dugald" had a very bad quarter of an hour with his stern chief.

During the hot weather the troops were, when practicable, cantoned in a town during the day, "but must invariably be out of the town at night." "Staff officers to make themselves acquainted with all the roads so as to be able to conduct troops in whatever direction they may be ordered to march without going through the town." Rice and barley were issued daily to the men for their use.

On August 16th, Brevet Major Ewen M'Pherson, 92nd, was appointed Town-Major of Lisbon.

At Villafranca the battalion remained for nearly a month, Hill having been directed to act on the defensive, but at the same time to prevent Drouet from succouring King Joseph. The French outposts were close by, and five companies of infantry bivouacked every night two miles from the town, with Light Companies in advance, and cavalry in advance of them, their orders being not to engage in anything serious, but merely to give time to the troops in the town to turn out in case of attack.

One evening, when the right wing of the 92nd was on this duty, irregular firing was heard in their rear, the troops stood to their arms, and Colonel Cameron, who was in command, took every precaution, thinking the French had eluded the vigilance of the outposts; but an officer with some men being sent to obtain intelligence, it turned out that the disturbance was caused by the Spaniards firing off muskets to show their loyalty to King Ferdinand, and their joy on hearing that King Joseph Bonaparte had fled, and that the Marquis of Wellington (this title having been conferred on him after Salamanca) had entered Madrid in triumph on the 12th of August. In consequence of Wellington's successful movement on Madrid, Soult raised the blockade of Cadiz, and moved towards Grenada. King Joseph had retreated to Valencia, and Hill received orders to follow and fight Drouet.

Accordingly, on the 28th of August, Sir Rowland's Corps, including the 92nd, left Villafranca and advanced to Usagre, and next night bivouacked near Villa Garcia. On the 30th the battalion proceeded to Maquilla, leaving Llerena to their right; on the 31st they entered El Campillo and spent the day in the houses, but bivouacked near it for the night. The enemy were retreating, and on the 1st of September the troops had a very fatiguing march of fourteen hours over a parched, waterless country, under a scorching sun, and entered Zalamea, where they took possession of great stores of grain left behind by the French. Here they met with a flattering reception from the inhabitants, and the wine casks of the town apologised to the thirsty soldiers for the want of wells in the country. Marching by Quintana and La Hava, the battalion occupied Don Benito on the 4th and halted. Here there were great festivities in honour of the raising of the siege of Cadiz, the town was illuminated, and the troops had a merry time.

I find in Regimental Orders about this time an expression of the commanding officer's "mortification at the disgrace attached to the regiment from the desertion of Kelly of the 6th Company and Wellbank of the 2nd Company," adding that "he does not dread a repetition of so infamous a crime."

A pensioner of the 92nd used to tell of two men of such bad character that none of their comrades would speak to them, who finally deserted to the enemy. Some time afterwards one of these men found himself in the front rank of a French battalion, within hail of the 92nd; for the French, not trusting the deserters they employed, were accustomed to place them in front, where, not daring to be taken, they were bound to fight.[1] It happened that the regiment had lately been served out with hose tartan; the deserter, in bravado, stuck on his bayonet the piece of red and white cloth which he still retained, and impudently waved it at his old corps. The colonel's attention being drawn to it, turning to the men he said—"Will some one send a pill with my compliments to cure that scoundrel's impertinence!" Norman Stewart and another crack shot knelt and fired, the "pill" went home, and the deserter's flag was lowered for ever.

Some men of the First Brigade had taken melons from the gardens, and "safe guards" were posted in the gardens and melon grounds near cantonments.

Colonel Cameron expected all his men to show an example of soldier-like bearing and good conduct, and considered the misbehaviour of even a few as a slur on the whole. He was "much concerned" at the misconduct of some men of the 1st Company at Don Benito, and tells the battalion that a continuance of such behaviour will ensure their being sent "to the rear, or to Gibraltar." He was a strict disciplinarian, and very particular that there should be no straggling or irregularity on the march, thereby ensuring the comfort of good soldiers and the character of the regiment, while the country people were encouraged by good treatment to bring in regular supplies.

[1] French deserters do not seem to have been allowed to enter British regiments.

The retreat of Marshal Soult from Cadiz, and of the Count d'Erlon (Drouet) to join him, rendered Hill's presence in Estremadura no longer necessary, and he was directed to move by Almaraz upon Toledo and Aranjuez, for the purpose of covering the capital while Wellington left it to lay siege to Burgos. Accordingly, the battalion left Don Benito on the 13th of September, forded the Guadiana, and marched with the First Brigade to Majados, and, moving by Villa Macia, arrived at Truxillo on the 15th. Here they halted till the 18th, when they bivouacked in the beautiful country near Jaraceijo, the place where they had remained concealed before the surprise of Almaraz. On the 19th they proceeded through the Pass of Mirabete, crossed the Tagus by a pontoon bridge placed near the site of the one they had destroyed in May, and encamped in front of Almaraz. The 50th being the rear battalion of the column of march this day, on crossing the bridge, Colonel Stewart halted and formed line fronting the *tête-de-pont* and Fort Napoleon, in the capture of which they had acted so conspicuous a part ; the whole gave three times three cheers, the band all the time playing " God save the King " ; they then followed the rest of the brigade, the band playing " The Downfall of Paris " and " The British Grenadiers."

The line of march continued by Naval Moral, Calzada de Oropesa, and La Gartera, where they halted a day ; marching at midnight on the 25th, they entered Talavera de la Reina next morning amidst the acclamation of the whole population, who had not forgotten the battle of Talavera in 1809, or the conduct of the British soldiers on that occasion ; and Sir Rowland Hill's desperate defence of the left of the position was spoken of by all with an admiration which was reflected on the troops now under his command. It was therefore with considerable regret that they took leave of the warm-hearted Talaverians on the 27th, and marched through miles of vineyards, the grapes, ripe and delicious, overhanging the narrow roads in beautiful and inviting clusters. These autumn marches were very different to the summer heat and arid plains of Estremadura, the peasants selling cheaply wine, peaches, plums and grapes, often giving fruit for nothing to the soldiers as they passed along, and at night the troops either lay in the village houses, or bivouacked round bright firwood fires ; where groups might be seen listening to old Highland tales, singing Gaelic or English choruses, such as the Skye soldiers' song[1] :

Tha mi'n duil	" My desire, my desire,
Tha mi'n duil	My desire is to return
Tha mi'n duil ri bhi tilleadh	To the land of MacLeod,
Dh' ionnsuidh Duthaich Mhic Leod	Where in childhood I
Far am b'og bha mi mireadh.	wandered."

[1] There were many Skye men in the regiment at this time.

Or—

"The wind blew the bonnie lassie's plaidie awa'."

Or a veteran of Egmont-op-Zee and Egypt would be heard reciting Corporal MacKinnon's poems. Nor were there wanting songs inspired by more recent adventures, notably one on Arroyo dos Molinos, to the tune of "Hey, Johnnie Cope," in which the French General Gérard took the place of his English prototype.

On arrival at their next halt, Cebollio, the Highlanders had, however, rather a cool reception. The reader may remember how in November 1808, when they entered Spain under Moore, some thoughtless people of the 71st had told the natives that the kilted men were cowards; this, like other scandalous tales, had spread further than the inventors of it intended, and had reached Cebollio, while the contradiction had failed to do so, and the villagers looked on them as men totally unworthy of notice; but the men could now talk Spanish, and they told the people that they were proud of their distinctive and honourable dress, while the 71st were even more anxious than themselves to remove the impression which they deeply regretted their thoughtlessness had created, so that the people were soon laughing heartily at their mistake, and the country was inundated with Highlanders proceeding by invitation to partake of the fruits of the surrounding vineyards.

Next day during the march they passed through five or six considerable towns, in each of which they were received with every demonstration of welcome, and at Torrejos, where they halted, Sir Rowland was received by the magistrates in their robes of office, and the town was illuminated in the evening. At Toledo, where they arrived on the 29th September, the Governor, the magistrates, the famous guerilla chief El Medico, and many of the nobility and gentry, met them a mile from the gates, congratulating their general on the favourable state of affairs, and cordially welcoming the first British troops who had visited them. When within the walls of this ancient city, they were received with enthusiasm; cheering crowds lined the streets, the balconies and windows were filled with ladies waving their handkerchiefs and throwing flowers as the soldiers passed, crying "Viva George III," "Viva Wellington," "Viva Hill," "Viva Ferdinand VII." The troops halted in the principal square, where every house was decorated with flags and gaily coloured cloth; and in the evening the city was brilliantly illuminated.

Splendid spectacles have generally their comic incident. It was furnished here by a vendor of brandy, who went along the ranks bawling the excellence of his liquor. Thinking Colonel Cameron a likely customer as he stood giving orders in front of his battalion in the square, he approached, roaring, "Aguardente,

Señor, aguardente," holding out for admiration an enormous bottle. Whack came the cane of the angry Highland chief, the big-bellied bottle was flying in a thousand pieces ; and the spirit merchant bounded into the crowd, in terror lest his brains should follow his brandy, while a titter ran along the ranks.

On the 30th the battalion left Toledo (famous for the manufacture of sword blades), crossed the Tagus by the bridge under the walls, and bivouacked in a rich valley twelve miles up the river ; on the 1st of October they occupied Aranjuez, a handsome town about twenty miles from Madrid, where, previous to the French invasion, the royal family resided for part of the year. Officers and soldiers were permitted to see the apartments of the palace, and three officers at a time were allowed four days' leave to Madrid.

During these campaigns, escorts were often sent to the rear with prisoners of war, or with sick and wounded, sometimes bringing back from the hospital men who had recovered, or ammunition, etc., on mules. The muleteer is a great character, a jovial, wine-loving, hardy fellow, sleeping in the open with his pack-saddle for a pillow, a favourite with all the countryside. The peasants were often hospitable to the soldiers. In the mountain districts the Spanish peasants wore sheepskin coats, woolly side out ; and kept enormous wolf-dogs to guard their sheep. In the plains they still tread out the corn with cattle or mares instead of threshing it. They live frugally, and in the evening sit on stone benches in front of the cottages, singing and playing the guitar. They are fond of dancing. These escort duties were often pleasant excursions. Sometimes officers and soldiers fell in love with the agreeable girls they met in their billets, and many romantic adventures or ridiculous situations ensued. At times when the regiment had changed its quarters, the returning escort might lose its way and meet an enemy, as happened once when, on a sultry day, one of these parties arrived at the summit of a ridge where a grassy hollow invited a halt. The wooded rocks which partly surrounded it offered shade to the tired soliders, while a spring of cold water quenched their thirst. Presently the glitter of helmets was seen at a corner of the winding road, up which a foraging party of French dragoons were advancing. Not knowing which might be the stronger, the officer quickly moved his men and lay in ambush in the wood above, hoping the French would pass, and that he might then resume his march in peace ; but the place was so tempting for a halt that the Frenchmen dismounted, watered their horses, and leaving them in charge of one or two of their number, the rest with the muleteers stretched themselves in the shade and proceeded to take a pull at the wine-skin. They were few in number, and the Highland officer, having crept to where he could see their arrangements, formed the design of capturing them and their spoil. He desired his little force to line the rocks above the unsuspecting marauders so as to cover them all. As he approached the rocks, a piper who was with him (probably returning convalescent), not understanding that stratagem rather than strength was to be used, was about to blow the onset, when a fierce, if whispered, rebuke prevented his play from spoiling the game. Then, as his men suddenly appeared with their firelocks presented at the astonished dragoons, the young officer called to them in their own tongue, "Surrender, or you are dead men!" Their carbines were on their saddles; there was nothing to be done, and they did it! The Highlanders had the satisfaction of rejoining their battalion with their prisoners, and the price of the horses and mules made a handsome lining to their sporrans.

These little adventures were not always, however, so much to their credit. An old soldier used to relate with great glee how, being one of an escort on a long, hot, and dusty march, they halted and sat down to rest in the shade of some trees by the road, while their officer retired some little distance out of sight in the wood. Soon there appeared coming up the road a procession of peasants leading asses and mule carts laden with grapes, corn, and wine, while a monk in his long robe and bare feet, carrying a cross, led the way. The soldiers supposed he had been collecting tithes; and thinking they would like a share, jumped up, and without saying a word began to help themselves; then they heard a peculiar sound from the front, and looking round saw the monk on his knees calling down curses on their heads; on which a wild sort of young fellow named Sandy M'Intosh, reckless of the clerical anathemas, ran across the road, lifted his musket, which was not loaded, and pointed it at the poor monk, who, to their great amusement, sprang to his feet, threw down the cross, and, kilting his coats, ran away as fast as his legs could carry him. Having taken their toll they let the carts go, and were much relieved to find that the officer, who had probably been taking a nap, knew nothing of their performance.

The battalion remained at Aranjuez in peace and plenty for three weeks, waiting and hoping for news of the capture of Burgos, but a very different experience was before it. In October the army under Sir Rowland Hill, consisting of five divisions of infantry and one of cavalry, with several divisions of Spanish infantry and cavalry—in all, about 40,000 men—were concentrating about Madrid.[1] Soult, having reinforced himself from the armies of King Joseph and Suchet, was advancing with 60,000

[1] Alison and Colonel Cameron's letters.

men, including 8000 cavalry and 84 guns, to Sir Rowland's head-quarters at Aranjuez; his advanced guard approaching the cantonments on the 22nd October. Wellington was before Burgos; but the French army near him, under Souham, was also powerfully reinforced, and now greatly outnumbered the Allies, especially in cavalry and artillery.

Hill began his movements in face of the enemy on the 23rd, on which day the 92nd marched with the First Brigade from Aranjuez, crossed the Tagus and halted at Colmanar de Orijo. On the 25th the First Brigade was ordered to protect the bridge of boats at Fuente Duena, nine miles south-eastward of Madrid, where they shifted the boats to the right bank, and the destruction of other bridges along the line was also commenced. On the morning of the 27th the enemy appeared on the opposite heights, and part of the First Brigade lined the bank of the Tagus, and kept up a smart fire across the stream, which is here narrower than below Aranjuez; but the shots were rather long, and little execution was done. In a letter of Colonel Cameron to Sir John Hope he says—"The enemy's cavalry began to make their appearance on the heights beyond the left bank in the most regular, cautious and beautiful manner I ever saw. Towards evening they dashed down to the water's edge, their skirmishers in the most superior style." Another officer of the 92nd mentions that some of the men were enjoying a bathe, when they were warned of the rapid approach of the French cavalry. One strong swimmer was on the opposite bank as the enemy came up, when, making a contemptuous gesture, he plunged in, and was striking out for his own side when an angry Frenchman galloped up, and deliberately levelled his carbine. To shoot at an unarmed man was so entirely contrary to the custom of civilised warfare, that a groan or shout of indignation burst from the British who were lining the bank, and the dragoon, ashamed of himself, lowered his weapon and rode off.

At ten o'clock that night the First Brigade broke up from Fuente Duena without beat of drum, marched the whole night and till one o'clock next day, when they occupied the village of Villa Conijos; the same evening at seven o'clock they resumed their march, crossed the Taramo branch of the Tagus at Ponte Largo[1] at midnight; and at four o'clock on the morning of the 29th, took up their ground on a strong position, where they found

[1] Hill had ordered two arches of the bridge of Ponte Largo to be destroyed, and the engineers were at work as the brigade crossed. In the darkness an officer's horse rammed his head against the end of the big drum of the 71st; the hollow sound of the drum, and the cries and noise occasioned by the kicking and plunging of the terrified steed, gave the idea to those in rear that the mine was exploding. A panic ensued, which, though it ended in nothing worse than some swearing and a good deal of laughter, was very near having serious results.

the rest of Hill's army concentrated, and where a fight was confidently expected.

The troops stood to their arms at daybreak, but as Soult did not attack, the men cooked and took what sleep they could get, so as to be ready for whatever might happen. It seems that Val de Moro was the position in which Hill intended to fight, and to it his army began to remove on the night of the 29th. "We were all in high spirits," says Colonel Cameron, "and I most certainly think had Joseph and Soult come on, the contest would not have been long doubtful, but fate had determined otherwise; and to the deep mortification (as it is said) of Sir Rowland Hill, he was ordered late that night to proceed forthwith to endeavour to effect a junction with Lord Wellington, who was on his retreat from Burgos."

Accordingly, next morning the 92nd passed close under the walls of Madrid, followed by the lamentations of the citizens. Rain poured on the downcast and wearied soldiers, who had no tents to shelter them from the storm when they halted for the night on the road to the Escurial, which they reached on the 31st, and where they were quartered in the Palace. On the 1st of November they got to the foot of the Guadarama Pass; and next morning had a stiff ascent of three hours, and descending the western slope at a quicker pace, bivouacked at Villa Custine as the sun set. Passing by Blasco Sancho, Fontiveras, Numo Sancho, and Penaranda de Bracanista, they crossed the Guadarama, and proceeding by Buttaya, Trabancos, and Ventosa, passed through Alba de Tormes and encamped near it on the 7th, having effected a close communication with Wellington about Cantaravella. The fields near their bivouac were still strewn with the wreck of the battle of Salamanca.[1] "Joseph and Soult," says Colonel Cameron, "followed us so closely that they did not take time to do much harm to Madrid. They constantly hung on our rear, but did not venture to press on us in the way the Northern Army did on Lord Wellington; they, however, followed the same plan when we joined his lordship."

Before sunset on the 8th November the First Brigade, supported by General Hamilton's Portuguese, was ordered to occupy Alba de Tormes and to defend it to the last extremity. An old castle commanded the bridge over the Tormes; a low dilapidated Moorish wall surrounded the town, which was divided into three districts, one of which was given to each British regiment (50th, 71st, 92nd),

[1] Soult, who was present at Queen Victoria's coronation in 1838, was delighted to meet Lord Hill. "What, have I found you at last?" he said; "you whom I followed so long without ever being able to overtake you," alluding to Hill's masterly retreat from Madrid to Alba de Tormes.

the Portuguese being distributed between them. The streets were barricaded. One wing of each battalion lined the walls, the several companies being told off to their respective posts, the other wing being held in reserve in the square, relieving each other at intervals day and night. Colonel Cameron writes—"We did what we could to improve our situation during the short time left to us; I threw an old door across the place where the gate once had been, and barricaded it with sticks and stones. We had not a single piece of ordnance. Just as the clock of Alba struck two, the French columns moved to the attack, and from that time till night we sustained a hurricane of shot and shell from twenty pieces of cannon. Their riflemen threw themselves into the ditches and ravines round the walls, but their masses never forsook the protection of their artillery, which was most dastardly for Soult with 10,000 men."

Before the attack began a French officer of rank rode near to the walls to examine the position; several of the 92nd were about to fire at him, but Cameron, whose chivalrous nature shrank from taking such an advantage, forbade them to shoot. It was Soult himself who was thus saved![1]

The pickets of General Long's British cavalry were driven in on the 9th, and on the 10th their main body retired over the Tormes. It was then the attack commenced. The town was completely commanded from the heights; 8000 French infantry were repeatedly formed for the assault of the place; "but," writes an officer of the 92nd, "notwithstanding the dreadful showers of shot and shell which plunged and danced in the streets in every direction, the bold and determined manner in which the soldiers who lined the walls did their duty, and the firmness of the officers commanding regiments, deterred them from making the attempt."

While part of the 92nd were lining the wall on the 10th, and continually exchanging shots with the French skirmishers, who were firing briskly, Private Norman Stewart, who was an extraordinarily good shot, and who was separated by some little way from his comrades, placed his bonnet on a stone so as to look as if it was still on his head, which, however, he carried to a safe distance; and proceeded to fire at the enemy with his "brown bess," which he lovingly called his "wife," and if a Frenchman advanced nearer than he liked, would say, "Dia, mar dean thu 'halt' gheibh thu pog o m'bhean" ("By G—d, if you'll no' halt, you'll get a kiss o' my 'wife'"). When the French retired

[1] This incident is mentioned in Grant's "Romance of War." It was, however, told to me by an officer long before that book was written, and is an instance of the general accuracy of the writer, whose stories of "The Highlanders in Spain" were told him by his father, who was at that time an officer in the 92nd.

Norman took up his bonnet, which was a hopeless wreck ; he tried it on amidst the laughter of his comrades, which attracted the attention of Colonel Cameron, who scolded him for destroying his necessaries ; but, hearing how it happened, gave him a kindly clap on the shoulder, saying in Gaelic that he would give him a new bonnet—" Aye, or two if you want them."

There was a nunnery in the town much exposed to the fire, and the poor inmates were running in all directions seeking shelter, and invoking all the saints for protection. The French having command of the road by which supplies should have come up, the men were badly off for bread ; but discovering a store of wheat, they ground it between stones for supper, and next day they found flour and set the bakers to work.

From the 10th to the 14th the brigade held Alba, during which time the enemy manœuvred along the banks of the Tormes, as if desirous to cross and bring matters to an issue on that side. An officer and a sergeant with a telescope were always on the castle to report the motions of the enemy, whose light troops kept up a skirmish the whole time ; but such was the steadiness of the troops that he dared not venture to assault a place assailable at all points, and with no other defence than the brave soldiers of the First Brigade.

R.O., *Alba de Tormes*, *November* 12*th*, 1812.—The pibroch will never sound except when it is for the whole regiment to get under arms ; when any portion of the regiment is ordered for duty and a pipe to sound, the first pipe will be the warning, and the second pipe for them to fall in. The pibroch only will, and is to be, considered, as invariably when sounded, for every person off duty to turn out without a moment's delay.

No. 2.—Captain MacPherson will have charge of the duties of the right wing, and Captain Seton those of the left wing ; the commanding officer being occasionally out with both.

R.O., *November* 13*th*, 1812.—For the working party in front of the line, Lieutenant Winchester. For the lookout at the castle, Lieutenant Hobbs (to report any movement of the enemy). The baggage to be loaded, and assemble in the square, agreeable to Brigade Orders. For baggage guard, Lieutenant Ross.

The main army having gained sufficient distance, and the enemy having crossed the Tormes above Alba, the troops there were recalled on the 14th. They withdrew in perfect order, blew up the bridge ; and as the colonel says in his letter, " The last sentinel of the 92nd knocked the Frenchman opposite to him heels over head."

During this trying and determined defence on the 10th, 11th, 12th, and 13th, the loss of the defenders was 100 in killed and wounded ; the 92nd had 8 men killed, and Lieutenant Andrew Will and 33 n.c. officers and men wounded, of whom 7 died of their wounds.

KILLED IN ACTION.	DIED OF THEIR WOUNDS.
Private Murdoch MacDonald.	Private W. Rennie.
" Thos. Macmillan.	" Murdoch MacKenzie.
" Donald Ross.	" A. MacPherson.
" David Shairp.	" W. MacDonald.
" George Simpson.	" Thos. Leitch.
" George Weston.	" Robert Brown.
" Walter Moir.	" David Philips.
" George Stewart.	

On the 14th of November, Major-General Howard was transferred to the First Brigade of Guards, and expressed in Orders his warmest thanks to Lieut.-Colonels Stewart of the 50th, the Hon. Cadogan, 71st, and Cameron, 92nd, for their zeal and attention, and to the officers, n.c. officers and soldiers for their general good conduct on every occasion ; "and it was probably owing to the steadiness of the troops that the enemy did not choose to pursue the attack on this town, which they commenced with so brisk a cannonade and such a superiority in point of numbers on the 10th inst." ; and he further expressed his regret at leaving the brigade, and his approbation of Captain Blassier in command of the 60th Rifle Company.

CHAPTER XVII

THE battalion, with the First Brigade now commanded by Lieut.-Colonel (Brigadier-General) Stewart of the 50th, rejoined the Second Division on the Arapiles, where Wellington concentrated his united force of 60,000 men, and prepared to give battle to Soult with 90,000. The weather in the evening was horrible ; the men were without tents, their clothing worn bare ; provisions were scarce, and of money they had none. The only thing that kept up their spirits was the chance of beating the enemy, whose infantry they could see on a height a mile away, while an immense body of his cavalry was drawn up on the plain below. The Second Division, not having been at Salamanca, were in hopes of taking part in a repetition of that victory on the same ground.

On the morning of the 15th November, the Allies were drawn up in order of battle. All forgot their hardships, and expectation was at its height ; but Soult declined the challenge and manœuvred to the British right, threatening to cut off their communication with Portugal, from whence their supplies came. In consequence of this, the order to retreat was given, causing gloom and dismay where before there were cheerfulness and confidence ; but every one felt satisfied that nothing but the most pressing necessity had compelled their Chief to retire before his adversary. Wellington, feeling too weak to attack, and seeing the enemy's cavalry pointing to the Ciudad Rodrigo road, suddenly formed three columns, and defiled in order of battle before the enemy at little more than cannon shot, carried his army round the French left, and crossed the Valmusa River. In this bold manœuvre he was favoured by fog and rain,which rendered the lanes and fields by which the enemy moved nearly impassable, while he had the use of the high roads. About 2 p.m. the Second Division marched on the high road to Ciudad Rodrigo, crossing a flooded plain, where their dress gave the Highlanders an advantage, but soon a stream swollen by the heavy rain took the men to the middle. On gaining the rising grounds the 92nd had entered an open wood, when the French cavalry made a furious charge right up to the bayonets, but were so roughly handled that they took themselves off without doing much damage, nor did they receive serious injury from the Highlanders, whose ammunition was wet, and many muskets would not go off, but they lost some men and horses by the fire of our artillery.

Although they are not noticed in "Cannon's Record," it appears that Private Donald MacLeod was killed on this day ; and Private Andrew Sibbie died of his wounds on the 18th, though it does not appear on what date he was wounded.

The Gordons bivouacked in a wood behind the Rio Valmusa.

They managed to light fires with the wet branches and thus got some large trees kindled, which gave out a good heat.[1] Bread was so scarce that at one place a staff officer asked a peasant journeying with his wife and children to Ciudad Rodrigo to sell him a loaf he was carrying, and offered a doubloon (about £4). The man refused, saying, "My little ones cannot eat gold."

On the 16th, 17th, 18th, and 19th the allied army continued its retreat. The 92nd bivouacked near Matilla on the 16th, and behind the Rio Cuebra on the 17th. During these days the rear guard was engaged more or less with the enemy. On the 18th, the battalion bivouacked near Moraes Verde. The French cavalry pressed on them, sometimes riding so close as to bandy wit in Spanish with the British, and any man who fell to the rear was sure to be taken prisoner. As a battalion was halted for rest, they might see a column of dragoons in their brass helmets and white cloaks appear upon the scene, ready to charge; but the infantry stood to their arms, and showed that, though retreating, it would not be safe to molest them. The enemy penetrated in the forest between the columns of march, and once made a dash at a party of officers, taking some prisoners, among them Lieut.-General Sir Edward Paget. The rivers were all in spate, and only passable at fords. At one of these a mule, laden with biscuit, with the proverbial stubbornness of his kind, stopped in mid-stream and declined to move. The 92nd had crossed, and part were left to guard the passage of commissariat from the pursuing French, who were coming up on the other side. The muleteer bolted, and the precious bread seemed sure to be lost, when a bugler, gathering up his kilt, waded back, mounted the mule behind the panniers, and, leaning forward, sounded the "advance" in its ear; the startled beast obeyed double quick, and as the plucky boy landed, a voice from the rocks above called, "Well done, my lad, you shall have all the biscuit!" It was Brigadier-General Stewart, commanding the brigade, who had been watching the performance, and the narrator, who was in the bugler's company, added, "We had a real good feed that night."

It is customary in Spain to drive the swine into the oak forest to feed on the acorns. One night a fusilade was heard as the hungry soldiers shot the pigs, causing the troops to stand to their arms, thinking the enemy was upon them, which afterwards drew a severe rebuke from the general. Some of the 92nd were about to join in this noisy proceeding, when a young officer said, "Don't shoot them; stick the beggars!" His advice saved both their

[1] There were no lucifer matches in those days. Flint and steel and tinder or gunpowder were used to light fires, a man kneeling to fan the flame with the apron of his kilt—the "Highlandman's bellows."

credit and their bacon, and that officer found a pork chop placed handy for his supper !

"In nine cases out of ten a bivouac is more or less enjoyable, but from the 15th to the 19th any one unacquainted with the service could hardly be persuaded to credit the sufferings we endured ; deluged with rain, roads deep and miry, repeatedly fording rivers and streams, some of them breast high. The ground of our bivouacs soaked, no dry wood to be had, fires smoky and cheerless ; miserably provided, having neither bread, biscuit, nor flour."[1] The officers and men were to be seen gathering acorns to eat instead of bread with the beef which they cooked in the ashes on the points of swords or ramrods, for the camp kettles were with the baggage and often a day's march ahead of the troops, but they had bullocks driven along, which, if they arrived in time, were slaughtered as they halted. The marches were not long, but from the circumstances and nature of the ground they took a long time, and the men had an immense weight to carry, for the ordinary marching order weighed about 60 lb., besides the blanket ; while the feeling of retreat has always a depressing effect, and the difficulty of bringing the various columns along without jostling, and if possible without being attacked, caused long and tiresome halts.

On the 18th November a handful of Indian corn was issued to each man, which they ground between large stones, passing them from one to another till far on in the night, while the noise banished sleep. On the 19th, the 92nd went into quarters in a miserable little village crowded to excess ; the people had little to spare, but the commissary managed to buy potatoes, of which each man got 2 lb. The baggage was here restored, and, miserable as were the cottages, officers and men recovered their gaiety as they roasted chestnuts and boiled their potatoes in the chimney corners.

The enemy had now given up the pursuit ; they had gathered a great spoil of baggage and over 3500 prisoners. In the whole retreat of Wellington's army from Burgos, together with that of Hill's from the Tagus, the loss in killed, wounded, and prisoners was not less than 9000 men—British, Portuguese, and Spaniards, including those lost in the siege of Burgos. The loss of Hill's army, however, between the Tagus and the Arapiles was only 500 in killed, wounded, and prisoners, including that of the First Brigade at Alba.

It being rumoured that Soult intended to invade Portugal by the valley of the Tagus, Hill's troops were moved as far as Robleda, to which place the 92nd marched on the 20th. An officer of the battalion describes its appearance on parade. He compares the jackets to "parti-coloured bedcovers," the "clothing literally

[1] Scherer.

composed of shreds and patches." Colonel Cameron, writing to Sir John Hope of the campaign from the 27th of October, says—"Since that time to the 20th November we have been exposed to more hardships than I thought the human frame could bear. Mine, I know, had very nearly yielded. In weather which would have been thought inclement in England, with the canopy of heaven for a covering, wet, cold, and hungry, we were marching day and night generally. During the days of the 16th, 17th, and 18th, fifteen poor fellows of the 92nd fell down and were lost. My heart bled for them."

Robleda was a dirty place, and the streets and lanes where they were cantoned were at once cleaned by the battalion.

R.O., Robleda, 22nd November 1812.—Officers commanding companies will please arrange with the batmen and servants of their companies to give up their kilts for the purpose of mending the kilts of the duty men ; such batmen and servants as give up their kilts to march in their pantaloons till the new clothing is issued. The shoemakers and tailors to be constantly employed in mending the shoes and clothes of their companies. The commanding officers of companies to report personally as to the state and appearance of their companies, and on the arrangement made with the batmen and servants respecting their kilts.

R.O., Robleda, 22nd November.—Sergeant-major Ewen Kenedy is appointed to do duty as quartermaster-sergeant, in room of Quartermaster-sergeant Cameron, promoted to a commission. Sergeant Duncan Macpherson is appointed to do duty as sergeant-major until further orders, in room of Sergeant-major Kenedy, appointed quartermaster-sergeant.

Men who had been guilty of any misconduct during the march were ordered for drill. The officers and men received their pay up to the 24th of June.

On November 22nd it was announced that a battalion, "to be called the 13th Royal Veteran Battalion, is to be formed at Lisbon. Ten companies of 100 rank and file of men from different regiments who are fit only for garrison duty. The lieutenants to be transferred from the line, or old quartermasters or ensigns who have been n.c. officers. The ensigns to be entirely deserving n.c. officers." Sergeant Symon and thirty-six rank and file of the 92nd were transferred to this battalion.

R.O.—Prisoners who are not to be tried by court-martial to be employed on all regimental fatigues, and all prisoners to parade, properly dressed, in rear of the battalion, and to be inspected by the officers of their companies.

On the 28th of November the 71st and 92nd, under Colonel Cameron, marched to San Payo, and thence across the Sierra de

Gate, in lovely weather and through splendid scenery, the road winding, rugged and dangerous, through thick woods overhanging steep precipices and round the boldest rocks, which re-echoed the sound of pipe tune or marching song. They lay that night in Perales, and on the 30th they occupied Casas de Don Gomez. During the march the usual measures were adopted to prevent men from falling out; one of these was stopping the allowance of wine; another, which seems a little hard, was not dismissing a company till any straggler belonging to it came in; but as an old Peninsular soldier said to the writer, "If the officers didna claw us, the general would claw them." They went into quarters on the 1st of December at Corea, a town of about 600 houses, prettily situated on the River Alagon. Here newspapers arrived, and there was plenty of game to shoot. There was a good market, and they were well supplied with food and wine, also with shirts and other necessaries from Lisbon; but the colonel, not wishing to put the men to the expense of new fatigue pantaloons, "will not at present order new ones, and hopes that by a little care and mending they may stand for some time, and such men as have none may make them out of old greatcoats brought by the quartermaster from Lisbon." The names of all n.c. officers and men who were present with the battalion between the 15th and 19th November are to be sent to the Orderly Room, agreeable to General Order, probably for them to get an allowance for clothing. Many men were in hospital at Corea; and Lieut.-General Sir W. Erskine, commanding the cavalry of Hill's Corps, Colonel Stewart of the 50th, and Colonel Wilson, 39th, died from the effects of exposure. A good many had died or been invalided from the 92nd during the past year, a return of Walcheren fever and ague being often the cause; with rare exceptions, the invalids were described as of "good character." At Corea the battalion had resumed its usual smartness. Any man seen walking in the streets in a slovenly manner, or not properly dressed, "to be ordered to his quarters by any officer who meets him."

At this time Wellington addressed a circular letter to the superior officers of the army. After saying that he will be able to keep the troops in cantonments for some time, and that clothing and necessaries are on their way to the different divisions, he draws particular attention to discipline, which, he remarks, becomes relaxed in every army after a long and active campaign. He goes on to declare "that discipline had deteriorated during the last campaign in a greater degree than he had ever witnessed or ever read of in any army, and this without any disaster, any unusual privation or hardship, save that of inclement weather; that officers had from the first lost all command over their men, hence outrages of all kinds and inexcusable losses had occurred. No army had ever made shorter marches in retreat, no army had been so little pressed by a pursuing enemy ; and that the true cause of this unhappy state of affairs was to be found in the habitual neglect of duty by regimental officers. He does not question their zeal, far less their gallantry, but their constant and minute attention to the conduct of the soldiers, as well as to their arms, ammunition and food ; and remarks the facility with which the French soldiers cooked in comparison with our army," etc.

These severe reproaches caused great dissatisfaction. It was felt that in some particulars they were founded on statements which were unintentionally inaccurate. The marches, though short as to distance, were long as to time : " the troops, ankle-deep in clay, mid-leg in water, lost their shoes, and with strained sinews heavily made their way, and withal they had but two rations in five days." [1] " Wellington knew not that the commissariat stores he had ordered up did not arrive regularly because of the extreme fatigue of the animals who carried them, and were often not available because the conductors, alarmed by reports of the enemy's cavalry, carried off or destroyed the field stores." [2] The destruction of the swine, though it was a serious loss to the unfortunate owners, may be in some measure palliated, because that evening the soldiers had nothing but acorns to eat ; but no doubt there had been great disorder in some parts of the retreat, and much of the suffering of all arose from these very disorders on the part of some. If soldiers break up the arrangements of their general by want of discipline, they have no right to complain of the misery which those arrangements were designed to prevent. The circular was not strictly just because it excepted none from blame ; and those brigades and battalions who knew that they had honestly done their duty felt that their hard working, if unobtrusive exertions had not been appreciated.

With regard to cooking, that most important part of military lore—for an army marches on its belly—a n.c. officer of the 92nd remarks that the camp kettles then in use were made of strong iron to hold sixteen quarts for twelve men, and were carried on mules ; and when the troops were engaged with the enemy or retreating before him, the baggage often did not arrive till midnight ; but in 1813 they got light kettles, one to six men, which they carried themselves, and then they could cope with the French in cooking as well as in fighting.

An officer of the 92nd admits that discipline was permitted to relax to a criminal degree in some regiments ; " but I cannot allow that the misconduct of one or of a dozen battalions could ever form a good apology for bestowing upon the others a sweeping censure

[1] Napier. [2] *Ibid.*

for offences which they knew they had never committed." He believes that if Wellington had taken a little time before giving vent to his indignation, he would have distinguished between the guilty and those who had done their duty under all circumstances. Since, however, this course was not followed, he affirms "publicly and unhesitatingly, that every officer in the regiment to which I belonged performed their several duties with credit to themselves and advantage to the service; and I fearlessly assert that throughout the whole of the retreat, the n.c. officers and privates obeyed the orders of their officers with the same cheerfulness and alacrity for which they have ever been distinguished." Colonel Cameron seems to have been of the same opinion, for in Regimental Orders he calls attention to the orders of the Commander of the Forces, and "in the most pointed manner" desires that they may be enforced by the captains of companies, and that they see that the subalterns and n.c. officers exert themselves "to prevent in the 92nd Regiment what has given occasion to the Commander of the Forces so strongly to animadvert upon the conduct of some other corps." Sergeant Robertson writes that after the imputation cast on the commanding officers, discipline was exercised with the utmost rigour. "The 92nd was not so severely dealt with as several regiments; but whether we deserved it or not, I cannot say."

R.O., December 24th.—The commanding officer requests officers commanding companies will endeavour to give their men as good dinners as they possibly can to-morrow and something (extra) for breakfast, as it is Christmas Day, and that they will see them at it.

A letter from the adjutant-general, Horse Guards, to Lord Wellington was published at Corea. His earnest attention is requested by command of H.R.H. the Commander-in-Chief to a branch of the service from which H.R.H. expects the most essential benefits will be derived; "I allude to the establishment of military chaplains and the duties which attach to them." The letter continues that the chaplains have "been selected with the utmost care and circumspection by the first prelates of this country." They are to have the pay and allowances of major, and are to receive from all persons that respect which is so justly due to their rank and profession. They are to visit the sick and the hospitals at least twice a week. At divine worship no more are to be assembled at one time than the voice can reach; service to close with a "short practical sermon suited to the habits and understandings of soldiers"—"in conformity with the custom of the Established Church."[1]

[1] Regimental chaplains had been done away with thirteen years before, because the duty was not properly performed. There were no paid Presbyterian or Roman Catholic army chaplains till a much later period.—Clode's "Military Forces of the Crown."

R.O., 28th December 1812.—The detachment joined this day, under Lieut. Macpherson, to be posted to companies.[1]

R.O., Corea, 31st December 1812.—As to-morrow is New Year's Day, the commanding officer recommends it to officers commanding companies to feed their men well by letting them have a breakfast and as good a dinner as they can upon that occasion, and that they will visit them at their meals.

No doubt their Spanish hosts entered into the festivities of the soldiers, who would keep up "first footing" and "Oidche Challuinn" with all the spirit of their country; the Spanish wine making the best of "plotty," and *aguardiente* doing duty for mountain dew as the dram which rewards the rhymers in the latter quaint old custom.[2]

R.O., Corea, 3rd January 1813.—It is intended to commence the regimental school to-morrow. Such n.c. officers and men as wish to attend it will be allowed leave from evening parade daily and from one day's duty each week; and the commanding officer strongly recommends the officers to advise their n.c. officers and the young men of their companies to attend the school as a thing which will tend so much to their own advantage hereafter.

Corea, 5th January 1813.—The battalion will march to-morrow morning at eight o'clock, agreeable to this day's brigade orders. First pipe to sound at half-past six.

This unexpected move was caused by an expedition of the French, under General Foy, for the capture of Bejer, a town about half-way between Salamanca and Corea, in order to ascertain if any large body was collected behind it, for the vivacity of Sir Rowland Hill troubled the French, on whom his successful enterprises had

[1] In monthly returns, dated 25th January 1813, mention is made of Lieut. Ronald Macdonald, two sergeants, and thirty-six rank and file as having joined from the 2nd Battalion then in Scotland, and two rank and file from Portugal.

[2] On New Year's Eve, known in the Lowlands as "Hogmanay," and as "Oidche Challuinn" in the Highlands, the people gathered at the residence of the principal person of the place and marched round the house three times from east to west (deiseil) chanting an invocation for a blessing on it and all within, such as—

Beannaich an tigh 's na bheil ann,	"May blessing on this house descend,
Eadar fhiodh 'us chlach 'us chrann,	Each stone and beam, each lock and chain;
Moran bidhe 'us pailteas eudaich,	May food and raiment plenteous be,
Slàinte dhaoine gun robh ann.	May health abide: throughout this year."

At the door each was admitted only after reciting a verse or sentiment (duan), which the mistress rewarded with a dram. A bit of singed wool from the breast of a sheep (caisein Challuinn) was handed round under the noses of the guests, who were supposed to pretend the smell delicious. After this, a man with a bull's or other hide drawn over his head and shoulders entered and ran through the house, the others shouting and laughing and beating him with sticks till he gained the outside. All were entertained by the master, and the night was spent in singing and dancing.

This ceremony, which I have seen several times in the West Highlands, is no doubt very ancient in its origin, and symbolical in character, but is seldom practised of late years. It is, however, still continued in the regiment, though in more modern fashion.

made a profound impression, so that his slightest change of quarters caused them to concentrate their troops, expecting one of his sudden blows. Also the enemy was aware that a large quantity of woollen cloth was in the possession of a manufacturer there, and, being badly off for clothing, they intended to seize it.

Sir Rowland Hill, on receiving notice of this intention, at once sent the First Brigade, then under command of Lieut.-Colonel Cameron of the 92nd, to prevent it. The 71st and 6th Caçadores moved from Monte Hermosa to aid the inhabitants of Bejer, while the 50th and 92nd were pushed forward from Corea to Monte Hermosa. These movements convinced Foy that the rich prize would not be tamely given up, and he thought it prudent to retire.

The Gordons regretted their agreeable quarters at Corea, for Monte Hermosa was exclusively occupied by farmers, who in Spain do not generally reside on their farms ; the streets were filthy and in bad repair, the houses had wooden shutters in place of glass in the windows, giving a choice between darkness and a draught, and were altogether rather comfortless. But if the farmers were slovenly in their habits, they had an example of the contrary in the Highlanders, whose commanding officer was most particular on that point ; the streets were cleaned, and the main guard had orders to confine any man seen in the market not properly dressed with side arms—as an old soldier said, " If you had but ae button on your coat it bode to be bricht."

The battalion was practised constantly in light infantry drill, and the men were employed in taking down the feathers from their old bonnets, and in dressing them and putting them in good order for the new bonnets. " They will likewise wash and clean their heckle feathers, as it was not in the commanding officer's power to procure new ones for them. Every man who has any knowledge of tailor work is to be employed with the clothing. The commanding officer expects that every man is now supplied with good gartering and rosettes."

Monte Hermosa, 21st January 1813.—Sergeant John M'Combie of the Grenadier Company is appointed to do duty and receive pay as quartermaster-sergeant until further orders, in the room of Quartermaster-sergeant Kenedy, promoted to a commission.

On the 24th, the commanding officer was so well pleased with the schoolmaster-sergeant's report that, unless particularly ordered to the contrary, the scholars are excused morning parade every second day.

The following prices of necessaries may interest soldiers of the present time.

R.O., Monte Hermosa, January 1813.—Prices of articles furnished by Quartermaster MacFarlane, fixed by a board, the tradesmen's bills being produced—Shirts, 6s. 8d. each ; black gaiters, 2s. 11d. a pair ; grey trousers, 10s. 6d. a pair ; shoe brushes, 2s. 3d. a pair ; gartering, 3d. a yard ; writing paper, 3s. 6d. a quire ; a paper of ink-powder, 1s. 6d. ; soap, 1s. 4d. a lb. ; tobacco, 4s. 6d. a lb. ; a razor, 10d. Hose tartan to be charged to officers commanding companies at 2s. 8d. a yard, and by them to the men at the rate of 2s. a pair.

On the 3rd of February the battalion was inspected by Sir Rowland Hill in their new clothing ; knapsacks neatly packed, greatcoats and blankets well folded and rolled, canteens well cleaned, etc. The haversacks, being so bad, were not to be worn, and, as the feathers had not been put up, they paraded in " humble bonnets." The men to wear purses, but, as all had not got them, " the front rank men and the flanks of the rear ranks will be completed with purses."

At Monte Hermosa the officers and soldiers of the battalion received their share of prize-money for the captures at Arroyo dos Molinos, and also their pay up to the 24th of September last, the captains to retain a sufficient sum to provide their companies with vegetables. The men were now rich, and " officers commanding companies are to cause all the wine houses in their cantonments to be shut up at eight o'clock, patrols to clear the houses after that hour " ; and the Alcalde issued a proclamation to prevent the inhabitants selling wine after that hour. Still, many a jovial evening was spent by the soldiers with their hosts and their wives and daughters—dancing, or singing in Spanish, English, and Gaelic. So the time passed, till the second week in February, when General Foy put a stop to their gay doings by making another dash at Bejer. Sir Rowland Hill, however, received timely notice, and ordered the brigade to advance. When the 71st and Caçadores occupied Bejer on the 12th of February, Foy was only a few miles from it, but on hearing of the British advance he again retired. To secure the place against future attacks, the 50th and Caçadores were thrown into Bejer as a permanent garrison ; the 71st occupied the villages of Puerta de Banos and Candaleria, and the 92nd the town of Banos. The 60th Rifle Company were quartered in Herbas.

A few days after this attempt had been frustrated, Foy, thinking to gain his object by surprise, advanced again at the head of 3000 picked troops, with so much secrecy that one of the outlying pickets was nearly surrounded before its officers knew that the enemy was nearer than Salamanca. Driving in the pickets, the French were before the gates of Bejer in a few minutes, expecting an easy conquest of the place and the stores it contained. The attack was made with all the bravery characteristic of their nation, but the entrances were guarded with unflinching courage by the old 50th ; and notwithstanding his superiority of numbers, Foy, after reiterated attempts, was unable to intrude himself into Bejer society, and retired with the loss of a hundred killed and wounded, his aide-de-camp being amongst the latter. On the first notice of Foy's attack, Colonel Cameron, commanding the brigade, advanced the 92nd and four companies of the 71st, but they were too late to take part in the action. The 92nd were now cantoned in Banos, which town is situated in a narrow valley surrounded by rugged mountains, on the borders of the provinces of Leon and Estremadura. People resorted to it from all parts of Spain on account of its hot mineral springs, and the houses were furnished in a superior manner for the accommodation of visitors, so the battalion found themselves in most comfortable quarters. In April a draft of forty-six n.c. officers and men joined from the 2nd Battalion. The people of Banos were famous for their industry, and for their loyalty to their sovereign and the cause of Spain, and the Highlanders and inhabitants were soon on most friendly terms. An officer of the 92nd describes the pleasant situation in which "our merry blades" found themselves on entering Banos, "where in every house the nut-brown knee and weather-beaten countenance met with nothing but smiles and the most marked attention." "The longer we remained among them the more friendly did the townspeople become, till the soldiers and the inhabitants of each house messed together as one family, the former furnishing beef, bread, etc., and the latter pork, oil, and all sorts of vegetables." The spring is a glorious season in Spain, flowers and creepers surround the houses, and the people sit smoking and singing in the evening air. The dancing parties were started again, Highland games were practised, and all went happily till a sad event cast its gloom over the battalion.

"In the latter end of March, a military execution of a most afflicting description took place. The crime for which the unfortunate young man was condemned to die was desertion to the enemy, and attempting to stab the n.c. officer who apprehended him. A little before the hour of execution, the regiment was drawn up so as to form three sides of a square, the other side was left open as the place of execution. The schoolmaster-sergeant accompanied the poor lad to the fatal spot, and all the way from the village read portions of Scripture. On their arrival at the point assigned them, the criminal joined very audibly in singing a few verses of a Psalm, and then, after spending a few minutes in prayer, the fatal cap was drawn over his eyes, and the provost-marshal with his party advanced from the rear to carry the sentence into effect. At this awfully affecting part of the scene the whole regiment, officers and men, knelt down, and, on behalf of him who then stood

IN BILLETS AT BANOS

on the verge of eternity, offered up humble supplications to the throne of mercy. In a few moments the party fired, and in an instant the world closed upon the culprit for ever."

The above is taken from the memoir of an officer of the regiment.[1]

Although there was no sympathy with the crime of desertion to the enemy, there seems to have been an impression that the man was not at heart a traitor to his country, but had been led to the rash act by a love affair.

While at Banos the men who had enlisted in 1806 for seven years were given the opportunity of renewing their engagements. Men not above thirty-five years of age were allowed to enlist for life, and received sixteen guineas of bounty ; those above thirty-five, for seven years only, and received eleven guineas. These were large sums in those days, sufficient to send a welcome help to the old folks at home, and leave enough to drink their healths in many a cup of Spanish wine. There seems to have been a charm about the constant variety and adventure characteristic of the Peninsular campaign, which appealed powerfully to manly natures and outweighed the occasional hardships and dangers ; and most of those who could do so re-engaged for unlimited service, which also entitled them to a pension.

The delay in opening the new campaign arose from various causes. At the close of the campaign of 1812, the cavalry and artillery had lost great numbers of horses, which had to be replaced. The clothing and accoutrements of the infantry had to be renewed ; the heavy camp kettles of the soldiers had been exchanged for light ones carried by the men, and the mules now carried tents, which were of great advantage at halts, though they did not always arrive in time to be of use on rapid marches. Wellington had to wait till men, horses, money, and military stores could be brought from England ; but these difficulties being removed, and having reorganised the Spanish troops at his disposal, he prepared to take the field with as fine an army as could be desired.

The relative strength of the contending forces was no longer in favour of the French, whose numbers had been reduced by drafts to Germany to 231,000 men and 29,000 horses ; 30,000 men were in hospital. Wellington had so well used the winter months that nearly 200,000 allied troops were ready to take the field.[2] Officers and men were filled with the warlike intelligence gained by the experience of several campaigns ; full of vigour and confidence, their bronzed and daring countenances looked danger

[1] Regimental Return, 1st Battalion 92nd, Banos, March 24th, 1813—"Private George Mackie shot by sentence of general Court-martial."

[2] Napier.

boldly in the face, as if they cared neither for man nor devil ; and their great leader, confident that they would grandly execute his well-planned design for driving the enemy to the Pyrenees, rose in his stirrups as he passed the stream which marks the boundary between Spain and Portugal, and, waving his hand, cried, " Farewell to Portugal ! " [1]

The various parts of the great military machine being prepared, an order of readiness to move was received in the middle of May ; and the Gordon Highlanders left Banos on the 20th, and encamped with the Second Division near Bejer, where they renewed their acquaintance with the regiments from whom they had been separated in winter quarters. To show the attachment of the inhabitants to the Highlanders, which he attributes to " the kindred spirit which seems to possess the breast of every mountaineer throughout the world," an officer mentions the following fact. When Sir Rowland Hill passed through the town a day or two after the Highlanders had left Banos, a deputation of the principal inhabitants waited on him to return thanks for the protection he had afforded them against the enemy. The General inquired if they had any complaints to prefer against the regiment that had just left them, to which the head of the deputation replied, " Sir Rowland Hill, had you been here when the Highlanders marched out of our village, our tears would have answered your question."

On the 23rd the Second Division, now commanded by Lieut.-General the Hon. William Stewart, was reviewed by Sir R. Hill ; and on the 24th the 92nd marched to Fuenteroble, and continuing, crossed the Tormes near Salamanca on the 26th, where Wellington and Hill joined forces, and where they had a brush with the enemy, who had retired from the city. The French, under Villatte, though not strong in numbers, were posted advantageously, and maintained their position against repeated assaults of our cavalry and horse artillery, till the Second Division forded the river in haste above the city, intending to cut them off ; but as soon as they gained footing on the other side, the French retreated in good order, to the admiration of their opponents. The troops, both British and French, suffered much from the intolerable heat. In this affair the enemy lost about 160 killed and wounded, 140 prisoners, and six guns.

The battalion then encamped with the division half a mile above Salamanca, where they were visited by numbers of the higher class of the citizens. Some of them were rather disconcerted by

[1] Napier.

AUTHOR'S NOTE.—During the winter the grumblers of the army wrote that Wellington had lost his prestige—that Soult had out-manœuvred him ; and the arm-chair critics at home even suggested his being recalled ! Such self-sufficiency is not unknown in the present day among those who are entirely ignorant of the circumstances which govern a general's actions.

the ladies, whose curiosity induced them to pull aside the doors of the tents without the smallest ceremony, in order to have a full view of "los Ingleses," who were not all at the moment in drawing-room costume.

On the morning of the 27th, Hill's troops were informed that the Marquis of Wellington would review them on the march. He took post with his numerous staff on a height about four miles from Salamanca, and each battalion marched past in ordinary time. The morning was extremely fine, and the appearance of the troops truly magnificent. As each corps passed, Wellington paid them a flattering compliment, and as the last company saluted, he turned and said, "Sir Rowland, I will take the gloss off your corps this campaign"—and he kept his word.

It happened that the commissariat had killed a lot of sheep the day before, and the heads, not being appreciated by the English troops, had been appropriated by the Highlanders, of whom many carried them slung on the backs of their knapsacks; when the battalion halted to clean and brush up before passing the Commander of the Forces, some officers just come out were shocked at the idea of men marching past with sheep's heads dangling at their backs; but Sergeant Robertson remarks that before the campaign was over, they would have carried them themselves if they could have got them! Wellington was evidently of the sergeant's opinion; for, after saying that he had seldom seen so fine-looking a regiment, he noticed the heads, and remarked, laughing, that the Highlanders were a day ahead with their rations, and he wished all regiments understood campaigning as well.

The morning state, 1st Battalion 92nd, on the 27th May 1813, shows:—

Officers Present

Lieut.-colonel, 1; major, 1; captains, 8; lieutenants, 15; ensigns, 5; staff, 6.

Sergeants

Present for duty, 45; sick absent, 3; on command, 6.—Total, 54.

Drummers

Present for duty, 15.—Total, 15.

Rank and File

Present for duty, 820; sick present, 16; sick absent, 70; on command, 16; prisoners of war, 14.—Total, 936.

Sir Thomas Graham's Corps, forming the left of the allied army, had crossed the Douro, ascended the Esla and advanced on Zamora, and now, the line of the Tormes being gained, Wellington pushed forward his troops so as to cover his communications with

Ciudad Rodrigo, and keep open communication with Graham. He now left Hill in command, while he himself went off to see after his combinations on the Esla—crossing the Douro in a basket slung on a rope stretched from rock to rock, the river foaming several hundred feet below ![1]

After the review of the 27th, the Gordons encamped at Orvada, where the regiments of the Second Division took the opportunity to practise brigade drill, which, from their separation during the winter and spring, had been impossible ; and during the few days they rested here, games and amusements were promoted, the officers taking an active part in the sports. At this place they were joined by the Life Guards and Blues.

Hill was now ordered to advance along the main road to Burgos, and make the French yield the castle or fight, and on the 3rd of June the 92nd marched from Orvada, crossed the Guarena, and encamped at Villa Buena on its bank. Here 200 French prisoners passed them on their way to England, having been captured in an action by the Hussar Brigade and Don Julian Sanchez' Spaniards. On the 4th of June they crossed the Toro by a bridge, of which two arches, which had been destroyed in 1812, had been rendered passable for one man only at a time by boards laid across ; but so elastic were they, that some men could hardly keep their footing, and General Stewart, who stood at the far end till all had passed, caught the hand of one of the 92nd just in time to save him. The mounted officers and baggage animals crossed by a ford, some of them swimming. That night they encamped at Morales, on the road to Valladolid ; a fifteen-mile tramp brought them to Villa Sexmil on the 5th, where, as there was neither wood nor long grass to boil the kettles, two houses were given by the Chief Magistrate to supply them with fuel, compensation being paid. Leaving Valladolid to their right, they encamped at Mueientes on the 6th, and, pursuing the retreating enemy, they encamped on the 7th at Duennas, which the French had just left. At Torquemada (which King Joseph left the previous day), on the 8th, it blew a hurricane and the rain poured. Next day they plodded on up to their knees in mud, and after crossing the Pisuerga, encamped on the bank of that river. On the 11th June they only moved forward three miles to Los Valbasas, and the following day they drew near to Burgos, through a country luxuriant in cornfields and vineyards. Here they expected warm work, and just as the division had taken up its ground and the usual order, " Make yourselves comfortable," had been given, the sound of artillery in front called them to arms, and in a few minutes they were on the road to the scene of action. Soon they were stopped by a deep little stream,

[1] Napier.

the enemy having destroyed the bridge, but, surmounting the obstacle, they advanced two miles and halted; for the little fight between our light troops, aided by the horse artillery, and the enemy's rear guard, was over; the latter losing some killed and wounded and one gun. The troops, wet and weary, retraced their steps to camp.

In the morning, about daylight, a tremendous explosion was heard, which made every man jump to his feet, and soon afterwards they learnt that Joseph had blown up the Castle of Burgos and evacuated the town. The destruction of this fortress was the first fruits of Wellington's admirable plan of operations; it was evident the enemy did not intend to join battle south of the Ebro. Sir T. Graham had been successful on the left, and continued to manœuvre and fight to the Ebro, which he crossed on the 14th, and by turning the enemy out of this position, opened a passage for the Centre Division on the 15th, and for Hill's Corps on the 16th.

On the 14th of June the battalion marched to Villa Toro, and on the 15th to Villa Esquiar; during these days they came up with some of the enemy and took them prisoners. The road on the 16th lay through a dreary wilderness of rocks and stones, till all at once they found themselves looking down on the valley of the Ebro, near Arenas, one of the richest, loveliest, and most romantic spots in Spain. The road led down a deep corrie, then in zig-zag down the face of a high and rugged mountain, the rocks re-echoing the wild notes of the bagpipe with very beautiful effect. Below they were all among fruit and flowers, and as the division crossed the bridge over the Ebro the music of each battalion struck up the "Downfall of Paris," and the happy band of British soldiers made hill and valley resound with their cheers. Such was the effect of the exhilarating scene that many of the men were seen dancing with delight across the bridge.[1]

The Gordons wended their way a short distance up the left bank of the river, with inaccessible mountains on their right, the road cut out of the rock which overhung it; then, turning up a lovely valley to the right, encamped at Pesquez. On the 17th, 18th, and 19th they continued marching through the grandest scenery, the road winding sometimes over steep mountain passes, sometimes through luxuriant orchards and flower gardens, while horse, foot and artillery appearing and disappearing as they moved along, with music playing and arms glittering in the sun, gave a lively interest to the scene. At the entrance of every village through which the troops passed, they were welcomed by the peasant girls with garlands of flowers, dancing before them, rewarded, no doubt, by many a compliment in Scots Spanish.

[1] Memoir of a 92nd officer and Captain Kincaid, 95th Rifles.

Though the country afforded many excellent defensive positions, the pressure both on his front and round his right flank by the British left was so great that the enemy dared not take advantage of them, but continued to concentrate his forces on Vittoria. The Light Division, however, came suddenly upon a French corps crossing their path, took their baggage and pursued for some little distance. On the 20th the army closed up to within twelve miles of Vittoria, where the French, under King Joseph Bonaparte and Marshal Jourdan, had taken position the preceding day in order of battle. The 92nd encamped that night at Robeo.[1] Wellington, with extraordinary determination and rapidity, had mastered the line of the Douro, and placed Graham's Corps upon the Esla before the enemy were aware of his intention. His admirable plans had been so well carried out by his generals and by the hardy, high-spirited and well-disciplined troops they commanded, that Joseph had been obliged to retire behind the Ebro ; and had gathered not only his army, but all his baggage, treasure, stores and encumbrances of all sorts, in the valley or basin of about ten miles by eight, at the farther end of which stands the city of Vittoria.

[1] Robeo is about 9 kilometres or nearly 6 miles, with rough intervening hills, from the village of Puebla, where Hill crossed the river. It lies towards Vittoria but on the other side of the Zadora. Though Robeo is the name in the Order Book it is possibly a clerical error for Pobes, a village less than 3 miles from Puebla.

CHAPTER XVIII

THE long-expected battle was now imminent, nor were the sides unequal. The Allies had present about 60,000 Anglo-Portuguese sabres and bayonets, of whom about 22,000 were British (for the Sixth Division, 6500 strong, was left at Medina-de-Pomar). They had 90 guns, and the Spanish auxiliaries numbered 20,000. The French had about 70,000 men, all veteran soldiers superior to the Allies except the British, and 150 pieces of cannon.

The plain of Vittoria is almost surrounded by hills, and is intersected from east to west by two rugged ridges, which formed strong positions for the French. There was also the River Zadora, which ran west of the city, and by the Pass of Puebla, to join the Ebro at Miranda, being joined by several mountain streams by the way. It was spanned by seven bridges, which, though the enemy had neglected to destroy them, were very susceptible of defence. The Bayas River flowed from the mountains to the left of the allied position into the Ebro above Miranda. Hill's Corps was, on the evening of the 20th, between these rivers near the village of Puebla.

The right of the French occupied the heights in front of the Zadora, above Abechuco, covering Vittoria from approach by the Bilbao Road ; their centre extended along the left bank of the Zadora, from Arinez to Puebla de Arlanzon, and fronted the defile of Puebla. A detached corps secured the road to Pampeluna, and Foy was stationed in the valley of Senorio towards Bilbao.

Wellington, having reconnoitred the enemy's position, made his dispositions for an attack.

Sir Thomas Graham, on the extreme left, with the First and Fifth Divisions, Lonja's Spaniards, and Anson's and Bock's cavalry—in all about 20,000 men, with 18 guns—was to make a circuit by the Bilbao Road, and fall upon the extreme right of the French under Reille, force the bridge at Gamarra Mayor, and intercept the enemy's line of retreat. The centre attack, directed by Wellington in person, consisted of the Third, Fourth, Seventh, and Light Divisions of infantry, with most of the artillery and cavalry, numbering in all about 30,000 men. Sir Rowland Hill's Corps, about 20,000 strong, was composed of the Second British Division, Morillo's Spaniards, Sylveira's Portuguese, with some cavalry and artillery. They were destined to force the passage of the Zadora at Puebla, to assail the troops on the heights beyond, to pass the defiles of Puebla, and to enter the basin of Vittoria ; thus turning the French left, and securing to Wellington the passage of the Zadora at the bridge of Nanclares.

Battle of Vittoria

Early on the 21st June the troops moved from their camps on the Bayas; the centre of the army passed the ridge of the Morillas in its front, and slowly approached the Zadora. Hill, having seized the village of Puebla, passed the river there. The First Brigade of Morilla's Spaniards led up the ascent, their Second Brigade ascended half-way so as to connect the first with the British below; a stubborn fight was maintained on the heights, and Hill, with the rest of his corps, threaded the long defile of Puebla, and won the village of Subijana de Alava on the other side. Having connected his right with the troops on the mountain, he maintained his position, in spite of the efforts of the enemy to dislodge him.

Meanwhile, Wellington had brought the centre divisions down to the Zadora, where they waited till all were ready to attack, when they crossed, not without fighting. By one o'clock, Hill's assault of Subijana de Alava was developed; the smoke and distant sound of artillery far off to the left showed that Graham's attack had also begun. The Third and Seventh Divisions were rapidly moving down to the bridge of Mendoza; the enemy's artillery opened on them, and his light troops commenced a vigorous fire of musketry, while the British riflemen, who had crossed by another bridge, were between the French cavalry and the river, and were engaging their light troops and gunners so closely that the British artillery, from the other side, were unable to distinguish them in their dark uniforms from their foes. This gallant episode enabled part of the Third Division to pass the bridge, while the rest, and the Seventh Division, forded the river higher up. The French abandoned the ground in front of Villodas, and the battle, which had slackened, revived with extreme violence. Hill pressed the enemy on the right, the Fourth Division passed the bridge of Nanclares; the sound of Graham's advance became more distinct, and the banks of the Zadora presented a continuous line of fire. King Joseph, finding both his flanks in danger, had given orders to retire by successive masses, and these orders had shaken their confidence, but the Allies were too close for a regular retrograde movement to be made. The Seventh Division and a brigade of the Third Division were engaged with the French right in front of Margarita and Hermandad, and at the same time Wellington sent forward the rest of the Third Division.

"Come on, my ragged rascals!" cried Picton, who commanded them, as they doubled across the front of both armies to seize an important hill in the centre. General Cole, with the Fourth Division, advanced, and the heavy cavalry galloped up and formed between Cole's right and Hill's left. The French threw out clouds of skirmishers, and fifty guns played with astonishing activity, being answered by several brigades of British artillery. Both sides were surrounded by smoke and dust, and under its cover the French retired to the second ridge, but still holding Arinez on the main road. Picton's troops plunged into that village under a heavy fire of artillery and musketry, and captured three guns; but fresh French troops arrived, and this important post was disputed with terrible obstinacy, till at last the British emerged victorious from the strife. At the same time a conflict was going on at Margarita, till the French guns were driven away and the village carried. The village of Hermandad was also won, and the whole army advanced fighting.

Meanwhile the French left, hard pressed in front and flank, retreated towards Vittoria; but the courage of the French soldiers was not quenched, they took advantage of the broken nature of the ground to renew the contest at every favourable point. Reille maintained his post on the Upper Zadora, while more than eighty guns massed together shook the hills and "streamed with fire and smoke, amidst which the dark figures of the French gunners were seen bounding with frantic energy."[1] This terrible cannonade kept the Allies in check, and the battle became stationary. The French generals commenced drawing off their infantry, covered by their resolute cavalry. Joseph, finding the royal road to Bayonne blocked by carriages, indicated the road of Salvatierra as the line of retreat, and the action resolved itself into a running fight and cannonade. The French reached the last defensible height, a mile in front of Vittoria, about six o'clock; "behind the city thousands of carriages and non-combatants, men, women, and children, were crowding together in all the madness of terror, and, as the British shot went booming overhead, the vast concourse started and swerved with a convulsive movement, while a dull and horrid sound of distress arose, but there was no hope, no stay, for army or multitude."[1]

The foregoing account of the general events of the battle is taken from Napier's History, but as this work is the history of the Gordon Highlanders, I will now relate more particularly the part taken by their First Battalion in this great victory, as described by an officer who commanded a company that day, a sergeant who was also present, and others.

It has been remarked that people seldom get their fill of food and fighting on the same day, and when the dawn ushered in the morning of the 21st of June, there was less appearance of breaking fast than of breaking heads, for the bread was hardly baked when the battalion was ordered to be ready to march at a moment's notice; but fortunately they were ordered to fall out for two hours,

[1] Napier ?

when they again stood to their arms, and marched along the road to Puebla. They still knew not whether they were to be engaged in pursuing or fighting the enemy, but now all doubt was removed, for the French held the heights above, and they found that they were within three miles of the main body. A smile of satisfaction played on the soldiers' faces—the veterans encouraged the juniors by telling of former fights, some calculated the numbers of the foe, and all made sure of victory. At a halt, arms and ammunition were closely inspected, and while the captain of the First Company was engaged in this duty, a little incident occurred which shows how the natural fear implanted in most natures may be overcome. A young lad, who had never been in action before, told him he was so unwell that he would be obliged to fall out. The captain asked what was the matter, and received for answer, "A sair wame, sir." The captain walked him up to Colonel Cameron, who, divining his real complaint, took the sick man by the shoulder, and causing him to face the heights of Puebla, pointed to some French sentries on their summit, and asked if he saw them. He replied in the affirmative. "Well, my man," said the colonel, "those fellows are the best doctors in the world for complaints like yours"; and raising his voice, he continued, "and, by Heaven, if I live *you shall consult them this day*!" The poor fellow rejoined his companions covered with shame, but during the action was one of the keenest spirits in the fray, and ever after his captain had more difficulty in restraining his courage than he had in rousing it at Vittoria. The officer remarks that young soldiers in going into action for the first time should be mixed with old ones, and taught to subdue bodily fear by moral courage, for had this lad been allowed to retire he would ever after have been an unhappy coward, instead of a gallant and respected soldier.

The battalion, having crossed the Zadora, followed the high road leading to Vittoria, till at a turn it first beheld the dark masses of the enemy in order of battle, and our advanced cavalry two or three hundred paces from those of the French. It was a grand spectacle that presented itself. On the right bank Wellington's troops moving to their stations, on the Highlanders' right hand the Spaniards climbing the steep heights of Puebla, on which the former could see the French posted, supporting strong bodies of infantry, and some artillery in a clump of trees near the base of the hill. Immediately on their front stood the village of Subijana de Alava, in which, and on a height to its right, were the centre divisions of the opposing army, which, with numbers of cannon, literally covered the country. As the French battalions successively appeared, the sight, instead of damping the courage of the soldiers, seemed to make them more and more delighted, as "with drums beating and music playing, we advanced as if we were going to a common parade or field-day." At about ten o'clock the sound of musketry on the heights announced to 150,000 warriors that the conflict had begun—first a few shots, then more continued firing, followed by volleys, accompanied by the British cheer or the French "Vive l'Empereur!" Morillo's Spaniards had acted for some time in conjunction with Hill's Corps, and were considered superior to almost all the other Spanish troops, and nearly equal to the Portuguese; they had, however, little confidence in many of their officers, but Morillo was a brave man and capable general, and when he commanded, as on this occasion, they behaved with great spirit; but the superiority in ground, and latterly in numbers, would have given the French a decided advantage, had not the 71st Highland Light Infantry and the Light Companies of the division, both under Cadogan, arrived to their support; yet the fight was doubtful. Morillo was wounded, but did not quit the field; and Colonel Cadogan was turning to cheer his followers, and had just repeated, "Well done, well done, brave Highlanders," when he fell from his horse, mortally wounded, into the arms of Captain Seton, commanding the Light Company of the 92nd. He was a brave officer, of high promise, and beloved by his regiment.

An' green be our Cadogan's grave
Upon thy field, Vittoria! [1]

The position being most important, Villatte's Division was sent to succour the French, and so well did they fight that the Allies could hardly hold their ground, till Hill sent the 50th and 92nd to their assistance; but when they had almost gained the top they were ordered back. They had descended half-way when they were again stopped, and the battalions separated; the 50th proceeded to its original destination on the summit, while the 92nd moved across the face of the hill to drive back a body of 1000 French infantry, which had advanced to form a link in the chain of communication between the troops at the clump of trees and those who were to capture the heights, about 7000 in number. The soldiers of the 50th and 92nd did not much relish this separation, but there was no help for it, and the former clambered up and proceeded along the ridge till they reached the brink of a ravine which ran across the hill, above and to the left of the column which the 92nd were to attack. Here the Spaniards and 71st lay along the face of the hill, engaged in exchanging fire with the enemy's light troops. The 92nd descended a few hundred yards, and directed their march towards their opponents posted on a ridge. They had

[1] "Battle of Vittoria."—W. Glen.

to pass through fields of wheat taller than the men, over ditches so thickly lined with briars and thorns that blood trickled down many of the soldiers' legs before they arrived at the base of the ridge. On crossing the last ditch at the foot of the hill, the battalion loaded, the colonel rode along the line, cautioning his followers both in Gaelic and English to be firm and steady, and on no account to throw away their fire. He urged them to be silent till the order to charge was given, and then to join in the good old British cheer.

Full of hope and joy, the line advanced slowly and firmly, every moment expecting to see the enemy. Not a whisper was heard, till on arriving a short distance from where the foe was originally posted, instead of the expected volley, they found that their antagonists had retired during their advance to another height at some little distance ; and with them the hopes of putting the prowess of Frenchman and Scotsman to the test before an audience of 150,000 men had vanished for the time.

Just as they crossed the ridge, General the Hon. W. Stewart, commanding the Second Division, arrived. He ordered the battalion to form column at quarter distance, and two Spanish guns to cannonade the enemy at the clump of trees. This drew the fire of a French battery, and a sergeant, corporal, and one or two privates were hit. The General said to Colonel Cameron, "Poor Cadogan is mortally wounded ; the French are pushing strong columns towards our right, they must be opposed." Then waving his hand to his front, "It is on the heights of Puebla the battle must be fought ; being now senior officer of the First Brigade, you will instantly proceed thither with your battalion, and assume command on the heights. Yield them to none without a written order from Sir Rowland or myself, and defend your position while you have a man remaining." Then, taking a pencil from his pocket, he wrote the order, the shot and shell flying about his head all the time.

Meanwhile Hill had attacked the French in front of Subijana de Alava with the rest of his corps. He was met with the greatest determination, and the Highlanders from the hill above witnessed the cool bravery of both sides. The British advancing to the muzzles of their opponents' pieces before giving fire, their repeated assaults were foiled by the devoted bravery of the defenders, till by praiseworthy perseverance and gallantry they carried their point. Before the village was taken, the 92nd had arrived on the heights, and touched the summit half a mile in rear of the post held by the 50th behind the ravine, of which the western slope was in their possession and that of the 71st, while the enemy were on the opposite bank. The enemy was in superior numbers and our position a strong one, but unfortunately the senior officer on the heights, after Cadogan's fall, ordered the 71st to leave it, to cross the ravine and attack the French on the opposite side ; and the gallant Light-bobs at once set out on their perilous undertaking. The enemy had two corps of infantry out of sight of the British, and as the 71st moved round the northern slope, these corps kept moving round the southern slope, till, being nearly in rear of the 71st, hoping to kill or capture the whole battalion, they poured volleys on them, making many officers and men bite the dust ; and it required all the good military qualities which that regiment is well known to possess, to extricate them from their dangerous situation. They were not sparing of their powder, but against such odds all they could hope for was to be able to rejoin the 50th, and this they endeavoured to do ; but the French were trying to prevent them, and had partially turned the left of the 71st, when their friends the 92nd arrived to take part in the struggle.

On reaching the heights, the Gordons halted a minute to close up and take breath, and then, in open column of companies right in front, hurried along at the double till within two hundred yards of the 50th, when they formed line on the right centre company without halting. As soon as the four leading companies had filed far enough to the right for the centre one to have a clear front, Colonel Cameron placed himself on its left, and, telling the men to "be steady and sure and to remember their country" ("Socair chinn-teach 'illean cuimhuichibh 'ur duthaich"), called to the piper to play "The Camerons' Gathering," and to the officer who led it, "Now push forward double quick and give it them sweetly !"

"During the advance," says the officer, "a dead silence reigned through the ranks, men's thoughts being employed in the business they were engaged in. Animated by the presence of the chief, and the warlike sounds of their favourite bagpipe, the men advanced with a front as firm as the rocks of their native mountains, to meet the foe flushed with a temporary success over their countrymen."

As they approached the 50th, the officers and men of that regiment joined in cheer after cheer, and the Highlanders arrived at the western brink of the ravine just as the French were ascending it. At once they poured down on them a shower of shot, then re-echoed the cheers of their friends, and rapidly loading, and being joined by the other companies, a second volley was sent into the thickest ranks of the enemy, making them fly precipitately down the brow, the living on their feet, the dead rolling over and over. This rapid movement secured a safe retreat to the 71st, who now slowly retired to a position on the right of the 92nd. But the enemy, determined if possible to gain the heights, made another trial ; his beaten troops were moved round the western shoulder of the hill, and a fresh body of infantry carried round the southern, to renew the assault. During the interval, rather a remarkable incident took place, a public auction amidst the roar of musketry and artillery ! A man named Walsh, whose character was so bad that not a soldier in the company would associate with him, had annoyed his comrades by the abominable language he used during the ascent, and when they arrived near the 50th had shocked them by his blasphemous profanity, when, as the words were in his mouth, and before any one had heard the sound of a bullet at that point, he fell, shot through the head. Prepared as they were for sudden but honourable death on the battlefield, his comrades were horrified at the idea of his being killed in the very act of uttering a torrent of blasphemy ; but though they disliked the man, they felt for his widow and children, and while the battalion was resting after repulsing the French, the pay-sergeant of his company, a corporal and private, asked permission to bring the man's knapsack from the rear, and dispose of its contents for their behoof. The colonel, pleased with the idea, sanctioned it ; the auction began, the bidding was brisk, and £1 11s. was added to his balance and remitted to the widow. The good feeling exhibited by this little interlude attracted the admiration of some officers of the 50th who witnessed it.

Shortly after, the head of the French column began to descend the opposite eminence ; some skirmishers kept up a smart fire on them, the rest remained in line behind the brink, sitting down with arms sloped to the rear, the colonel explaining that they were to remain in that posture till the enemy were within twenty paces, then to stand up and give them pepper. His orders were admirably obeyed ; not a whisper was heard while the enemy was crossing the ravine or till they arrived close up, when "the silence was broken on our lads resuming their standing position, and giving their first fire." The scene which followed was an animated one, and, after a rough encounter, in which the 50th, 71st, and 92nd took part, the French were driven back with considerable loss. A French colonel commanding part of the attack had dismounted before crossing the ravine ; he was a very corpulent man, and when they retreated, his pace downhill was like the waddle of a duck, which could not be expected to carry him out of the clutches of a Highlander. He was taken, puffing, panting, and perspiring, "and our lads were ill-mannered enough to indulge in a hearty laugh at his expense. Seeing he was the butt of the group, he good-humouredly joined in the laugh, saying to a 92nd officer as he surrendered his sword, 'Mon Dieu, mon Dieu ! what a fool I was to part with my horse ! For want of it I am now your merryman.' "

Again the fugitives were withdrawn, and a third column of attack was formed, in order to recover the position which they should never have lost. Although this column was even stronger than the others, masses of infantry being collected to support them, they did not show the same spirit as in the two former cases, their efforts were comparatively feeble, and they were soon repulsed. This the British attributed to the state of affairs in the centre, which had now begun to take an active part in the business of the day.

The efforts of the First Brigade and their Allies having been successful on the extreme right, where they now held undisputed possession, and where their presence had a material effect on the issue of the battle, they had leisure to look down on the magnificent panorama presented to their view not far below. They could clearly see every eminence bristling with the artillery on both sides vomiting fire and death ; thousands of infantry marching against each other ; they watched with excited confidence the lines of red coats cheering as they charged, the French firmly waiting with the cry of "Vive l'Empereur !" till the British steel was close, when they sometimes, but not always, gave way. They could see the curling smoke far away to the left, where Graham's Corps was slowly and with difficulty forcing back the French right, and hear the salvos of their distant guns ; with the spy-glasses which some carried, they watched the hussars cross the river, and could distinguish the combatants and the flashing swords in the terrible charges of the heavy dragoons against the equally brilliant French horse. Gradually they observed, with pride and pleasure, the enemy giving way. The fields were intersected by ditches and hedges, behind which they extended their line. "Often," says the 92nd officer, "during that awful struggle did I witness the British soldiers walk up to the brink of a fence, behind which their opponents were arrayed, and in the most cool and determined manner cross their pieces with the latter before they gave their fire. On these occasions the combat invariably assumed a sanguinary aspect, for the ditches were generally too deep for our men to cross, and the French kept up a smart fire till artillery and cavalry came to dislodge them."

In this manner they were spectators of the later battle raging beneath, till, between four and five o'clock, the French divisions whom they had driven back earlier in the day to a new position on the heights, seeing that their troops below were falling back, began a retrograde movement. The First Brigade and two brigades of Portuguese then stood to their arms and advanced, the enemy quickened their pace till it degenerated into a race downhill, the brigade firing and the men shouting to them to stop, as they gave chase ; but so superior were the French at that game (for the Allies kept their knapsacks, while many Frenchmen threw theirs away), that in an hour and a half the pursuers entirely lost sight of them. As a French officer who was taken said to one of the 92nd, " I will back my countrymen against any soldiers in the world in a race of that kind."[1]

The 92nd continued the pursuit past Vittoria till eleven o'clock at night, when they bivouacked in a wood near the road, four or five miles beyond the town. The question of supper is the most important one even at the end of so glorious a day. The road was covered with carriages, baggage waggons, and all the *impedimenta* of King Joseph's Court and army. Hams and provisions of all sorts were to be had for the taking, nor was good wine wanting for a sleeping draught, ere the weary Highlanders lay down that summer night to enjoy their well-earned repose.

Thus ended the great battle of Vittoria; and never was victory more complete. Never had such an enormous quantity of military stores and private wealth fallen to the lot of an army. The accumulated plunder of Spain during the five years the French had occupied the country was here collected, and its amount exceeded anything witnessed in modern times. More than five millions of dollars were taken in the military chest, but the amount of private wealth cannot be estimated, and Napier, who was present, says that for miles the pursuers may be almost said to have marched upon gold and silver without stooping to pick it up ; but the regiments which followed, and were not so warmed in the fight, were less distinterested, and camp-followers and non-combatants, as well as many soldiers, took enormous spoil, a dishonourable action on their part, depriving others of their fair share of prize-money ; and as a Gordon's journal remarks of a hussar who showed his boots full of gold coin—" He got it without much risk—a soldier should be able to say when he gets home, ' The wars are o'er and I 'm returned, my hands unstained with plunder.' "

Vast numbers of carriages with ladies belonging to the French army, nuns, wives of the generals and officers, actresses—arrayed in the height of luxury and fashion, with their poodles, poll-parrots, and monkeys—filled the roads, which were blocked by guns and waggons whose drivers had escaped with the horses. Laces and velvets, silks and satins, valuable pictures, jewels, gold and silver plate, cases of claret and champagne, lay scattered in all directions. Other vehicles were stuck fast in the fields, their occupants in helpless terror ; but the British officers and soldiers were very con-

[1] "Français plus qu'hommes au venir, moins que femmes à la retraite" (The French, more than men in the advance, less than women in retreat) is an old French saying.—" 1815," by H. Houssaye.

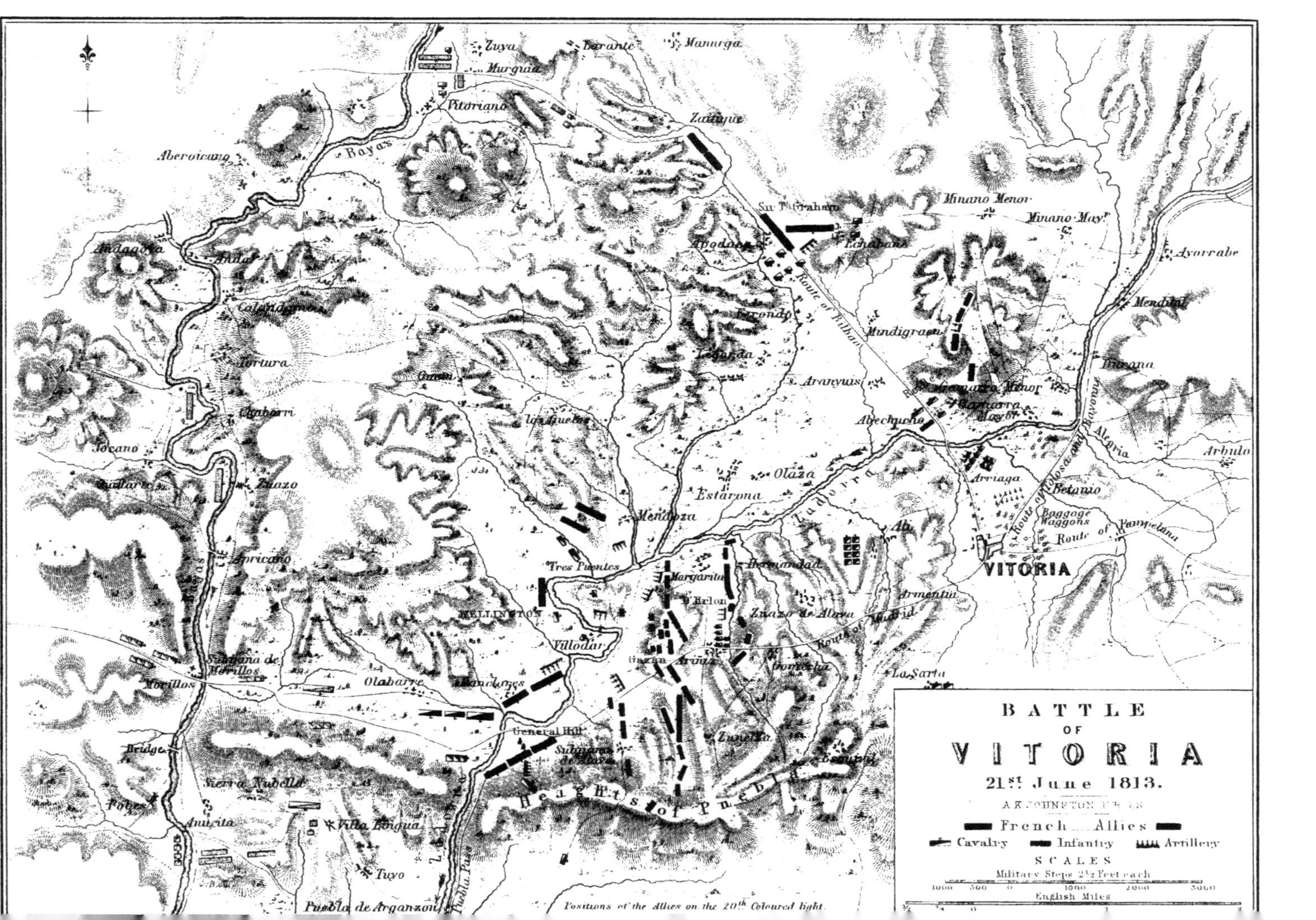
BATTLE
OF
VITORIA
21st June 1813.
French Allies
Cavalry Infantry Artillery
SCALES
Military Steps 2½ Feet each
English Miles
VITORIA
Heights of Puebla
Positions of the Allies on the 20th Coloured light.
Route of Bilbao
Route of Madrid
Route of Pampelana
Baggage Waggons
Puebla de Arganzon
Puebla Pass
General Hill
Tres Puentes
Villodas
Wellington
Bridge
Sierra Nubella
Boyas
Zadorra
Vitoriano
Murguia
Zuya
Mendoza
Estarona
Olaza
Arriaga
Betonio
Alegria
Arbulo
Ayorrabe
Mendibil
Minano Menor
Minano May.r
Echabarri
Abechucho
Margarita
D'Erlon
Hermandad
Zuazo de Alava
Armentia
La Sarta
Gomecha
Zumelza
Esquibel
Subijana de Alava
Olabarre
Subijana de Morillos
Morillos
Jocano
Zuazo
Apricano
Tortura
Andagoya
Aberoicano
Zaitique
Mandigrana
Arangus
Lagarda
Ali
Villa Bngua
Tuyo
Anucita
Pobes

siderate, and though they took a share of the provision baskets which most carriages contained, the ladies were treated with respect, many of them being next day forwarded under flag of truce to Pampeluna. King Joseph's carriage and sword of state, Marshal Jourdan's bâton, the papers and accounts of the army, were taken, with the whole of their artillery except two guns. They lost 6000 men killed and wounded, and nearly 1000 prisoners.

The Allies lost 5176 men in killed, wounded, and missing; of these the British were more than double the number of Spaniards and Portuguese put together. The loss of the 92nd was small, having regard to the very important effect of the part taken by their brigade on the issue of the battle. The Gordons had 4 privates killed and 1 sergeant and 15 rank and file wounded, of whom 2 privates died of their wounds.

KILLED

Private Robert Grant. Private John Macdonald.
,, Jas. Graham. ,, W Walsh.

DIED OF WOUNDS

Corporal T. Watson. Private George Gow.

Their conduct was commemorated by the word *Vittoria* on their colours, and a medal was conferred on their commanding officer, Colonel Cameron [1] (who also commanded the brigade after the fall of Colonel Cadogan). Officers and soldiers afterwards received their share of prize-money.

In the afternoon of the 22nd, the 92nd moved with the Second Division from their bivouac in front of Vittoria. The road, which lay through a corn-covered plain, had a wall on each side, the march being constantly delayed by broken-down waggons and carriages, and all the debris of the French flight; so it was after midnight when they encamped a few miles beyond Salvatierra. After a very short sleep the bugles sounded, but a good breakfast made the men start as gay as larks, till their gaiety was damped by a pouring rain during five hours spent standing under arms, and marching a few miles to Arbeniz, where they encamped. On the 24th they followed the Third, Fourth, and Light Divisions towards Pampeluna; a fearful thunderstorm overtook them on the march, and an officer of the 34th was killed, his watch being melted and his sword belt cut into little bits; nine men of the same regiment were struck, but recovered. They encamped that night at Araquil, the 25th at Stormanda, the 26th at Zuaza, and the 27th at Orcayen,

[1] On the thanks of Parliament being given personally to Sir William Stewart, he said, "I should be ungrateful for the services rendered me by Colonel Cameron and by General Byng on that, as on all occasions, if I were not to advert to them in my present place."

three miles from Pampeluna, where they were well supplied in every way. During these marches they saw nothing worthy of notice but the smoking ruins of houses burned by the retreating French, and one of the two guns they had brought from Vittoria, which had been dismounted by our troops in front. In this district is great plenty of good red wine, which no doubt helped them to sustain the fatigues of the march.

It had been Wellington's first intention to reduce Pampeluna, but being now so far from Portugal, a new base of operations was required, such as was to be found on the coast of the Western Pyrenees, at Bilbao, Santander, and San Sebastian ; and the capture of the latter fortress was of more immediate importance than the reduction of Pampeluna, which he now decided to invest, entrusting the blockade to O'Donnel's Spaniards. The French, on quitting Pampeluna, which they left strongly garrisoned, retired across the Pyrenees towards their own country in two columns, one by the Pass of Roncesvalles, the other by that of Maya. The Third and Fourth Divisions, and the Second Brigade of the Second Division and some Portuguese and Spanish troops, were pushed forward towards the former ; and the rest of the Second Division with a brigade of Hamilton's Portuguese were ordered to proceed against the enemy at Maya. Accordingly, on the 2nd of July, the First, Third, and Fourth Brigades, Second Division, a Portuguese brigade, some artillery and cavalry moved from the camp in front of Pampeluna, and encamped in the afternoon near La Zarza. Next day, after a wet and weary walk, they camped on a mountain near the village of Lanz, and early on the 4th crossed the Pass of Lanz, arriving about one o'clock at Almandos, where the column closed up preparatory to an attack on the enemy's position at Irueta, three miles off. At two o'clock, the 92nd with the First Brigade, led by General Stewart, entered Almandos, turned sharp to the right and filed through the fields by a footpath, crossed a ravine and the stream running through it, and, scrambling up a steep place covered with loose stones, gained the summit, surprised that the enemy had allowed them to do so without resistance.

On seeing them established on the left of the enemy's advanced posts, Sir R. Hill moved the other brigades through Almandos to attack the French outposts ; these, however, retired on their main body, of which the left occupied a high mountain, the centre the village of Irueta, while their right extended to the Bidassoa. The Portuguese and French skirmishers began an animated fire, during which the 50th attempted to carry the village in their usual gallant style, but the French brought up a very superior force, and drove them back. The left wing of the 92nd was pushed forward to their support, the right wing and the 71st being kept in reserve ;

but before the Gordons reached the scene of action, the 50th had been ordered to retire, Sir Rowland not wishing to bring on a more general affair that evening.

The troops cooked early next morning, and, on the arrival of Wellington at noon, moved against the enemy, who remained in the same position. The First Brigade, forming the right of the little army, ascended a high, steep, and slippery mountain on their right, in order to throw itself in rear of the enemy's left. It was a stiff climb, but from the top they got a glimpse of France and of the sea, which the men called the high road to home, and they saluted it with deafening cheers, which made the French bestir themselves, attributing the demonstration to a different cause. The Light Companies were now detached towards the left flank of the enemy, supported by the rest of the brigade ; their centre was attacked at the same time by the other brigades and forced to retire from Irueta ; then the whole (7000) retired towards Elizondo, partly by the narrow road and partly by the fields, occasionally turning and firing, constant skirmishing going on from walls, hedges, or houses; our men often flanking them, and giving them a volley as they left their temporary strongholds. Here they came on a regiment in red coats, which the men took to be British, till finding they were Swiss in the French service, they soon drove them off. Thus the retreat and pursuit were conducted to Elizondo, where a wall had been thrown up across the entrance of the town, from which the French annoyed our troops as they closed upon them ; but soon they were forced to yield the entry, and the British pursued through the streets amidst the loud acclamations of the townspeople, while the church bells rang a merry peal in honour of their deliverance. The enemy then took up a position on a high ridge at the head of the valley of the Bastan, and prepared to fight for his last foothold in Spain. Wellington reconnoitred this position on the 6th and on the morning of the 7th, when orders were given to attack it.

The Second Brigade got under arms about 11 a.m. and ascended by a narrow way to attack the enemy's right on the rock of Maya, five miles distant. Before arriving, a thick fog came on, which concealed their approach till they were close to the French, who fired down on them ; but after a short fight, mostly with the bayonet, the British cleared the mountain of their opponents. Meanwhile, in order to deceive the French general, the other brigades remained quiet in camp; and it was only when they heard the firing on the rock, and these brigades advanced, that the enemy slowly retired. The First Brigade, followed by Ashworth's Portuguese, advanced rapidly up the valley, passed Maya village at the double, and the Light Companies soon came into contact with the enemy, who tried to prevent communication between the main body of the British and their Second Brigade. A smart fire was kept up, till, after various movements, a considerable body of combatants were brought to close quarters ; the 50th were more than once on the point of crossing bayonets, and being hard pressed, the 92nd were sent to help them, but the "Old Half Hundred," with their usual determined pluck, repelled the assault before their Highland comrades arrived. The French General Gazan now

ENTRANCE TO ELIZONDO, ON THE BIDASSOA. (*From an original sketch.*)

made desperate attempts to regain the rock ; nothing could excel the conduct of the Second Brigade, who held it, and the affair was only ended by the night, which was so dark that when the skirmishers were being called in, many could not find their battalions, and some French soldiers passed our pickets and were made prisoners. The French were only about 200 yards from our advance posts, and for hours voices could be heard calling out in various languages.

Next day General Gazan made another attempt, but failing as before, he kept up an irregular fire for some time and then took advantage of another fog to retire, pursued by the Portuguese, who skirmished with his rear guard till evening. During the fight of the 7th, three Spanish peasants of the village of Maya joined our light troops, and fought splendidly in the heat of the conflict, one being killed and another wounded. Sergeant Robertson, in his Journal, mentions that in these various affairs " we had a number wounded," but whether he alluded to his battalion or his brigade does not appear certain. The Order Book of that time is lost, and in such returns as are preserved, I find the name of only one 92nd man wounded at Elizondo. Nor does the Journal of an officer allude to loss, though he mentions the coolness of a young lad named M'Ewen, and his facetious rebuke to the bullet which had cut his bonnet just over his ear, before burying itself in the earth behind him.

The pursuers gave the French a parting volley and cheer as they descended the mountains to the frontier of France, the light troops following them to near Urdax. The Gordon Highlanders encamped on the heights of Maya, near the road leading from Pampeluna to Bayonne. The surrounding scenery is exactly like the West Highlands of Scotland, and from their camp on the Pyrenean Mountains, they looked down on the plains of France, and beyond them could see the white sails of the British fleet on the Bay of Biscay. " We fondly hoped," says the sergeant, " soon to descend like a mountain torrent sweeping all opposition before it." They little knew how stern and strong that opposition would prove.

CHAPTER XIX

NAPOLEON was in Germany (where, and in Poland, he had about 700,000 French soldiers) when he heard of the battle of Vittoria; he immediately appointed Marshal Soult [1] to command the army of Spain in place of his brother, King Joseph. There were still in Catalonia and Aragon over 60,000 French, and the whole force employed against Spain amounted to 180,000 men and 20,000 horses, of which 156,000 were present under arms; but when the garrisons of Pampeluna, San Sebastian, Bayonne, etc., were deducted, Soult had only 77,500 men present under arms, of whom 7000 were cavalry. This army was divided into three corps of battle and a reserve. Clauzel commanded the left at St. Jean Pied de Port; Drouet, Count d'Erlon, commanded the centre at Espelette and Ainhoa, with an advanced guard at Urdax; Reille, with the right, was in position overlooking Vera; while the reserve under Villatte guarded the Bidassoa from its mouth up to Irun, where he destroyed the bridge. The cavalry were on the banks of the Nive and Adour.[2]

The position of Wellington's army was as follows:—Byng's Brigade of the Second Division and Morillo's Spaniards were on the right, in front of the Pass of Roncesvalles near St. Jean Pied de Port. On Byng's left was Campbell's Portuguese Brigade, supported by the Fourth Division under Cole, at Viscayret. On Campbell's left, Hill defended the Bastan with the rest of the Second Division and Hamilton's Portuguese. Picton, with the Third Division, was at Olague as a reserve both to Hill and Cole. On Hill's left, the Seventh and Light Divisions occupied the mountains from Echellar to Vera, and behind them at San Estevan was the Sixth Division. Lonja's Spaniards continued the line of defence from Vera to Giron's position, which extended along the Bidassoa to the sea, and behind Giron was Sir Thomas Graham with the army besieging San Sebastian. Thirty-six pieces of artillery and some cavalry were with the right and centre, but most of the cavalry were left behind the mountains about Tafalla. The great hospitals were at Vittoria and Bilbao. The whole force under Wellington in Navarre and Guipuscoa, including British, Spanish, and Portuguese, was about 100,000, of whom about 82,000 were present under arms.

The 92nd, under Major Mitchell (for Colonel Cameron was acting as Brigadier to the First Brigade), remained a fortnight in camp, 200 yards to the left of the road where the Atchiola Mountain rises from the pass. They were disturbed only by a tremendous storm of rain and wind, which blew down most of the tents one night, and damaged the arms and accoutrements greatly, besides

[1] Soult had left Spain some months before Vittoria. [2] Napier.

destroying a good deal of ammunition. The 71st was 300 yards to the left of the 92nd; the 82nd, from Barnes' Brigade, Seventh Division, were about a mile up the mountain to the left of the 71st; the 50th lay half a mile from the 92nd down the slope on the Spanish side, and three Portuguese guns were between the 92nd and the road. The Second Brigade, Second Division, were behind, near the village of Maya, having the 34th advanced towards the heights on the right, where strong pickets were posted.

It is remarkable how many battles have been fought on Sunday. The chaplain attached to the First Brigade had more than once lost his congregation owing to the movements of the enemy, and on Saturday the 24th of July he visited the camp, and on leaving said, "Gentlemen, we shall have divine service to-morrow, God and the French being willing." With what fearful effect the French interposed will now be described.

On the 25th, the 92nd had a foraging party with the regimental mules out early for firewood, which had not yet returned; the rest were preparing, in the sunshine of the forenoon, for the expected divine service. Unaware that Soult, wishing to relieve Pampeluna, had been concentrating troops for a simultaneous attack on Byng at Roncesvalles, on Campbell and Cole near Viscayret, and on Maya, General Stewart was at Elizondo on that Sunday morning thinking all quiet in front.

The Col de Maya is three miles broad. There were three passes to defend—at Aretesque on the right, Lassessa in the centre, and Maya on the left. General Pringle's Brigade was to defend Aretesque, and Colonel Cameron's Brigade the Maya and Lassessa Passes. The officer relieving the picket at Aretesque was told by his predecessor that a glimpse had been obtained at dawn of troops in movement in front. Captain Armstrong of the 71st, who was on picket at Maya, also observed troops in motion far away, and reported the circumstance to Colonel Cameron, who, with a number of officers, proceeded to satisfy himself. After looking through a spy-glass, a young officer remarked that what Armstrong took for French troops was nothing more dangerous than a drove of bullocks. "By J——s, young man, if they 're bullocks they have bayonets on their horns!" said the indignant captain. "Few," adds the officer who relates the incident, "had greater reason to remember the correctness of the captain's remark than the individual who provoked it, for he was severely wounded in the action that followed by one of 'Armstrong's bullocks,' as Frenchmen were afterwards denominated by us." [1]

[1] On the first alarm the soldiers packed their knapsacks and accoutred, ready for orders.—Sergeant Robertson.

"General Stewart was surprised; his troops were not."—Napier.

The Light Companies of the Second Brigade were ordered to support the pickets, and had just formed at the Rock of Aretesque when the Corps d'Armée under d'Erlon advanced. D'Armagnac's Division mounted the great hill in front, Abbé followed, and Maransin, with a third division, advanced from Urdax against the principal pass of Maya, meaning also to turn it by a path leading up the Atchiola Mountain. D'Armagnac's men came on in several columns and forced back the picket with great loss upon the light companies, who with difficulty sustained the assault. The alarm guns at Maya were now heard, and General Pringle hastened to the front; but his regiments, moving hurriedly from different camps, could only come into action one after the other and by companies, breathless from the steep ascent, first the 34th, followed by the 39th and 28th. But before the two latter could arrive, and notwithstanding the desperate fighting of the picket, the light companies, and the 34th, d'Armagnac's troops, supported by Abbé, had, by numbers and valour combined, established themselves on the ridge of the position. Colonel Cameron then sent the 50th to the assistance of the overmatched troops, and that "fierce and formidable" old regiment, charging the head of an advancing column, drove it clear out of the pass of Lassessa. But the French were so many that though checked at one point, they assembled with increased force at another; nor could General Pringle restore the battle with the 39th and 28th Regiments, which, cut off from the others, were forced back, fighting desperately, to a lower ridge crossing the road to Elizondo. They were followed by d'Armagnac, but Abbé continued to press the 50th and 34th, whose line of retreat was towards the Atchiola, where Cameron's Brigade was. "And that officer, still holding the pass of Maya with the left wings of the 71st and 92nd Regiments, brought their right wings and the Portuguese guns into action, and thus maintained the fight; but so dreadful was the slaughter, especially of the 92nd, that it is said the advancing enemy was actually stopped by the heaped mass of dead and dying." [1]

It was at this moment that General Stewart reached the field of battle. The passes of Lassessa and Aretesque were lost, that of Maya was still held by the left wing of the 71st; but Stewart, seeing Maransin's division gathering on one side and Abbé's on the other, abandoned it for a new position covering the road to Atchiola, called down the 82nd from that mountain, and sent messengers to the Seventh Division for help. Stewart, although wounded, continued his resistance; but General Maransin suddenly thrust the head of his division across the front of the British, and

[1] It was as much by their own dead as by those of the 92nd that the French were stopped, as appears from the account of a 92nd officer. (See page 295.)

connected his left with Abbé, throwing as he passed a destructive fire into the wasted remnant of the 92nd, which even then gave way sullenly. The men fell till two-thirds of the whole had gone to the ground, and the left wing of the 71st came into action ; but one after another all the regiments were forced back, and the first position was lost, together with the Portuguese guns.[1] Abbé's division now followed d'Armagnac's towards the town of Maya, leaving Maransin's division to deal with Stewart in his new position, which was held (though the small British force had shrunk in numbers and was short of ammunition) till a brigade of the seventh division under General Barnes arrived from Echellar and drove the French back to Maya ridge, when d'Erlon, probably thinking larger reinforcements had arrived, recalled his other divisions and united his whole corps. His loss was 1500 men and a general.

The above account is taken from Napier's History, who gives, in Appendix No. 4, the following extract from a MS. memoir by Captain Norton, 34th Regiment, showing the situation of the 92nd as it appeared from a distance. "The 39th Regiment then immediately engaged the French, and after a severe contest also retired. The 50th was next in succession, and they also, after a gallant stand, retired, making way for the 92nd, which met the advancing French column first with its right wing drawn up in line, and after a most destructive fire and heavy loss on both sides, the remnant of the right wing retired, leaving a line of killed and wounded that appeared to have no interval. The French column advanced up to this line and then halted, the killed and wounded of the 92nd forming a sort of rampart ; the left wing then opened its fire upon the column, and as I was but a little to the right of the 92nd, I could not help reflecting painfully how many of the wounded of their right wing must have unavoidably suffered from the fire of their comrades. The left wing, after doing good service and sustaining a loss equal to the first line, retired."

The following more detailed account of the movements of the 92nd, particularly of the right wing, is taken from the memoirs of the subaltern officer who brought it out of the first part of the action.

It was after eleven o'clock when the attack on the pickets began. As we have seen, the charge of the 50th, following the 34th and 39th, checked for a moment the career of the French ; but d'Erlon, availing himself of his great superiority in numbers, charged these corps in front, at the same time sending strong columns on each

1 "The force of the enemy in our front yesterday is generally estimated at 14,000 men." "I am sorry to say we were so pressed on the height that it was impossible to bring away the four guns which were there ; they were spiked and thrown over the cliff."—*Sir Rowland Hill to the Quartermaster-General, dated 26th July 1813.*

flank to surround them. At this critical period the right wing of the 92nd, nearly 400 strong, under Major John MacPherson, entered the field and took part in the fray. On their arrival, the Highlanders were a good deal blown, having advanced a mile and a half mostly at the double. The situation of their friends, however, was such that they formed line without a moment's delay and at once advanced. The enemy, seeing their intention to charge, halted, and thereby afforded the 34th and 50th an opportunity of retiring to re-form their ranks. Enraged at the failure of his attempt to capture these two battalions, the French general now turned his fury against the Highlanders and tried to annihilate them by showers of musketry. They, however, nothing intimidated, returned the fire with admirable steadiness and effect. Perceiving that the enemy was acting cautiously, Cameron (Brigadier), wishing to draw him to ground where he could charge him, retired the Highlanders, when the French general, mistaking the reason for the retrograde movement, pushed forward over 3000 troops, who advanced making the air ring with their shouts of "Vive l'Empereur!" Conceiving that the enemy had made up his mind to meet the steel, Cameron ordered the Highlanders to halt—front—and prepare to charge. On seeing them halt, the French did the same, and instantly opened a terrific fire of musketry. At this time the space between the combatants was not more than 120 paces, while the enemy numbered about eight to one. From the 92nd to the French front line the ground was almost level, but in rear of their foremost troops was a narrow ravine, behind which rose abruptly a considerable eminence, from the face of which the French fired over the heads of their comrades on the small body of Highlanders. These did not, however, return it, but directed the whole of their fire on that part of the enemy's force stationed on the brow of the ravine nearest to themselves, and so coolly and admirably was it given, that in ten minutes the French dead lay literally in heaps. The slaughter was so appalling, indeed, that the utmost efforts of the officers failed to make their men advance beyond their slain. At times they prevailed on a section or two to follow them, but the sight of their comrades' mangled corpses was too much, they invariably gave way. One officer rendered himself conspicuous by his repeated and gallant attempts to induce his men to charge the diminishing band of feathered bonnets. He advanced alone about fifteen paces before them, struck his sword into the ground, and crossing his arms on his breast, stood facing our men. His noble conduct might have had the effect he desired, a spark from his spirit might have fired the increasing mass of men in the ravine behind him, who might, by sheer weight of numbers, have overwhelmed the Highlanders, had not Private Archibald M'Lean, stepping to the front and kneeling down, taken deliberate aim and shot him.

The officer from whose memoirs I take this account, and who was present, says :—" To have killed this officer under any other circumstances than those in which the 92nd were placed would have been considered by us as an act of deliberate cruelty, but when the respective numbers of the combatants are kept in view, every impartial man will admit that the death of the officer was indispensably necessary to our safety. I never felt so much for any individual as for that truly brave man."

During the hottest of the fire, Sergeant Cattanach,[1] whose place was in rear of the officer commanding his company, tapped him on the shoulder, saying he wished to speak to him. On his officer turning, he said, " Oh, sir, this is terrible work, let me change places with you for a few minutes." Respecting the generous motive of the sergeant in offering to place himself between the bullets and his officer, the latter thanked him, but desired him to attend to his duty in the rear ; in a few moments the sergeant repeated his proposal, but being told not to mention the subject again, he resumed his post with a look of disappointment, and shortly after was shot in the groin. His brother, a private, ran to his assistance, when he begged to be laid in rear of the company, but had scarcely spoken when a second bullet killed the brave and warm-hearted sergeant on the spot. William Bisset, a private in the same company, was wounded in the thigh. He quitted the scene of action leaning on his musket, blood flowing copiously as he hobbled away ; but halting at a little distance to look back, and seeing his comrades still supporting the unequal conflict, he returned and took his place in the ranks. His officer advised him to retire and at least get his wound bandaged, when Bisset said, " I must have another shot at the rascals, sir, before I leave you." He fired once, and was about to have a second shot when a ball broke his arm, and he was compelled finally to retire from the field. Many similar instances of devotion no doubt occurred.

This sanguinary combat was sustained for over twenty minutes, during which time more than half the men had been killed and wounded, and all the officers wounded and carried from the field except two lieutenants, while of the soldiers still standing many were short of ammunition. No help was in sight, and 3000 against 200 is long odds ; therefore the senior of the two subalterns decided " under all the circumstances "[2] to retire, which they accordingly did in perfect order, pursued slowly by the French, who did not take a prisoner but such as fell by the musketry they poured on them during the retreat. This determined and un-

[1] Sergeant Cattanach belonged to Kingussie. [2] " Military Memoirs."

flinching stand remains an unsurpassed example of the perfection of steady fire-discipline, dauntless courage, and devoted bravery.

On getting behind the height in front of which they had been engaged, they found the 28th and right wing of the 71st hastening to their relief. The former attacked the enemy's leading columns, but soon after, moving down the hill, they joined the 34th and 39th in the valley, and the rear of the shattered band of Highlanders was completely unprotected. Under these circumstances they, with the 50th, retired to the pass, where General Stewart was trying to retard the progress of his opponent. He detached the right wing of the 71st, and part of the 50th, to a position in rear, and at the head of the left wings of the 71st and 92nd awaited the enemy. During this interval Pipe-major Cameron, thinking a little music would be grateful to his comrades, set his drones in order, and made the hills re-echo the "Pibroch of Dhonuil Dhu." The effect was electrical. The weary Highlanders were on their legs in an instant anxiously looking to their wounded General (Stewart), who was a few paces in rear, for the order to advance. He at once ordered the piper to stop, and warned them of the fatal consequences that might follow a forward movement at that particular moment. Meanwhile the French below were increasing in numbers, and in ten minutes the piper, probably impatient of the general's tactics, tuned up again, and again his comrades jumped up eager for action. The angry General peremptorily ordered him not to play without orders, on peril of his life. He obeyed, but was heard muttering, with a sublime confidence in the power of his own music, "Mur leig e leom a phiob a chluich cha'n eil Frangeach 's an duthaich nach bi nuas oirn" (If he 'll not let me play, every man in the land of France will be here).[1]

The French, after a little skirmishing, brought forward a mass of infantry to overpower all opposition, and Stewart, after a few rounds, withdrew the advanced wings, and marching them through the intervals of the 50th and 71st right wing, placed them in position 200 yards in their rear. The enemy followed, and were warmly received by the 50th and 71st, who then retreated through the intervals of the 71st and 92nd left wings, and thus relieving each other with the utmost regularity, and disputing every inch of ground, they fell back fully a mile, when, being reinforced by the 82nd, they halted in a new position, from which they had the mortification of seeing their camp plundered by the enemy ; the batmen and mules which had been out in the morning had been unable to remove all the baggage, and some of them had been taken. An officer mentions how his faithful servant, Hugh

[1] Told by Private John Downie and by an officer.

PASS OF MAYA FROM THE NORTH OR FRENCH SIDE, TAKEN FROM ABOVE URDAX

Johnstone, had saved his most valuable articles and money by emptying his knapsack of his own kit, and filling it with his master's valuables.

At the beginning of the action Colonel Cameron detached Captain Campbell of the 92nd with 150 men to the summit of the rock of Maya, which was the key of the position. From this post its garrison rendered considerable assistance, for the face of the hill was covered with whinstone rocks, which Campbell and his men hurled down on the pursuers, frequently with great effect. But neither bullets, bayonets, nor boulders could check the advance of the enemy, who had separated the two brigades (the Second having retired across the valley of the Bastan).

Fresh troops were advancing from Urdax. Our troops were from 2000 to 2500 men, their opponents 8000 to 9000 ; and about 7 p.m. General Stewart, in order to stop further bloodshed, proposed to retire, and sent an order to the troops on the rock to abandon it ; but, before the bearer of the message could deliver it, the cheers of the troops at the base of the hill reached the summit. These were occasioned by the arrival of General Barnes with the 6th Regiment and some Brunswick infantry. A more seasonable reinforcement was never received. The tired soldiers were resting when it arrived, but rose to cheer—" Our lads were perfectly frantic with joy." General Stewart, having regard to the extraordinary loss and fatigue sustained by them, desired that the 92nd should not join in the charge of Barnes' troops. But this time the pipe-major was not to be denied. He struck up the charging tune of " The Haughs of Cromdale,"[1] his comrades, seized with what in the Highlands is called " mire chath "—the frenzy of battle—without either asking or obtaining permission, not only charged, but led the charge, and rushed down on the enemy with irresistible force, driving back their opponents in the most splendid style. The power of the national music over the minds of Scottish soldiers was never more conspicuous.

D'Erlon, taking it for granted that such temerity must be backed by stronger reinforcements than appeared, retired about a mile, and General Stewart made such dispositions as would convey the impression that he intended to renew the combat in the morning.

Intelligence was, however, received that Generals Cole, Picton, and Byng had been compelled by the overwhelming forces of Soult

[1] This tune is older than the battle of Cromdale (1689) ; the old words have more to do with weddings than warfare—

Tha banais aig na Graintaich uile, Air mullach Beinn a' Chromdail.	" There 's a wedding for the Grants On the heights of Cromdale."

Rev. W. Forsyth, Abernethy.

to yield their positions, by which the right of the army was turned. This rendered the recovery of the Col de Maya useless, and Hill, withdrawing the troops during the night, posted them on the heights in rear of Irueta, fifteen miles from the scene of action. It now became necessary for Wellington to concentrate part of his army in front of Pampeluna, to prevent the relief of that fortress.

Such was the disastrous conflict of Maya, in which 2600 British troops not only retained the key of their position for nine hours, despite the utmost efforts of about 11,000 of Bonaparte's best infantry, but, on receiving a reinforcement of only 1000 men, actually recaptured about a mile of ground which the enemy had acquired earlier in the day. It was certainly one of the most brilliant achievements of the whole Peninsular War.

The following anecdote shows the impression made by the conduct of the Gordon Highlanders on the commander of the French army corps engaged. An officer relates that a wounded British Colonel had to be left behind in the valley of Bastan. In passing through the village, the French General, Count d'Erlon, called on him, and after condoling with him said, "Pray, Colonel, how many *sans-culottes*[1] have you in your division?" "One battalion," answered the Colonel. "One regiment of several battalions I presume you mean, Colonel," retorted d'Erlon. "No, General, only one battalion, I assure you," replied Colonel H——. The Count, with a smile of incredulity, then said, "Come now, Colonel, don't quiz me; do tell me candidly how many Highlanders you had in action on the right of your position on the 25th?" Colonel H—— then said earnestly, "I give you my word of honour, General, there was only half a battalion, not exceeding 400 men." D'Erlon fixed his eyes on Colonel H——, and after a pause said with emotion, "Then, Colonel, they were more than men, for before that body of troops I lost one thousand killed and wounded."

In commemoration of the noble and devoted conduct of his regiment, Lieutenant-Colonel Cameron was permitted by His Majesty to bear the word "Maya" on the shield of his coat of arms.

Napier, in his account of the battle, says, "Never did soldiers fight better, seldom so well." "The stern valour of the Ninety-second would have graced Thermopylae."[2]

[1] "Men without breeches," the name used by the French for the Highlanders.

[2] In his first edition, published 1840, Napier added after the word "Ninety-second" the words "principally composed of Irishmen." This statement gave great offence to the regiment, not because they in any way disliked or undervalued Irishmen, but because the statement was inconsistent with fact and with their nationality. Colonel J. M'Donald, then commanding the 92nd, wrote to the author drawing his attention to the error in his History,

Many officers and men of the battalion were absent on July 25th on command, sick, prisoners of war, so that there were present and fit for duty on that day :--Field-officers, 3 ; captains, 7 ; subalterns, 20 ; staff, 6 ; sergeants, 40 ; drummers, 15 ; rank and file, 762—Total of all ranks, 853. Of whom Ensign Ewen Kenedy was killed in action, and Captain S. Bevan and Lieutenant Alexander MacDonald died of their wounds. The other officers wounded were—Lieutenant-Colonel John Cameron,[1] Majors Jas. Mitchell and John MacPherson, Captains G. W. Holmes and Ronald MacDonald, Lieutenants William Fyfe, Donald MacPherson, John A. Durie, James John Chisholm, Robert Winchester, Donald M'Donald, James Ker Ross, George Gordon, John Grant ; Ensigns Thomas Mitchell and George Mitchell.

It was evidently difficult to make an exact return of the killed in action, as the dead had to be left to be buried by the French, and several corrections seem to have been made. Altogether I find the names of sixty-seven n.c. officers and men killed. In many cases the rank is torn off the original return, and these I have entered as privates.

Names of n.c. officers and soldiers killed in action, 25th

"which you yourself will be anxious to rectify in any further editions of that valuable work. I do not deem any apology necessary for requesting that you will have the goodness to admit this mistake. Should you, however, be under the impression that the 92nd was not principally composed of Scotchmen, I shall be happy to furnish you with ample proof" ; which he afterwards did, and the objectionable paragraph was left out in later editions ; and in "Battles of the Peninsula," note, page 300, Napier writes, "In my original work, misled by false information, I said the soldiers of the 92nd were all Irish, but their Colonel (M'Donald) afterwards gave me irrefragable proof by a list of names that they were Scotchmen."

The Monthly Return, dated Heights of Maya, 25th July 1813, gives the nationality of he n.c. officers and rank and file of the First Battalion as under :—

	Sergeants.	Drummers.	Corporals.	Privates.	Total.	Per Cent.
English	0	3	0	33	36	3·49
Scotch	58	13	44	810	925	90
Irish	0	1	3	64	68	6·60
	58	17	47	907	1029	

It has been often said that Highland regiments were composed of Irishmen, and generally with as little reason as on this occasion. As an instance, I remember when serving with the Gordon Highlanders in 1855, an old gentleman told me he had, when a boy, been visiting where the 92nd were quartered in Essex in 1805 ; that he particularly remembered the tall men of the Grenadier Company, adding, "They were mostly Irishmen, for I used to hear them talking in the Irish language." I find the nationality of the n.c. officers and men of the Grenadier Company, First Battalion 92nd Regiment (Gordon Highlanders), on January 1st, 1805, was :—English, 7 ; Scottish, 87 ; Irish, 3—Total, 97 ; and of the Scots, the great majority were from the northern counties, and many would naturally speak Scots Gaelic, which is sufficiently like Irish Gaelic to account for the mistake. A list of the officers of both battalions of the regiment in 1813 is given in Appendix VIII. It shows nearly all Scots, most of them Highlanders, and nearly half from Inverness-shire.

[1] Colonel Cameron was wounded in two places and his horse was killed under him.

July 1813; taken from non-effective return of 25th August 1813:—

Sergeant William Cattanach.
,, James Moore.
,, Alex. Gordon.
Corporal John Mackenzie.
,, William Raymond.
Private William Allison.
,, Jas. Allardyce.
,, Jas. Archer.
,, Jas. Burnet.
,, John Campbell.
,, John Campbell.
,, Jas. Campbell.
,, John Cameron.
,, Robt. Copland.
,, Donald Currie.
,, Willm. Cunningham.
,, Hugh Cunningham.
,, John Coutts.
,, Alex. Davidson.
,, Roderick Dunbar.
,, Andrew Dunlop.
,, Willm. Dorrit.
,, Alex. Elder.
,, Angus Fraser.
,, Charles Fraser.
,, John Gordon, 1st.
,, George Gordon.
,, George Gould.
,, Donald Gunn.
,, Jas. Glyn.
,, Robert Hardie.
,, Alexr. Kermack.
,, Jas. Leishman.
,, John M'Kay, 1st.
Private Alexr. M'Kay, 4th Co.
,, Alexr. M'Kay, 7th Co.
,, Alexr. Mackenzie.
,, Lewis Mackenzie.
,, Donald Mackenzie.
,, Jas. Mackenzie.
,, Donald M'Innes.
,, John M'Rae.
,, Dougald Macdougald.
,, Ewen Macpherson.
,, Alex. Macpherson.
,, John Macintyre.
,, Wm. M'Donald, 1st.
,, Wm. M'Donald, 4th.
,, Neil M'Aulay.
,, Peter M'Millan.
,, Willm. Mathieson.
,, Peter Matthew.
,, Alexr. Morrison.
,, John Petrie.
,, John Primrose.
,, Arthur Reid.
,, Donald Ross, 2nd.
,, Willm. Ross.
,, David Rennie.
,, Norman Stewart.
,, Jas. Smith.
,, George Thomson.
,, Matthew Trolley.
,, John Williamson.
,, Jas. Watson.
,, Jas. Wright.
,, Jas. Wilson.

Of whom fourteen are given as "supposed killed on July 25th."

In the morning state of July 26th, 1813, are entered as "sent wounded to the rear"—Field officers, 2; captains, 1; subalterns, 9; staff, 0; sergeants, 5; rank and file, 261.

It appears that Major MacPherson and three other wounded officers and several men were able to remain with the battalion. The others went to Vittoria, Bilbao, and Santander, and this journey of several days by ox-carts must have been a trying one. The names are given in the return of twenty men as taken prisoners between the 25th and 27th, so that the battalion lost in killed, wounded, and missing about 367 officers and men.[1]

Among the veterans who died at Maya was Private Norman Stewart, the best shot in the battalion, with which he had served in all its many fights. He was a favourite both with his officers and fellow soldiers, a great character, and, as one of his old comrades said, "had not as much English as would put the pot on the fire." When an officer complimented him on his soldier-like behaviour in his first engagement, implying, however, that he had hardly expected it, Norman replied, "Cha'n'eil fhios de'n claidheamh a bhios 's an truaill gus an tairnear e" (It is not known what sword is in the scabbard till it is drawn). He did not like his first musket, because it kicked, and because it was a "widow," for its former owner had died; but when he got a new one he called it his "wife," and woe betide the French skirmisher who came within reach of her kiss![2]

Though Lieutenant Fyfe was wounded, his life was saved by the bullet striking his watch, which, however, he carried till he died at a good old age, near Fochabers.

An instance is recorded by an officer commanding a company, of the good feeling which has always prevailed between officers and men of the Gordon Highlanders. Fatigued in body by their tremendous exertions on the 25th, followed by a long night march, and distressed in mind by the loss of their comrades, they halted on a hillside near Irueta. All were without shelter, but in the course of the day his company erected a beautiful hut, and on their inviting him to enter, he found a neatly-made table groaning under a load of soup, beefsteaks, young potatoes, and a bottle of brandy. He was naturally much gratified by such a mark of attention at such a time, "showing the dispositions of the men I had the honour to command."

[1] Cannon's "Historical Record" gives Ensign E. Kenedy killed, Lieutenant A. MacDonald (died), and 17 officers wounded; but Captain Bevan is mentioned in the original return as died of wounds, 21st August. The other casualties given by Cannon are—34 rank and file killed, 268 wounded, and 22 missing. Ensign Ewen Kenedy had lately been promoted from quartermaster-sergeant. He belonged to Moy, in Lochaber, and was the support of a widowed mother, to whom he left considerable savings. Lieutenant Alexander MacDonald was a son of Dalchosnie. Most of the killed had sums of from £2 to £12, besides prize-money, at their credit, and the address of their heir is generally given.

[2] See page 255.

CHAPTER XX

D'ERLON remained inactive at Maya on the 26th July ; but, stimulated by Soult, he advanced on the afternoon of the 27th by Elizondo, and in the evening the 92nd, with Hill's troops, retired. Each division of the allied army was ignorant of what had happened to the other ; alarming reports were spread ; the narrow glens and roads were blocked with the baggage, stores, and artillery of the Sixth and Seventh Divisions, which had preceded them, and by the fugitive families of the country people ; so that when it got dark the troops were compelled to halt on one of the lower heights of the pass of Lanz. At daybreak on the 28th they resumed their march, and at noon halted near La Zarza. Thinking themselves secure of a rest for the night, the beef was boiling in the kettles, and officers and men were washing and shaving the four days' growth on their chins. All was going on quietly, when suddenly the bugles sounded the " Fall in ! " In a moment the ground was covered with soup, the half-shaved soldiers stood laughing at each other in the ranks, and in ten minutes they were on the road to Pampeluna, near which the first battle of Sauroren was being fought, with great determination and loss on both sides. Here the French, 25,000 strong, under Soult, with great gallantry attacked Wellington in a strong position, which he successfully defended.

Being too late to take part in the battle of the 28th, the 92nd halted near Lizasso and bivouacked, and next day advanced about a mile nearer to Pampeluna. Both armies rested on the 29th without firing a shot, while the wandering divisions on both sides were being brought into the line. Wellington had vindicated his position with 16,000 combatants ; he had now, including the troops blockading Pampeluna, 50,000, of whom 20,000 were British. Hill occupied strong ground between Lizasso and Arestegui, and was well placed for retaking the offensive.

Soult had been reinforced by 18,000 men, but finding it impossible to penetrate to Pampeluna, he proposed to make such dispositions as might enable him to relieve San Sebastian. On the morning of the 30th July, Wellington attacked and drove back the enemy in the second battle of Sauroren ; and Soult, seeing his retreat endangered, determined to prevent his flank being turned by crushing Hill's troops with superior numbers. For that purpose he took d'Erlon's Corps, the same troops who fought at Maya, and the division of cavalry in support, not less than 20,000 sabres and bayonets, Hill having barely half that number.

When Soult's attack became obvious, the 92nd, commanded by Major John MacPherson, moved with the First Brigade to line the brow of a ridge facing the plain on which the enemy was forming his columns ; the Second Brigade was in support, and two brigades of Portuguese occupied a height on the right of the road. The 8th and Light Companies of the 92nd formed a guard to Sir Rowland Hill on a height to the left. A strong body of French infantry was pushed along the base of the hill with the view of turning Hill's left, and the First Brigade, screened by trees and brushwood, made a corresponding movement along the summit. When the French attempted to ascend at an easier part of the ridge, the 50th halted, the 71st extended in skirmishing order, and the 92nd was formed into two divisions ; that on the right as a sort of movable column to support where required, while the left division was pushed along the hill-top to watch the enemy on the left. One of the officers was sent down alone, with orders to give notice of the first appearance of the enemy. After advancing some way, their approach was announced to him by a rustling noise in the bushes, and he immediately gave the alarm ; he was fired at, but escaped. As soon as the Highlanders arrived, they at once attacked the French Light Infantry, and drove them back with great loss. After this our men continued to skirmish in extended order, till a battalion of grenadiers issued from a wood on their left, and with drums beating and shouts of "Vive l'Empereur," advanced to the charge. The Highland skirmishers were at once called in. "We had four small companies, while the French were from 500 to 600 strong." Thinking it best to meet them half-way, Captain Seton, who commanded, stepped in front of his little corps and, with his bonnet in the left hand and his sword in the other, called, "Ninety-second, follow me !" After proceeding about twenty paces he fell to the rear, as is usual on such occasions, and gave the word "Charge."

"Our lads moved forward with great spirit to measure bayonets with their opponents. But from such an unequal trial of strength we were most unexpectedly relieved by the 34th Regiment, who, coming in sight of us just as we were moving forward, gave three hearty cheers and joined us in our offensive movement. Being still superior in numbers, the enemy seemed for a little quite determined to wait our assault ; but somehow, when we arrived within thirty paces, they wheeled about and retired, hotly pursued by our two little corps. Their loss was considerable ; their commandant, a fine young man with two or three decorations, fell mortally wounded." [1]

The enemy, reinforced, again advanced. They were opposed with determined bravery, till increased numbers enabled them to outflank Hill's left, and his troops were reluctantly obliged to

[1] "Military Memoirs."

retire about a mile from the ridge to some high rocks, from which they hurled defiance at the French below.

The Second Brigade and other troops were also hotly engaged, and the Highlanders, from their rocky fortress, admired the courage of some of the Portuguese troops, who, though forced to retire fighting to a height two miles in rear, on being reinforced by a brigade of their countrymen, in turn became the assailants, and drove the French down the ridge at the point of the bayonet. Hill, having been joined by Campbell and Morillo, again offered battle; but Soult, having gained his object by turning Hill's left and securing a line of retreat, declined it, and thus ended the serious operations in the combat of Buenza, though an irregular fire was kept up till sunset. The Highlanders could see the bonfires kindled at night by the garrison of Pampeluna, rejoicing in the hope of relief.

In the two battles of the 30th July, the Allies lost in killed and wounded 1900 men, of whom nearly 1200 were Portuguese. On the French side the loss was enormous; two divisions were completely disorganised. Foy's Division was entirely separated from the main body; at the lowest computation, 2000 men had been killed and wounded, and 3000 prisoners were taken.

The Gordon Highlanders killed in action were:—Sergeant Jas. Allan, Corporal Robert Anderson, Privates John Brookes, William Dougald, Hugh Johnstone, David M'Intosh, Symon MacKenzie, Donald Maclean, Murdoch Ross, John Wright. Captain G. W. Holmes and twenty-six rank and file were wounded.[1]

An officer mentions that an order was issued early in the morning of the 30th for all men who could not keep up with their battalion to be sent to the baggage. He selected three from his company; two went with apparent goodwill, but William Dougald respectfully told him he would rather die than leave his comrades. He had been hit three times by spent bullets on the 25th, and though not much minded at the time, the wounds had become so inflamed by subsequent exertion that on the 30th he could scarcely drag his right leg after him. "I shall never forget the exertions he made to keep up with his companions, and the admirable manner in which he performed his duty in action till stretched a lifeless corpse on the heights of La Zarza."

John Brookes, one of the two who quitted the company agreeable to order, had also been struck by a musket ball on the 25th;

[1] Cannon's Record says nine rank and file were killed, and Captain G. W. Holmes and twenty-six rank and file wounded. The morning state, 31st July, gives one captain, twenty-six rank and file to rear wounded. The above ten names of killed are taken from the Regimental Return.

it had been turned aside by his leather-stock, but his throat had become so inflamed that by the 30th he could hardly speak intelligibly. The brave fellow having obeyed orders with apparent alacrity, his officer was astonished, on going into action, to see him only a few paces in rear on his way to rejoin his company, but had no time to take notice of his disobedience at the moment; and Brookes conducted himself with his usual spirit and gallantry, till another bullet struck him on the same place and killed him on the spot.[1] The third man, Hugh Johnstone (the same who saved his master's property at the expense of his own at Maya), had rejoined the company along with Brookes, and soon after was very severely wounded. He was carried to the rear, but subsequent movements placed him in the enemy's hands, where he remained without medical aid till the French retreated next day, when his master sent a party to carry him to La Zarza. They attended to him, but he was exhausted from loss of blood, and expired in his comrades' arms, with a smile on his countenance. "Such was the premature fate of as good a soldier and faithful servant as ever graced the ranks of the British army." No doubt examples of chivalrous bravery occurred in the other companies similar to those recorded by this officer in his Memoir.

When the 92nd retired from the ridge where they had been so long engaged, a Portuguese battalion was ordered to cover the retreat, but their conduct proved an exception to the generally conspicuous gallantry of their countrymen in this campaign. In the valley, between the ridge and the rocks to which the 92nd retired, there were some houses which should have been held by the Portuguese, but of which they allowed the French to get possession, taking the shortest road to safety. Enraged to see this post lost by the bad behaviour of his men, their colonel rode up to the standard-bearer, snatched the flag from his hands, and galloped to within one hundred yards of the houses, where he remained for a considerable time with the enemy's shot flying about his ears, while he waved the colour round his head to induce his men to follow.[2] The gallant young colonel (he was only captain in the British army) was a Highlander, he had both an uncle and a brother in the 92nd, and it was with the greatest difficulty their officers could prevent the Highlanders from breaking away to render their countryman the aid which his Portuguese refused; but the orders were peremptory, and soon they had the pleasure of congratulating the young commander on his hairbreadth escape from the danger in which his gallantry had placed him, for his cloak and body-clothes were pierced in several places by musket balls. His

[1] Brookes belonged to Old Deer, Aberdeenshire.

[2] The Portuguese troops were always led by British officers.

name was John MacDonald of Dalchosnie, and he many years afterwards commanded the Gordon Highlanders.

On the evening of the 29th, the officers of the First Brigade were enjoying that restful day in the shade of a wide-spreading tree, when an officer of the 50th, fresh from Lisbon, passed in all the smartness of new equipment and clean clothes. When he dismounted to report his arrival to the commanding officer, Colonel Fitzgerald of the 60th Rifles, then commanding the Light Companies of the Second Division, offered to bet twenty dollars "that the officer just arrived is either killed or a prisoner within twenty-four hours." The wager was taken and won, for the officer, horse and all, was in the hands of the French before the period named ; but Fitzgerald was never paid, for a few hours later he was himself led captive to France.[1]

Hill's force was now 15,000 men, and Wellington's dispositions were such that the French marshal was placed between two fires, his only line of retreat being by the Pass of Donna Maria. Towards it he began his march soon after midnight, giving the rear guard to d'Erlon, whose divisions, having been hitherto successful, were in good order. They occupied their ground opposite Hill's Corps till about eight o'clock on the morning of the 31st, thus giving time for the rest of Soult's army to be well on their way. Hill immediately followed, the pursuit being led by the First Brigade, Second Division. They came up with the rear guard about noon, when serious skirmishing took place in the woods. The main body of d'Erlon's troops had by this time gained the pass, the road to which leads up a steep rocky hill covered with trees and brushwood. On getting through the wood where they were first engaged, our troops found themselves within three hundred yards of the enemy crowded together on the road. Now was the time for artillery. One field-piece and one howitzer were coming up, but the officer was at a loss for a road through the difficult ground, when Mr Firth, the chaplain, who was as good a soldier as a preacher, and had been up with the leading troops, acted as guide to ground from which the guns could open on the enemy with effect. The men were amused at seeing the minister in this new character ; "Gude guide us," cried one of them, "see to the minister leading the artillery !" "I 'm sure *he 's* nae business runnin' himsel' into danger," said another. "Haud yer tongue, ye gowk," answered a third. "He 's the very man that should be there ; *he 's prepared.*"[2]

The first shell knocked down a number of men, and almost every shot took fatal effect, throwing the enemy's rear columns into

[1] "Military Memoirs." [2] *Ibid.*

confusion, so that they scattered through the woods on each side. The infantry was now called on ; the 50th ascended the hill to the left of the road ; the 71st, in extended order, skirmished with the French along the hillside between the right of the 50th and the road ; while the 92nd took the high road, and attacked the main body of the enemy in their front ; the Second and Fourth Brigades, Second Division, were in support, and the Seventh Division, under the Earl of Dalhousie, moved by a parallel road to attack the enemy's left flank. Their skirmishers being driven in, the main body, several thousand strong, faced about and made good battle, opening on our troops with a heavy fire of musketry. The 92nd, under Major MacPherson, charged in defiance of shot and numbers, but were repulsed with loss. A captain was taken prisoner in the *mêlée* and disarmed, but he knocked down with his fist the French officer who took him, ran off, and succeeded in rejoining his company, being, however, shot through the arm as he did so. Again the 92nd charged with a like result. A third time they led the charge, this time joined by the Second Brigade, and the enemy was at last driven over the Pass. The loss of the Allies was under 400 men, that of the French is unknown, and was probably less. Lieut.-General Sir Rowland Hill, under whose eye the battalion was engaged, bestowed most flattering encomiums on its gallantry, " and the enemy's defeat was particularly ascribed to the persevering bravery individually displayed in the ranks of the 92nd."[1]

An officer relates, " At the close of this day's engagement, I could only muster thirteen privates out of eighty-two n.c. officers and privates which I carried into action six days before " ; and he was himself wounded.

Major MacPherson being severely wounded (his horse was killed under him), the command of the battalion devolved upon Captain Seton (who, though it is not mentioned in Cannon's Record, is entered in the Regimental Return as " slightly wounded "). The other officers wounded were Captains Jas. Lee and Dugald Campbell ; Lieutenant Jas. Hope and Ensign Thomas Mitchell. Cannon's Record gives ten rank and file killed and 69 wounded, but I can only find the names of the following eight n.c. officers and men killed, viz. Sergeant Charles MacGregor,[2] Privates Angus Campbell, Alexr. Gow, Ewen Gordon, John Murray (1st), Roderick Ross, Jas. Shaw, Alexr. Watt. The morning state of August 1st gives 3 sergeants, 76 rank and file, " sent to the rear wounded."

Besides the three officers already mentioned, the following

[1] Cannon's Record.

[2] The sergeant who lost his wife and children in the retreat to Corunna. " Thus," says Sergeant Robertson, " putting an end to the whole family."

n.c. officers and privates, wounded on the 25th, 30th, and 31st, died of their wounds between July 27th and August 13th inclusive, viz. Sergeant John Cumming; Corporals Willm. Leighton and Willm. Craig; Privates Alexr. Clarke, Graham Cattanach, Willm. Bain, Willm. Davidson, Thos. Gowrie, Jonathan M'Donald, John MacPherson, and Jas. Strachan. Though a large proportion of the wounded afterwards rejoined, the battalion did not recover during the war from the appalling losses of seasoned soldiers it sustained from the 25th to the 31st July 1813.

D'Erlon was not routed, but retreated slowly and in good order, and pursuit was prevented by a thick fog. That evening a picket of the Light Company was taken rather cleverly. A body of troops approached, and when challenged, answered "Españoles." The Highlanders, completely deceived, were surrounded before they discovered their mistake, disarmed, and carried off, each guarded by two men with fixed bayonets. They had to ford a river, just above a place where it rushed down, dark and deep, between high rocks wooded on either hand; as they waded across some of the men agreed to chance the bayonets and dive for it, and, while their guards were taken up with their own footing, these hardy Highlanders plunged down the rapid, were lost to view among the trees, and regained their battalion. The narrator, J. Ferguson, however, said he did not like the look of the bayonets at his side well enough to risk getting one in his inside, and with two or three others was kept prisoner.

No two generals ever commanded both the respect and affection of their troops more than Sir Rowland Hill, and the Hon. Sir William Stewart who commanded the Second Division of Hill's Corps, and who followed his superior's example both in professional enthusiasm and in consideration for the comfort of his men. To see them contented and happy was the delight of both. Sir William, wounded on the 25th, had been obliged to leave his command; but on hearing of the battle of the 30th, he had a pillow fastened to his saddle so as to keep the wounded leg from coming in contact with the horse's side, and rejoined his troops before they came up with the enemy on the 31st. The enthusiastic cheers of the soldiers welcomed him back, but he was soon again severely wounded, and compelled once more to go to the rear. On his rejoining the division a few weeks afterwards, a party of the Gordon Highlanders placed themselves on the road by which he had to pass. When he arrived, one of them stepped forward, and saluting, said, "Oh, General, ye maun drink wi' us." "With all my heart, my men," replied the general to this unexpected request, and, taking the cup, drank to them, expressing his gratification at their attention, and at finding them in such good health and spirits. He then rode off to the camp, where the rest of the regiment were standing in front of their tents, and received their general and friend with joyous acclamations, every bonnet flying into the air as high as its owner could heave it. Gratifying as the demonstration no doubt was to Sir William, it nearly deprived him of an aide-de-camp, whose horse, unprepared for this Highland welcome, whisked to the right about, and galloped down the hill, to the amusement of the spectators, but the hazard of its rider's neck.[1]

The evening of the 31st was spent in carrying in the wounded and in the sad duty of burying the dead, a mournful time for the remaining Gordons, who had lately lost so many with whom they had been united by bonds of closest friendship. But though, says Sergeant Robertson, they wished these scenes of carnage might be ended, they were none the less ready to do their duty when, on the morning of the 1st of August, they once more entered the valley of the Bastan and encamped at Elizondo, where they were visited by one of those terrific thunderstorms so common in that mountainous region. On the 2nd, the First Brigade took possession of its old ground on the heights of Maya, and the opposing armies reoccupied much the same position held by them on the 25th of July. One can imagine the feelings of the British, which I have heard expressed by some who witnessed the heartrending spectacle, when they found the unburied corpses of many of their comrades killed on the 25th, and left by the French a prey to the eagles and vultures. During those nine days several combats had taken place besides those in which the 92nd were engaged ; they were all fiercely contested. The encounters were often at the point of the bayonet, charges up and down the rocky mountain sides gave constant opportunity for individual prowess, while, with the exception of the 29th of July, the troops were in constant motion between the fighting, exposed in the valleys to the excessive heat of a Spanish midsummer which made the cold fogs and heavy rains of the high mountain passes the more trying.

In memory of this short but bloody campaign, the 92nd afterwards received the royal authority to bear the word "Pyrenees" on their colours, and when, after the lapse of many years, the Peninsular medal was given to officers and men, all the battles were represented by one clasp. Gold medals were, however, conferred at the time on Major MacPherson, who was promoted

1 "Military Memoirs."

NOTE.—Sir William Stewart, on being personally thanked by Parliament for his conduct in the Pyrenees, said—"Supported as I was by my gallant friend on my right (Major-general Pringle), by such corps as the 92nd Highlanders and the 50th British Infantry, I should have been without excuse if a less firm stand had been made on the positions of the Pyrenees than was made."

Brevet Lieut.-Colonel, and on Captain Seton,[1] who now commanded the battalion, which was reduced to about 300 rank and file.[2]

On the 8th of August the Second Division was relieved at Maya by the Sixth, and the 92nd encamped that evening at Los Alduides on the French frontier. Next day, after a long stiff pull, they encamped on one of the tremendous ridges near Roncesvalles on which the division on the right of the British line had been engaged on the 25th of July. During August the battalion was engaged in making blockhouses and breastworks on the commanding points of the position.

It may be here mentioned that in the monthly return for July, one private had been discharged by authority of the Prince Regent on account of his being promoted in the 87th Regiment, name not given ; and in September, Volunteer John Clarke, 92nd, was promoted ensign in the regiment.

At this time a colour-sergeant was appointed by War Office circular to each company of infantry, a rank which had not hitherto existed.

A few men kept rejoining from hospital or duty in Spain. On September 13th, 8 sergeants joined from the Second Battalion, and 2 sergeants and 22 rank and file on the 25th.

Meanwhile, on the 31st of August, San Sebastian was taken, where Lieutenant John Ross of the 92nd, serving with a Portuguese regiment, was killed ; and later, Lieutenant W. Fyfe was promoted captain "*vice* Arnot, dead of his wounds," but where wounded does not appear.

At Roncesvalles the British outposts were near those of the enemy, but although kept continually on the *qui vive*, the opponents exchanged many little courtesies. The French would offer wine or cigars, while they valued more than any other attention a London or Edinburgh newspaper, for their own were so unreliable that "He lies like a *Moniteur*" or "He lies like a bulletin" were common sayings with them. And the French General Foy actually sent an aide-de-camp with a flag of truce asking for a London newspaper with the details of the late actions.

Pampeluna capitulated on the 31st of October, which not only lessened Soult's inducements to attack, but added the troops employed in the blockade to Wellington's effective field force. The 92nd had moved twice during October for a few days to the heights of Don Carlos, returning, however, to Roncesvalles, where they now had leisure for amusement. This they found in weekly horse and

[1] In Cannon's Record and other places the name is spelt Seaton.

[2] Morning state, August 5th :—Field officers, 0 ; captains, 4 ; lieutenants, 7 ; ensigns, 4 ; staff, 3. Sergeants—present, 24 ; sick absent, 14 ; on command, 10. Drummers—present, 13 ; sick absent, 1 ; on command, 1. Rank and file—present for duty, 305 ; sick present, 4 ; sick absent, 463 ; on command, 51 ; prisoners of war, 39.

foot races, bull-fights, and in the wine-houses of the village of Roncesvalles. They made beds of branches covered with ferns and grass to keep them off the damp ground, where "we reposed as comfortably as on the best bed in England." When the weather at last became very bad, the troops were recalled from the heights, save the outlying pickets, and an in-lying picket of 500 men. These suffered greatly from a snowstorm, followed by severe frost on the 28th. "Snow fell to a greater depth than I had ever seen it in Scotland": the drifts were twelve feet deep, some of the outlying pickets were covered and had to be dug out, says an officer of the 92nd; and Sergeant Robertson relates that a picket of the 57th perished, and were only discovered by the top of the sergeant's halberd appearing above the snow. He also mentions that it was remarked by the general officers "that though the 92nd wore the Highland dress, none of them were frost-bitten." The First Brigade now marched into the village of Roncesvalles, where General Byng (afterwards Earl of Strafford) gave up the greater part of his house to the officers of the 92nd, and invited them all to dine with him.

The changed aspect of affairs now enabled Wellington to carry the war into the enemy's country. He had, by the famous passage of the Bidassoa at Fuentarabia and Hendaye, invaded French territory, and he now prepared to attack their fortified positions on the Nivelle from St Jean de Luz on the sea, to Ainhoa below the Pass of Maya, a distance of about twelve miles. Accordingly, the Second, Third, and Fourth Brigades, Second Division, marched from Roncesvalles on the 6th of November. The enemy on that day attacked a Spanish picket a mile in front of the village; the Spaniards defended themselves bravely, but were nearly all killed or wounded. The advanced post being withdrawn from the right of the position to the blockhouse between it and Roncesvalles, the enemy pushed forward a strong reconnoitring party on the afternoon of the 7th, and attacked the blockhouse occupied by Captain Holmes and a picket of the 92nd. On the alarm, the First Brigade came to their assistance, General Mina and his Spaniards being the first to draw the attentions of the French, who retired hotly pursued. The 92nd had a private wounded, and two taken prisoners in the skirmish. Next day at 4 p.m. they moved with the First Brigade, and after a march in the dark over rough mountain roads, arrived in the village of Los Alduides after midnight. At daybreak on the 9th they started to join the division in rear of Maya, where the whole country seemed covered with men, horses, cannon, and baggage. Here they rested till 9 p.m., when the advance was resumed; they passed their comrades' graves in the memorable pass of Maya, and stumbling through rough grass, stones, and heather, hardly able to keep themselves awake, they reached the Nivelle at 6.30 a.m. on the 10th November, and the brigade was in the act of crossing that little river which here forms the boundary between France and Spain, when the sound of artillery on the left of the army announced the commencement of the great battle. Here they were allowed to rest before being called into action.

On this occasion Wellington had the superiority in numbers, and he used it with great ability against Soult's very strong position, which extended along a range of heights from near St Jean de Luz on his right to Ainhoa on his left, which heights were defended by entrenchments and redoubts. The First and Fifth Divisions engaged the enemy's right, the Third, Fourth, Sixth, and Light Divisions were directed against his centre, and the Second Division, Hamilton's Portuguese and Morillo's Spaniards, attacked the redoubts on his left, the Sixth Division being also sent up the right bank of the Nivelle to their assistance. There were two strong redoubts on the French extreme left beyond Ainhoa, occupied by a considerable body of infantry ; lower down, and behind Ainhoa, were three redoubts, each mounting several guns of large calibre. It was against this part of the French line that the 92nd, again commanded by Lieut.-Colonel Cameron, was led. The battle on the British centre and left had been hid from their view, but now the Highlanders could see, and took part in, the interesting operations of 26,000 men, under Generals Stewart, Clinton, and Hamilton, pending the assault of the heights.

The country was difficult, and it was eleven o'clock before they got within cannon shot of the redoubts : but Clinton and Hamilton having turned the right of the enemy's position at Ainhoa, the French immediately opposed to the 92nd made little resistance, running out of the redoubts in confusion ; many prisoners were taken, and their own guns were turned to give them a parting salute. The Highlanders took possession of the Frenchmen's huts, where they passed the night very comfortably, being employed during the afternoon in collecting and attending to the enemy's wounded.

When, in after years, the regiments engaged were given authority to bear *Nivelle* on their colours and appointments, the 92nd and other regiments of the First Brigade were not included in the grant. It is true that the principal action was fought, and the great loss sustained, in the centre and on a space of about seven miles ; but the action of Hill's Corps and the Second Division, in which was the 92nd, had an important effect on the success of the combinations, which failed in no point,[1] and the fact that the battalion captured the redoubt with the loss of only one man

[1] Napier, Vol. VI. pp. 343-344.

La Rhune Mountain, from near St Jean de Luz, showing the Field of Battle of the Nivelle.

wounded, in no way detracts from the credit of their bold attack.

The Allies had about 90,000 men ; the French 79,000, of whom, however, Foy's Division acted separately. The former had 2690 officers and men killed and wounded, and Generals Knight and Byng were wounded. The loss of the French was 4265 officers and men, of whom about 1400 were prisoners, and one general was killed ; the field magazines and fifty-one pieces of artillery were taken.

On the day following the battle of the Nivelle, the army advanced in order of battle. Sir John Hope, on the left, marched by St Jean de Luz on Bidart ; Marshal Beresford, in the centre, moved upon Arbonne, and General Hill, communicating by his right with Morillo, who was on the rocks of Mondarain, brought his left forward into communication with Beresford, and with his centre took possession of Suraide and Espelette, facing towards Cambo. The time taken by these movements enabled Soult to rally his army upon a line of fortified camps, his right resting on the coast at Bidart, and the left at Ustaritz on the Nive. Foy had driven back Mina's Spaniards on the right of Hill's troops on the 10th, and taken a quantity of baggage, but finding that d'Erlon was giving way before Hill, he retreated during the night, and on the 11th reached Cambo and Ustaritz, ready to defend them against Hill.

About 10 a.m. on the 11th, the 92nd, with the Second Division, quitted the heights from which they had driven the enemy the preceding day, but after advancing three or four miles they halted. The rain, which fell in torrents, rendered the country roads impassable for artillery, and no attempt was made to drive the enemy further back. They, however, retired at sunset. The 92nd encamped on a heathery hill near Espelette, where they spent a most uncomfortable night, for even cooking was difficult, as the wood was too wet to burn. On the 12th, Hill tried to pass the fords of the Nive, and also made a demonstration against Cambo, but the floods rendered the fords impassable, and both places were successfully defended by Foy. In these operations the 92nd, with the First Brigade, moved by Espelette along the road to Cambo, drove back the enemy's pickets to within a short distance of his works, and then ascended a height which overlooked the town and its defences from the right. Sir Rowland reconnoitred, and finding the enemy better prepared than he expected, contented himself with driving in their light troops and cannonading the garrison. The 92nd passed the night in the neighbourhood. Their baggage had not yet arrived, and the fires they made roasted one side of a man's body while the other was iced by the winter wind. The battalion had one man wounded this day at Espelette.

The baggage arrived on the 13th, but the rain ceased not from the evening of the 12th to the 16th, when the enemy retired over the Nive, destroying two arches of the bridge to prevent his being followed; and the battalion marched into Cambo, after a skirmish with the enemy's rear guard.[1] It appeared that the French had told their countrymen that the British would murder them, and they found no one in Cambo but one dragoon, who had been left behind, a bedridden old man and his wife, and a pig—for the head of the latter a Highland officer paid four dollars. The soldiers did not regret the absence of the inhabitants, as they had left their beds, which were a treat to men, many of whom had only been under the roof of a house thirteen nights out of the last one hundred and eighty-one days.[2]

The order books and morning states of this period are lost, but it appears from monthly returns that a private was discharged, November 25th, by authority of the Prince Regent, on account of being promoted to ensign in the York Light Infantry Volunteers, name not given. A draft of 100 rank and file, from the 2nd Battalion in Scotland, joined at Cambo on November 27th, "the greater portion of them fine-looking stout young fellows, and proved a great acquisition." At the same time four officers, who were doing duty with the First Battalion, but who belonged to the Second, were ordered, much to their regret, to join it in Glasgow.

Notwithstanding the hardships, the British army was at this time exceptionally healthy, there being hardly any sick except the wounded, and of these numbers were rejoining the battalion; one private joined at Cambo "from reported dead." The troops felt a just pride in contrasting their present position, quartered as victors in a French town, with their situation at the same date in the previous year, when retreating before the French into Portugal.

As soon as his army entered France, Wellington's first care was to establish the system of regular payment for supplies, which had so greatly contributed to his previous success. He issued a proclamation to the allied troops, in which, after recounting the miseries which the exactions of the French troops had brought upon Spain and upon themselves, he added that it would be unworthy to retaliate on the innocent inhabitants of France. "The officers and soldiers of the army must recollect that their nations

[1] Sergeant Robertson says two of the Light Company were killed on this occasion, but I do not find them in the existing returns.

[2] "Military Memoirs": the author of which was one of them.

are at war with France, solely because the ruler of the French nation will not allow them to be at peace, and is desirous of forcing them to submit to his yoke." A proclamation to the same effect was published among the inhabitants, who at first would not believe in such moderation. The Spanish and Portuguese soldiers had in many cases suffered great wrongs, which they

French Peasants of the Pyrenees. *(From original sketches.)*

revenged on the French people, but these acts were punished with death, and when Mina's Spaniards continued to murder and pillage, Wellington sent them back into Spain, retaining only those commanded by Morillo. The confidence of the peasantry became so great that they brought their produce to the British army as being the best market. As an old soldier told me, "Wellington was awfu' agin' plunder ; I 've seen the son of a bailie o' the toun o' Montrose get 300 lashes for stealing a hen ! But he was richt, for a plundering army is whiles fu' an' whiles fasting."

From the 16th of November to the 8th of December the two armies remained in cantonments. The floods had rendered both roads and rivers impassable, and the troops at Cambo interchanged civilities with those on the opposite bank of the river ; the French bands played, and conscripts were drilled within sight and hearing. Newspapers were constantly exchanged, till intercourse was interrupted by active operations. One exception occurred to the general politeness of the Frenchmen. A lieutenant of the 92nd was on picket on the broken bridge, and, as he spoke French fluently, thought he would beguile the time by a chat with the officer opposite, and desired the sentry to tell him so, with his compliments; but the Frenchman, instead of agreeing to the proposal, threatened with violent language to have him shot, and when the discomfited Highlander walked back to his picket-house, a ball struck the door as he entered ; but in general a French officer, if he considered that one of his opponents was on forbidden ground, would courteously salute him, and explain that he must retire.

Quartermaster-sergeant M'Combie, 92nd, in his Journal, refers to the absence of personal ill-feeling between the British and the French, and mentions one occasion on which they met while foraging and helped each other in the most friendly manner. He also says of a French family on whom he was billeted, "We were canty an' crouse thegither." Lieut. Gleig of the 68th (afterwards chaplain-general of the army) mentions that once when the advanced sentries were separated only by a hedge, and the French were about to retire, the British soldiers helped them to put on their packs, and on another occasion the field officer, going his rounds at night, found the British and French n.c. officers and men drinking together.[1]

[1] "The Subaltern."

CHAPTER XXI

THE rivers having subsided, the army was put in motion, and along with Hill's Corps, the 92nd under Lieut.-Colonel Cameron left their quarters in Cambo on the 8th of December, and forded the Nive under cover of artillery fire. The current was still strong ; some men were carried off their feet, and one, a very short man, William M'Kenzie, would have been drowned had not Colonel Cameron ridden in and saved him ; but the enemy were surprised and offered but slight opposition. General Hill, having won the passage of the Nive, placed the brigade to which the 92nd belonged at Urcaray, to cover the bridge of Cambo, and support the cavalry which were scouring the neighbouring roads ; with the rest of his troops he marched to the heights of Lormenthoa, where the Sixth Division joined him. It was now one o'clock. Soult had extended his line to bar the road to Bayonne and offered battle, but though a heavy skirmish took place along the front, no general attack was made, the state of the roads having retarded the rear of Hill's Column. On this day the 92nd had one man wounded. The battalion bivouacked in a wood ; officers and soldiers, having been wet to the middle in the river, rose cold and stiff on the morning of the 9th of December, and were put into the neighbouring farm-houses, where the inmates did everything in their power to make their unbidden guests comfortable. They remained two days with these kind people, who live in much the same style as small farmers in Scotland. The Highlanders noticed that the men wore the same flat blue bonnet then general at home (Scotland has given it up, but in this district of France it is still universal). It is evident that the Gordons had ingratiated themselves with their entertainers, whose children told me that they had often heard their parents speak of the warlike dress and of the mild manners and good conduct of " Les Ecossais."

On the 10th, 11th, and 12th, the French attacked the British on the other side (the left) of the river, but were repulsed on each occasion after very severe fighting. On the 11th the 92nd advanced to Petite Moguerre, a village near Bayonne, situated between the rivers Nive and Adour. Wellington, feeling the want of numbers, brought back the Spaniards who had been sent to the rear, at the same time taking steps to prevent their plundering ; and Soult, finding he could make no impression on Wellington's left, marched with seven divisions to fall upon Hill on the right. That general, who had observed a movement of the enemy on the 12th, took a position on a front of about two miles. His left, composed of the 28th, 34th, and 39th Regiments, under General Pringle, occupied a wooded hill at the Château of La Ralde near Villefranque. It covered the lately constructed pontoon bridge over the Nive, but it was separated by a stream forming two ponds and a morass from the centre, which was placed on both sides of the high road near the hamlet of St Pierre,[1] and occupied a crescent-shaped height having thick hedges on the right side, one of which was almost impassable for 100 yards. Here Ashworth's Portuguese and Barnes' British Brigade were posted, the 71st on the left, the 50th in the centre, the 92nd on the right. The Portuguese were in front of St Pierre, their skirmishers in a small wood covering their right ; twelve guns under Colonels Ross and Tulloch were in front of the centre looking down the high road, and half a mile in rear Le Cor's Portuguese Division with two guns was held in reserve. The right under Byng consisted of the 3rd, 57th, 31st, and 66th Regiments. The 3rd was on a height running parallel with the Adour in advance of Vieux Moguerre, where it could only be approached by crossing the lower part of a narrow swampy valley which separates Moguerre from the heights of St Pierre. The upper part of the valley was held by the remainder of Byng's Brigade, and his post was covered by a large millpond which nearly filled the valley. A mile in front of St Pierre was a range of heights belonging to the French, but the basin between was open, and commanded by the fire of the Allies. The country was too heavy and enclosed for the action of cavalry. The enemy could only approach on a narrow front and by the high road, till within cannon shot, when one by-road branched off to Pringle's position, the other to where the 3rd Regiment was posted. A flood on the night of the 12th had carried away the bridge over the Nive ; it was soon restored, but on the morning of the 13th, Hill was completely cut off from the rest of the army ; and while seven French divisions, 35,000 strong, with twenty-two guns, approached him in front, an eighth under General Paris, and Pierre Soult's Division of cavalry menaced him in rear. To meet the enemy in front he had less than 14,000 officers and men with fourteen guns (for the Sixth Division was now on the left of the Nive), and there were only 4000 Spaniards, with Vivian's Brigade of cavalry, to protect his rear. It was a situation to try the nerve of the bravest general.

On the morning of the 13th of December, the 92nd marched from Petite Moguerre, and on arriving at the high road in rear of St Pierre, were ordered to turn to the right and halt at that hamlet. Soult had formed his order of battle in front of Bayonne under cover of a thick mist. D'Erlon, having d'Armagnac's, Abbé's and Darican's Divisions, Sharré's cavalry and twenty-two guns, marched in front, followed by Foy and Maransin ; the remainder

[1] Not St Pierre d'Irube, which is close to Bayonne, but the hamlet of St Pierre, properly called Lostenia, over two miles from the city.

of d'Erlon's command was in reserve. This formidable array was sometimes shrouded by the mist, sometimes dimly seen by the dauntless little army of Allies who waited their attack. At 8.30 the sun burst forth, just as Soult's light troops pushed back the British pickets in the centre, and the sparkling fire of the skirmishers crept up the hills on either flank, while nearly forty pieces of artillery shook the banks of the Nive and the Adour. Darican directed his division against Pringle ; d'Armagnac was ordered to force Byng's Brigade ; Abbé assailed the centre at St Pierre. Sir William Stewart commanded the troops stationed in front. Sir Rowland Hill took his station on Mount Horlope in rear of St Pierre, from whence he could see and direct the movements of all.

Abbé pushed his attack with great violence, and gained ground so rapidly with his light troops on the left of Ashworth's Portuguese, that Stewart sent the 71st and two guns to aid the latter. The French also won the small wood on Ashworth's right, and half the 50th was detached to that quarter. The wood was retaken and Stewart's flank secured, but his centre was much weakened, and the French artillery fire was concentrated on it. Abbé then pushed on a column of attack with such power that, in spite of musketry on his flanks and artillery in his front, he gained the top of the position, and drove back the remaining Portuguese and that half of the 50th which had remained in reserve. The British were now in imminent danger, when General Barnes brought forward the 92nd.

As the battalion cleared the houses of St Pierre, one of the pipers was killed at its head. The right wing quickly extended on the moor to the left ; the French skirmishers fell back before them, while the left wing, led by Colonel Cameron, charged down the road on the two regiments composing the column. "The charge," says Napier, "was rough and pushed home ; the French mass wavered and gave way." The Highlanders pursued and took many prisoners. Abbé immediately replaced the beaten column with fresh troops, and Soult, redoubling the play of his heavy guns, sent forward a battery of horse artillery, which opened its fire at close range with destructive activity. Cannonade and musketry rolled like one long peal of thunder, and the second French column advanced with admirable steadiness, regardless of their loss by Ross's guns. The Highlanders, unable to resist this accumulation of foes, were borne back, fighting desperately hand to hand, and even charging again and again with most determined fortitude and audacity, till General Barnes ordered them to retire,[1] while all the time the pibroch "Cogadh na sith"[2] rang in

[1] "General Barnes had just sent me word not to advance further."—*Letter from Colonel Cameron.*

[2] "War or Peace."

their ears.[1] Slowly, but with broken ranks, they regained their position behind St Pierre. The Portuguese guns, their British commanding officer having fallen, then limbered up and retired; and the French skirmishers reached the impenetrable hedge in front of Ashworth's right, when Barnes, seeing that hard fighting was the only chance of saving the day, made the Portuguese gunners resume their fire. The wing of the 50th and the Caçadores gallantly held the wood on the right, but Barnes was soon wounded, the greater part of his and of General Stewart's staff were disabled, and the matter seemed desperate. The light troops were overpowered by numbers and driven in; the gunners were falling at their guns, Ashworth's line crumbled before the fire; the ground was strewn with dead in front, and the wounded crawling to the rear were many. If their skirmishers could have penetrated the thick hedge in front of the Portuguese, the success of the French would have been certain at this point, for the main column of attack still steadily advanced up the causeway, and a second column on its right was already victorious, because the colonel of the 71st had withdrawn that gallant regiment, and abandoned the Portuguese. Pringle was bravely holding the hill of Villefranque against Darican's numbers, but on the extreme right, the colonel of the 3rd Regiment had also abandoned his strong post to d'Armagnac,[2] whose leading brigade was rapidly turning Byng's other regiments; and now Foy's and Maransin's Divisions were coming up to Abbé's support, at the very moment when the troops opposed to him were deprived of their reserve.

For Hill, with admirable decision, when he beheld the retreat of the 3rd and 71st Regiments, galloped down from his eminence, leaped his horse over the fence of the road, met the latter regiment indignant at being withdrawn from the fight, and turned them back delighted to renew the combat; then leading in person one brigade of Le Cor's Reserve to the same quarter, he sent the other against d'Armagnac on the hill by Vieux Moguerre. Thus at the decisive moment the French reserve was augmented, and that of the Allies was thrown into action as a last resource. Meanwhile Ashworth's Caçadores and the wing of the 50th held the little wood with unflinching courage, never doubting that their tried comrades of the Gordon Highlanders would soon return to their assistance; and their confidence was not misplaced, for as soon as the 92nd had time to re-form its shattered ranks, its gallant colonel, on foot—

[1] A letter from Archibald Campbell, Esq., suggests that Sir John Sinclair (President) "ought to communicate to the Highland Society the fact that two out of the three pipers of the 92nd Regiment were killed while playing 'Cogag na shee' to encourage their comrades"; as one fell another took it up; and that this "should be made known all over the Highlands." Mr Campbell was informed of this by a letter from Colonel Cameron.

[2] The colonels commanding the 3rd and 71st were both dismissed the service.—Napier.

THE CENTRE OF HILL'S POSITION FROM THE BAYONNE SIDE

Showing the Road down which the 92nd charged, and the Coppice Wood defended by the 50th and Portuguese

BATTLE
OF
ST PIERRE
December 13th
1813.

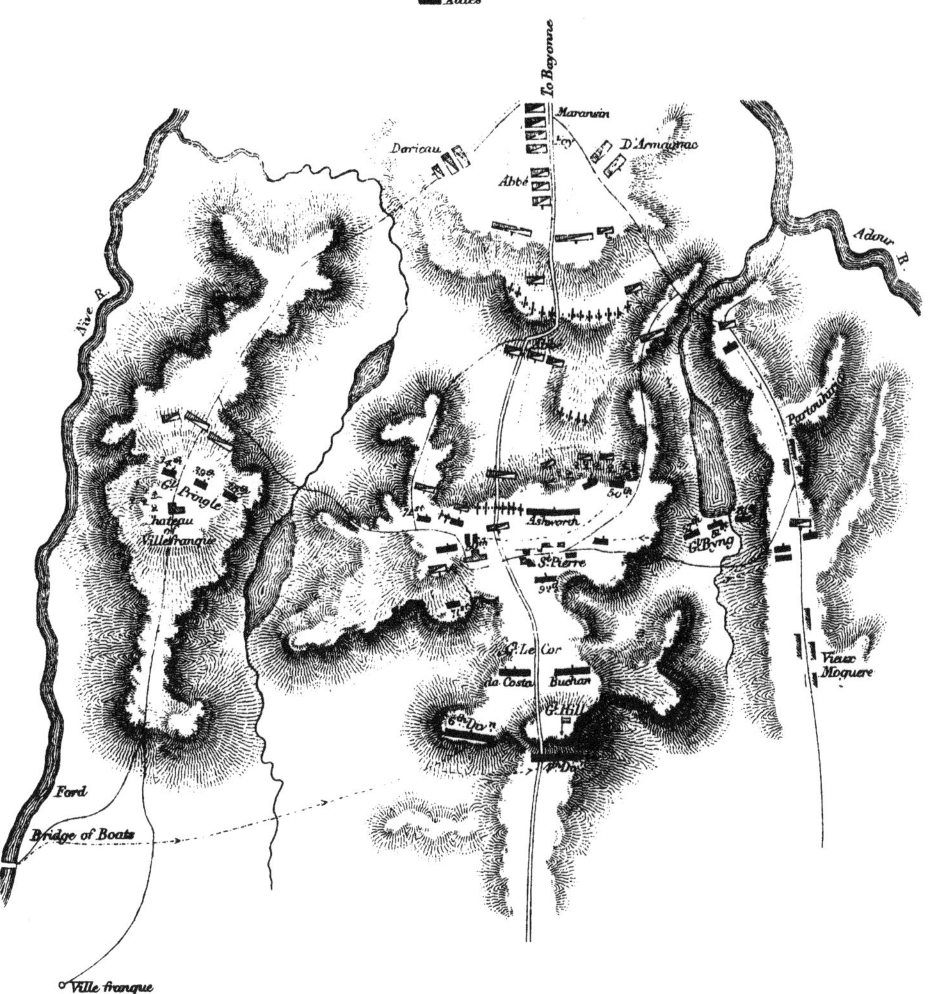

W. & A. K. Johnston, Limited, Edinburgh

for his horse had been killed under him—again led the battalion down the road, with colours flying and bagpipes playing, "resolved to give the shock to whatever stood in the way."[1] At this sight the British skirmishers on the flanks suddenly changed from retreat to attack, rushed forward and drove the enemy back on each side; yet the battle seemed hopeless; Ashworth was badly wounded, his line was shattered to atoms, and Barnes, who had not quitted the field for his former hurt, was now shot through the body. The 92nd was but a very small body compared with the mass in its front, for some of their companies were extended as light infantry,[2] and the French soldiers seemed willing enough to close with the bayonet; but an officer riding at their head suddenly turned his horse, waved his sword, and appeared to order a retreat; then they faced about and retired across the valley in good order, scarcely pursued by the victorious Highlanders, so exhausted were they by their several desperate encounters. The enemy's retreat was produced by the bold advance of the 92nd, and the returning rush of the skirmishers thus encouraged; but the 71st had returned with such alacrity to the combat, and were so well aided by Le Cor's Portuguese, Generals Hill and Stewart leading them in person, that the French were overthrown there also, at the same time that the 92nd came down the causeway in such gallant style. These deeds were witnessed by the French at Villefranque, where Pringle had given them enough to do, and now, disheartened, they fell back in confusion.

Meanwhile the Portuguese, detached by Hill to recover the Moguerre ridge, had ascended under heavy fire from Soult's guns, rallied the 3rd Regiment in time to stop the first brigade of French, who had passed Byng's flank; and while the fire of musketry and cannon continued, the contending generals restored their lines of battle. It was now noon. Soult had still Foy's and Maransin's Divisions with which to renew the fight. The ranks of the Allies were wasted with fire, nearly all the staff had been killed or wounded, and three generals had been obliged by wounds to quit the field. In this crisis Hill drew the 57th from Byng's position to reinforce his centre; at the same time, the bridge at Villefranque having been restored, the Sixth Division appeared on the mount from which Hill had descended to rally the 71st. It was followed by the Fourth Division, and that by the brigades of the Third Division; with the Sixth came Wellington, who had hurried from Barrouilhet when the first sound of cannon reached him, but only arrived to witness the close of the battle. Taking in the situation at a glance,

[1] Napier.

[2] Two companies of the 92nd had also been detached before noon to support the troops in the left centre.

he grasped Sir Rowland's hand, exclaiming, "The day is your own." Hill's day of glory was indeed complete; he had played the part of both general and soldier, rallying the 71st and leading the Reserve, "trusting," says Napier, "meanwhile, with a noble and well-placed confidence to the courage of the 92nd and 50th to sustain the fight at St Pierre. He knew, indeed, that the Sixth Division was close at hand, and that the battle might be fought over again; but, like a thorough soldier, he resolved to win his own fight with his own troops if he could. He did so after a manner that in less eventful times would have rendered him the hero of a nation."

Colonel Cameron, in a letter dated 14th January 1814, says—"General Barnes' Brigade fought without support for eight hours. Five companies of the 92nd, with General Barnes and myself at their head, charged three different columns of the enemy in succession; no sooner had we driven one before us than they were relieved by a fresh one." [1]

Napier, in speaking of Colonel Cameron and his regiment at the critical moment when they were forced back to St Pierre, says—"How desperately did the 50th and Portuguese fight to give time for the 92nd to rally and re-form behind St Pierre; how gloriously did that regiment come forth again to charge, with their colours flying and their national music playing as if going to a review. This was to understand war. The man who in that moment and immediately after a repulse thought of such military pomp was by nature a soldier."

Wellington in his dispatch, writing of the First Brigade, says, "They were particularly engaged with the enemy at that point, and these troops conducted themselves admirably." Next evening Captain Kincaid heard him say that "it was the most glorious affair he had ever seen."

General Sir R. Hill says in Orders, "Headquarters near Petite Mouguerre, 14th December 1813.—The Second Division has greatly distinguished itself, and its gallantry in yesterday's action is avowed by the Commander of the Forces and the allied army."

Lieut.-General the Hon. Sir W. Stewart, K.B., in his report of the action states, "Several brilliant charges were made by the Portuguese and British, more particularly by the 92nd Highlanders and 18th Portuguese Regiment, gallantly directed by Major-General Barnes." [2]

[1] Cannon's Record mentions four distinct charges with the bayonet.

[2] Sir William mentions four names as distinguished among the regimental officers of the British infantry engaged: Lieut.-Colonel Cameron, 92nd Highlanders; Lieut.-Colonel Leith, 31st Regiment; Major W. A. Gordon, 50th Regiment (commanding the light troops of the First Brigade), and Captain Cameron, 3rd Buffs (commanding the light troops of the Third Brigade). Major Gordon had served in the 92nd in Holland, etc. Captain Cameron had been promoted from the ranks of the 92nd.

In a letter from Sir R. Hill, dated January 8th, 1814, to Sir John Sinclair, President of the Highland Society,[1] acknowledging the receipt of the report of the competition of Highland pipers held at Edinburgh, for the information of the 71st and 92nd Regiments, he continues—"In justice to these distinguished corps, it affords me the highest satisfaction to state that they have on all occasions imitated the example of their warlike ancestors. The conduct of the officers and men throughout these campaigns has been so uniformly good as to render it almost unnecessary for me to select particular individuals for praise. Lieut.-Colonel Cameron of the 92nd does, however, demand that distinction. During the greater part of the battle of Vittoria he commanded my First Brigade, and also at Maya and other severe operations in the Pyrenees. I am also much indebted to him for leading the gallant 92nd in several successful charges against very superior numbers in the battle of the 13th of last month near Bayonne."

Extract from letter to Colonel Cameron, dated Edinburgh, 24th February 1814, signed R. M'Donald, Secretary of the Highland Society of Scotland—"The Directors cannot refrain from expressing their admiration of the conduct of the 92nd."

Cameron had just received General Barnes' message to fall back on St Pierre, which he was gradually doing, when his horse was killed, and he fell under it. He was close to the French, and in the press was not at first observed, till his orderly corporal ran back to his aid, calling out that the colonel was killed. He was helping him to struggle out from under the horse, when a Frenchman laid hold of the corporal by the neck ; a private ran up and shot him dead, but was himself killed immediately after. "The corporal, however, with extraordinary gallantry," persisted in taking his colonel's cloak and saddle with him, exclaiming as he rejoined his comrades, "Fagaidh sinn a chlosach aca, ach cha n'faigh iad an diollaid anns an do shuid fear an Fhasaidhfhearne" (We will leave them the carcase, but they shan't have the saddle on which Fassiefern sat).[2]

[1] Now the Highland and Agricultural Society of Scotland. Colonel Cameron gave great encouragement to the regimental pipers, some of whom were among the best in the Highlands, and were successful at various times in the competitions held by the Highland Society of Scotland, in the Theatre or Music Hall, Edinburgh, where the competitors were the most celebrated performers, some coming many days' journey from the distant Highlands,* before such competitions were common in Scotland. He particularly encouraged the specially *Highland* airs, and the high-class music (Ceol Mor). To this may probably be attributed the fact that *all* pipers in the Gordon Highlanders are still taught to play "Piobaireachd," and that this ancient and characteristically *Highland* class of pipe music is still played every day under the windows of the officers' quarters before dinner, according to the custom of the old Highland gentry.

[2] This incident is still told in Lochaber, where the credit is given to Private Ewen ban M'Millan, Cameron's servant, but the colonel's letter only mentions the corporal.

* Carr's "Caledonian Sketches," and an old officer of the 92nd.

Alison writes : " This battle, one of the most bloody and hard-fought on both sides which had occurred in the whole course of the Peninsular War, cost the British 2500 and the French 3000 men." Napier says—" Nor can the vigour of the combatants be denied where 5000 men were killed and wounded on a space of one mile square." The French account[1] agrees ; and Wellington declared that he had never seen a field so thickly strewn with dead. The Gordon Highlanders lost Major John MacPherson (mortally wounded), and Lieutenants Duncan MacPherson, Thomas Mitchell and Alan MacDonald were killed in action. The other officers wounded were—Captains G. W. Holmes, Ronald MacDonald, and Donald MacPherson ; Lieutenants John Catenach, Ronald MacDonald, James John Chisholm, Robert Winchester, and George Mitchell ; and Ensign William Fraser. I find the following twenty-six names of N.-C. officers and men killed in action:—

Sergeant	Jas. M'Haffie.	Private	Archd. MacLean.
Corporal	Ewen M'Pherson.	,,	Donald MacLeod.
,,	George M'Pherson.	,,	Donald M'Kinnon, 2nd.
Private	Adam Burnet.	,,	John Patterson.
,,	Jas. Cheetham.	,,	John Ross.
,,	Francis Elliot.	,,	William Ross.
,,	Kenneth Fraser.	,,	James Roy.
,,	James Glass.	,,	James Reid.
,,	Peter Gordon.	,,	John Smith.
,,	William Johnston.	,,	William Swan.
,,	John Lumsden.	,,	John Stephen.
,,	John MacDonald.	,,	Michael Taylor.
,,	William MacDougall.	,,	Michael Trimble.

Sergeant-major Duncan MacPherson (who was taken prisoner), six sergeants and 136 rank and file were wounded, of whom the following died of their wounds:—

Private	Robert Allan.	Private	Duncan Johnston.
,,	Patrick Burke.	,,	William Lyon.
,,	James Cairns.	,,	George Mills.
,,	Archibald Currie.	,,	George MacIntosh.
,,	Murdoch Davidson.	,,	Alexr. MacIntosh.
,,	William Dunbar.	,,	Donald MacDonald.
,,	Kenneth Fraser.	,,	Alexr. Ross.
,,	Alexr. Hardie.	,,	Alexr. Shaw.
,,	William Jamieson.	,,	James Wright.

The wounded were taken to Cambo ; one sergeant and four rank and file, who had also been reported dead, were found to be prisoners

[1] Pellot's " Mémoires sur la campagne de l'Armée française dite des Pyrénées en 1813-14."

THE THREE PIPERS AT ST PIERRE

of war.[1] Major MacPherson was a great loss to the regiment; Colonel Cameron placed great confidence in his tact and judgment, and often consulted him when he felt his own fiery temper might lead him astray.

In commemoration of this action, the Gordon Highlanders bear the word *Nive* on their regimental colour; an honorary badge with *Nive* was conferred on Lieut.-Colonel Cameron, and the senior captain, James Seton, was promoted brevet-major. When Peninsula medals were afterwards given, a clasp "Nive" represented all the engagements on both sides of that river.[2]

In the evening the enemy retired to their fortified camp at Bayonne, and the 92nd went into the quarters they had left in the morning; but before leaving the field, the Marquis of Wellington personally thanked them for their manly bearing in the action, and ordered them a double allowance of rum.[3]

Sergeant-major Duncan MacPherson was promoted ensign in the 92nd, December 23rd, and Sergeant David Taylor was made sergeant-major.

The battle of St Pierre had important results. It cut Soult's communications with St Jean Pied de Port, was a step towards forcing him to abandon Bayonne, and menaced the navigation of the Adour, rendering it difficult for the French to get supplies. For the latter purpose the 92nd marched and encamped at Urt, on the bank of the Adour, on the 4th of January 1814, and went into quarters there on the 8th. Here they had a troublesome time, not a night passing without some disturbance. The French supplies were guarded by gunboats, which returned the fire of the infantry pickets on the river bank. A frost at length hardened the roads and enabled some light artillery to be brought up, which sank some of their boats and rendered the passage very hazardous. There is an island in the river, on which the French attempted to establish a fortified post. Sir Rowland Hill sent a flag of truce to say that he could not permit them to do so, and, wishing to prevent bloodshed, desired that they should leave it. On their refusal, he opened on them with his guns, and the musketry of some Portuguese infantry and a wing of the 92nd, when they abandoned the island, which was occupied by a British picket.

On the 20th January the battalion marched for St Jean de

[1] Cannon's Record gives 28 rank and file killed and 143 wounded, but never mentions those died of wounds or prisoners of war.

[2] The battle of St Pierre is called "Moguerre" by the French, and is alluded to in the regimental documents of the time as "the action near Bayonne." It was the great action which terminated the five days of fighting, chiefly on the left bank of the Nive; but it was distinct from the rest, fought by a separate portion of Wellington's forces, and is well worthy of special remembrance in the records of the army and of the regiments engaged.

[3] Sergeant Robertson.

Luz, where they received their new clothing on the 23rd, and returned to their quarters in Urt on the 26th.[1] The 71st, who had relieved the 92nd, now went in their turn for clothing. The enemy had seen the 71st go, and thought it a good opportunity to make a descent on the British. They did so early on the morning of the 27th, when they were warmly received by the 92nd, whose return after dark they had not observed; their attempt proved abortive, and they were obliged to recross the river. In this affair Private W. Charles was killed.[2] After this, all remained quiet till the 12th of February, when a frost enabled Wellington to advance, and the 92nd, under Colonel Cameron, marched with Hill's Corps, 20,000 strong, for Urcuray, in order to force the passage of the Gaves, rivers which run into the Adour.

Hill's force was in two columns, and that to which the 92nd belonged attacked the enemy under General Harispe at Helette on the 14th. In this affair the battalion had Private Alexander Fairweather killed, and Lieutenant Richard MacDonnell and several men wounded; Cannon's Record says seven, Sergeant Robertson says ten. The morning state of the 18th mentions ten as "sick gone to the rear." The sergeant also says eight were killed, and particularises his "old and beloved comrade and bed-fellow" as "shot through the heart and died almost instantly," whereas Fairweather, according to the Regimental State, died of his wound next day. On the 15th the battalion marched in pursuit of the enemy, who took up a position on the heights of Garris, near the town of St Palais, closely followed by the Light Companies of the Second Division, who gained a parallel counter ridge, on which Hill's Corps was at once established, while his skirmishers descended into the ravine as the evening closed in, and two guns fired over them on Harispe's troops. At this moment Wellington arrived and, addressing the troops near him, said, "You must take the hill before dark";[3] and take it they did. The 92nd, particularly the Light Company, were employed along with other regiments, and the French being advantageously posted, though many of them were conscripts, not only stood the British fire bravely, but charged them with the bayonet, many being wounded in the contest with that weapon. They were beaten off, but twice returned to the attack, till Harispe, seeing the rest of the Allies coming on, retreated by St Palais, and blew up the bridge of the Bidouze. When the combat ceased, it was so dark that men could hardly see each other. Next morning they found many knapsacks thrown away by the French.

[1] Wellington, owing to want of transport, could not send the clothing to the troops.

[2] Sergeant Robertson says three men were killed and ten wounded, but they are not in any return I can find.

[3] Napier.

The enemy lost this day 500 men, of whom 200 were prisoners; the Allies' loss amounted to 160. Of the 92nd, Major James Seton [1] was mortally wounded (he died at Cambo), and three rank and file were wounded.

Major Seton had served at the age of fifteen with the 12th Regiment at the storming of Seringapatam in 1799, and was promoted in the Gordon Highlanders, where he was captain of the Light Company. His letters, which I have quoted, show him to have been an enthusiastic soldier. He was son of Sir W. Seton, of Pitmedden and Cushnie, and great-grandfather of the present baronet (1928). A monument was put up some years ago in the English Church at Biarritz, near Bayonne, to the memory of the British officers and men killed in France. As there were errors in the names of the officers of the 92nd, their tablet was replaced in 1893 by a new and correct one at the cost of past and present officers.

On the 16th February Wellington repaired the broken bridges at St Palais after a skirmish, and in the afternoon the 92nd crossed with Hill's troops and occupied a position in advance, the enemy retiring before them, but in good order. On the 17th, the battalion resumed its march, and about 2 p.m. came up with the enemy strongly posted in the village of Arriverete, on the right bank of the Gave de Mauléon, where they were endeavouring to destroy the bridge, to which they had set fire. A ford was, however, discovered higher up, by which the 92nd crossed under cover of the artillery and at once charged the troops in the village, after considerable resistance drove them out, and secured the bridge by which the rest of Hill's troops were enabled to cross. The enemy retired over the Gave d'Oleron, and the battalion, which had ten rank and file wounded in this gallant enterprise, was cantoned in the neighbouring villages. During the action a shell fell among the 92nd while in close column, but did no damage beyond breaking a bayonet! Colonel Cameron's horse was shot on entering the river, and the tradition is that Sergeant Angus Henderson and a Private M'Intosh carried him over.[2]

In memory of this occasion, Colonel Cameron was granted by Royal warrant the right to bear as his crest a Highlander of the 92nd, up to the middle in water, grasping a broadsword in his right hand, and in his left a banner inscribed "92nd" within a wreath of laurel, with the motto "Arriverete."

Extract from the Marquis of Wellington's dispatch, 20th February 1814:—"A ford being discovered above the bridge, the 92nd Regiment under Lieut.-Colonel Cameron, supported by the

[1] This officer's name is more properly "Seton," but it is "Seaton" in the Army List of 1813, and also on the memorial at Biarritz.

[2] Henderson was long a pensioner at Fort-William, and M'Intosh in Badenoch.

fire of Captain Beane's troop of horse artillery, crossed the ford and made a most gallant attack upon two battalions of French infantry posted in the village, from which the latter were driven with considerable loss."

The following letter relative to an application to be permitted to bear "Arriverete" on the regimental colours was addressed April 13th, 1816, by the Duke of Wellington to Lord Niddry,[1] colonel of the 92nd :—

My dear Lord,—I have received your letter of the 2nd regarding the desire of the 92nd Regiment to bear the word "Arriverete" on their colour, etc., to which I have no objection, and I will apply for the distinction, if after this explanation they should still desire it.

Arriverete is a village on the Gave de Mauléon, at which there is a wooden bridge. We had passed the river at other points, but our communication across it was difficult, and the enemy was of such a force at Sauveterre, in the neighbourhood, that we could not venture to move along it, and I wished to get possession of the bridge before the enemy could destroy it. The 92nd forded the river, and attacked and took the village against a very superior force of the enemy in the most gallant style (in the manner in which they have always performed every service in which they have been employed), but without much loss ; there the affair ended ; we were not prepared at that time to do more, and we held that village as a *tête-de-pont* till our means were in readiness for our further operations.

There is no doubt but the troops behaved as gallantly in this affair as they could in any of greater importance, but the result was not of that consequence to the ulterior operations of the army to have rendered it notorious to the army at large ; and although I reported it as I ought, I know there are many belonging to the army, some even who were present, who have no recollection of the name of the place which was the scene of the action, and some not even of the action itself.

It appears to me to be beneath the reputation of the 92nd to have to explain for what cause the name of a particular place has been inserted in their colours ; and notwithstanding that on no occasion could they or any other troops behave better than they did upon that, I acknowledge that I am anxious that they should not press the request. But if after this explanation they continue to wish it, I will take care it shall be granted.—I have, etc.

(Signed) Wellington.

Lieut.-General Lord Niddry, K.B., etc. etc.

It is generally considered that while Wellington had the entire confidence of his army, he had not Napoleon's knack of gaining the affection of his soldiers by personal intercourse with them. There are, however, many instances to the contrary. While the Gordons were marching at ease one day over one of the grassy commons of that part of France, the left hand man of the rear section, one Jock Webster, from Kingsmuir, Forfar, who, as his

[1] Sir John Hope, at that date Lord Niddry, and afterwards Earl of Hopetoun.

friend told me, "wasna blate," hearing the sound of a horse's hoof on the turf, looked over his shoulder, "an' wha but Wellington ridin' cannily by." Jock brought his firelock smartly to the shoulder, calling out, "Hoo far will ye tak' us, my Lord?" "How far do you want to go, my lad?" "To Paris, by G—d!" and Wellington rode on laughing. A 92nd pensioner at Strone, Kingussie, named M'Gillivray, delighted to relate how, at Quatre-Bras, where, though wounded, he was still trying to do his duty, Wellington spoke kindly to him, saying, "You have done enough, my man; make for the rear while your blood is warm."

The battalion remained some days in or about Arriverete, taking part, however, in an attack on the bridgehead occupied by the enemy at Sauveterre. On the 24th of February the battalion forded the Gave d'Oleron near Villeneuve, and on the 25th marched on the road to Orthes and encamped at Laas, near that town.[1] The allied army under Wellington crossed the Gave de Pau near Peyrehorade below Orthes on the 26th and at daybreak on the 27th of February, except Hill's Corps, which remained on the left bank of the river and menaced the bridge of Orthes and the ford of Souars above.

The battle of Orthes began with skirmishing and artillery fire on the right bank of the river. At nine o'clock Wellington commenced the real attack on Soult's position on a ridge of partly wooded hills presenting a concave to the Allies, and having long narrow slopes pushed out towards a marshy ravine which covered most of his front. The battle was continued on the left and centre with unabated fury for three hours, and at one time victory seemed to declare for the French. It is said that in the exultation of the moment Soult smote his thigh, exclaiming, "At last I have him!"[2] It was a most dangerous moment for the Allies, but Wellington, seeing the critical state of matters, suddenly changed his plan of battle: he assailed the flank and rear of the victorious French. Neither fire nor the marsh, though it took them above the knees, could stop the veterans of the Light Division. At the critical moment Wellington thrust the Fourth and Seventh Divisions, Vivian's cavalry, and two batteries of artillery through St Böes, which changed the aspect of affairs.

Meanwhile Hill, who was with 12,000 cavalry and infantry before the bridge of Orthes, received orders, when Wellington

[1] Sergeant Robertson mentions that a man of another regiment was hung on the 26th for plundering and fighting with French farmers, and Napier also notices this. Wellington was not at this time uneasy as to the power of Soult's army in the field, but he dreaded the danger of a popular rising against his troops. "Maintain the strictest discipline; *without that we are lost*," he told his generals. He made a British colonel quit the army for allowing his soldiers to destroy the archives of a small town.—Napier.

[2] Napier.

changed his plan, to force the passage of the river. Hill, though unable to force the bridge, forded the river at Souars just above the town, which he did under a heavy fire of artillery and musketry. The First Brigade, in which was the 92nd, led the way. The French posted on the opposite side were driven back, the heights above were seized, the French were cut off from the road to Pau, and the town of Orthes was turned, and this at the very time when the fortune of war had changed at the other end of the British line. Soult, arriving at the moment, saw that the loss of Souars rendered the whole position untenable, and gave orders for a general retreat. The 92nd then moved on the road to St Sever along the rear of the enemy, whose retreat, at first orderly, now became a confused flight ; and the battalion was conspicuous in the pursuit till it halted at Sault de Navailles, and the cavalry finished the business. The loss of the French was six guns and nearly 4000 men killed, wounded, and prisoners ; several thousands of the conscripts, however, threw away their arms and escaped over the country. The loss of the Allies was 2300, of which 50 with 3 officers were prisoners.[1] Among the wounded were Wellington (slightly), Generals Walker and Ross. The 92nd, though they had played an important part in the warlike game, escaped with the loss of Private Findlay Munro killed, and three rank and file wounded; one of them, Malcolm Nicolson, who lost an arm, had served with the Gordons from Corunna. Afterwards, as "Calum na rightaig" ("Stump-armed Malcolm"), he was a well-known pensioner in Skye, where he was a tenant in Uig, and died in 1863 at the age of ninety.[2]

In the pursuit they met the two other kilted regiments, the 42nd and the 79th, "and such a joyful meeting I have seldom witnessed. As we were almost all from Scotland, and having a great many friends in all the regiments, such a shaking of hands took place ; the one hand held the firelock, while the other was extended in the friendly Highland grasp, and then—three cheers to go forward. Lord Wellington was so much pleased with the scene, that he ordered the three regiments to be encamped beside each other for the night, as we had been separated for some years, that we might have the pleasure of spending a few hours together."[3]

The Gordon Highlanders bear the word "Orthes" on their regimental colour in memory of this victory. The King conferred honorary badges with the word "Orthes" on Lieut.-Colonel Cameron, who commanded the battalion, and on Major Mitchell, who commanded the Light Companies of the brigade, and the latter was promoted Brevet Lieut.-Colonel.

[1] Napier. [2] "Brave Sons of Skye." [3] Sergeant Robertson.

Next day the 92nd marched to Samadet, and proceeded to Grenade on the 1st of March ; at two o'clock on the 2nd, Hill came up with Villatte's and Harispe's Divisions drawn up on a strong ridge in front of the town of Aire, and General Stewart at once fell upon the French right with the First and another British brigade. A Portuguese brigade attacked the centre, and the other brigades followed in columns of march. The Portuguese gave way on the summit of the height before the charge of the French under Harispe, and the battle was like to be lost ; but Stewart, having won the heights on the right, immediately sent General Barnes with the 50th and 92nd to help the Portuguese, telling them he was sure they would take the hill or die there. Forward they went, and on gaining the top were received with a heavy fire. "Charge !" was the word ; "Forward, my heroes !" cried the General, "and the town will be ours in ten minutes." The vehement charge of these regiments turned the stream of battle ; the French were broken ; yet they rallied and renewed the fight with the greatest courage and obstinacy, till Byng's Brigade came up, when Harispe was driven towards the River Lees and General Villatte quite through the town of Aire. The 92nd pursued, and parties of both sides crossed each other at the street corners, where some fell. It was now dark, and the 92nd were quartered in the town. The French lost many men ; two generals were wounded, a colonel of Engineers was killed, 100 prisoners were taken, and the great magazines fell into the hands of the conquerors. The loss of the British was 150 killed and wounded, that of the Portuguese about the same ; General Barnes was wounded, and Colonel Hood killed.

The casualties of the battalion were—Captain W. Fyfe, Lieutenants J. A. Durie and Richard MacDonnell wounded ; one sergeant and two rank and file were killed, and one sergeant and twenty-eight rank and file wounded.[1] His Majesty George IV granted permission to Lieut.-Colonel Cameron to bear on his shield the word "Aire" and a view of the town.

DIVISION ORDERS

AIRE, *March 3rd*, 1814.

Lieutenant-General Sir William Stewart congratulates the division on its further advance and success against the enemy. To the admirable conduct of the Fiftieth and Ninety-second Regiments, led by their gallant commanders and by Major-General Barnes, the good fortune of yesterday's action is decidedly attributed, which the lieutenant-general has to state to Lieutenant-General Sir Rowland Hill for the information of the Commander of the Forces.

[1] The nominal list of killed and wounded at this period is lost.

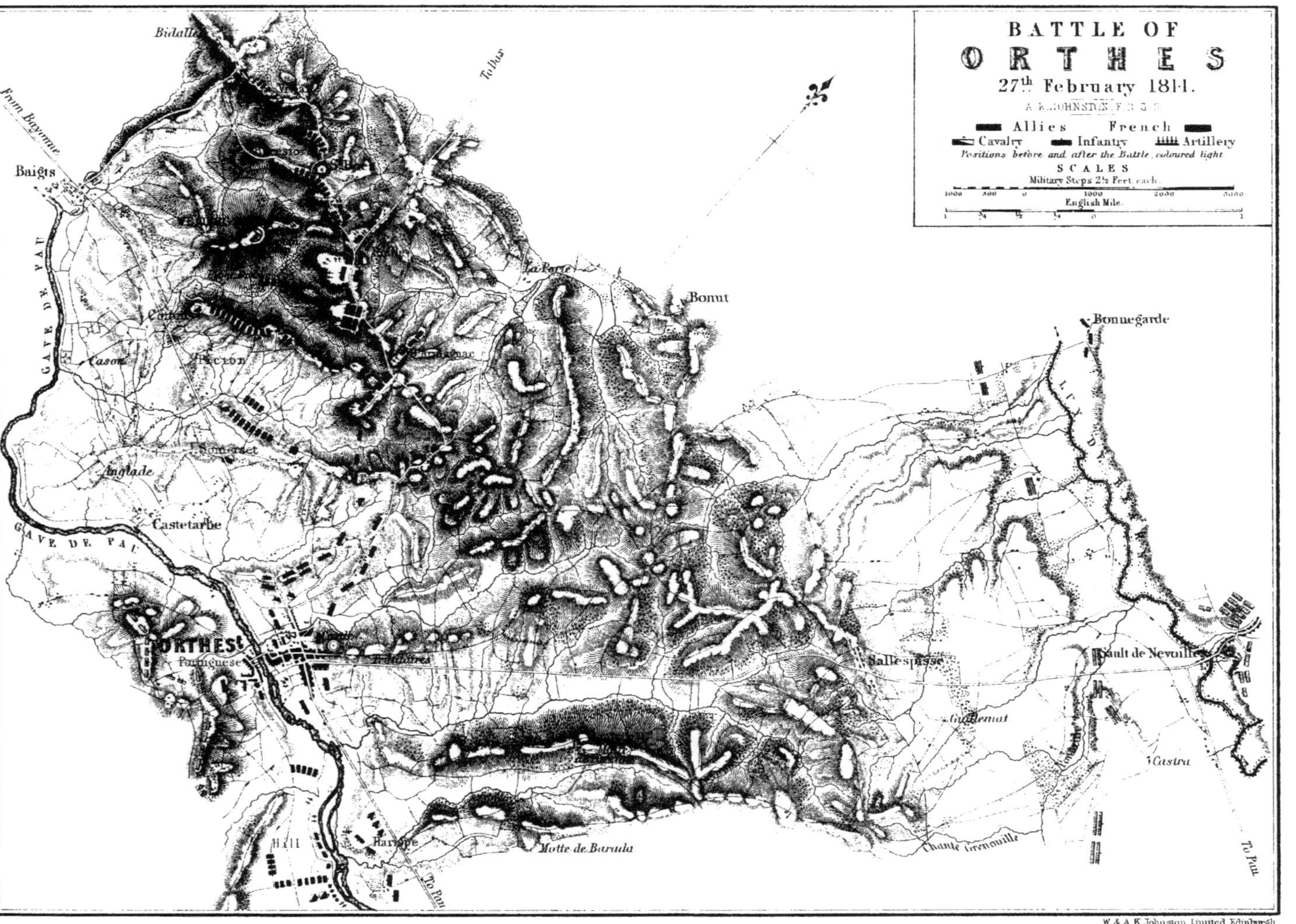

W. & A. K. Johnston, Limited, Edinburgh

General Orders

Aire, *March 5th*, 1814.

Lieutenant-General Sir Rowland Hill congratulates Lieutenant-General Sir William Stewart, Major-General Barnes, and Major-General Byng, on the brilliant part which they bore in the action of the 2nd instant.

The gallant and steady conduct of the Fiftieth Regiment under Lieutenant-Colonel Harrison, and of the Ninety-second Regiment under Lieutenant-Colonel Cameron, excited the admiration of all who were witnesses of it.

Aire, *March 10th*, 1814.

Dear Sir,—The Mayor and inhabitants of this town having requested me to convey to you, with their sentiments of gratitude, the accompanying address, I feel infinite pleasure in so doing, as I deem the sentiments expressed in it justly due towards you and the distinguished regiment under your command.—I have, etc. (Signed) William Stewart,
Lieut.-General.

Lieut.-Colonel Cameron,
Commanding Ninety-second Highlanders.

(*Translation*)

Aire, *9th March* 1814.

Sir,—The inhabitants of the town of Aire are not ignorant that if they were preserved from pillage and destruction at the close of the obstinate and sanguinary conflict of the 2nd of March, they are indebted for such preservation from that calamity to your honourable conduct, and the strict discipline which you have maintained amongst the troops under your command.

Penetrated by sentiments of the liveliest gratitude towards a commander distinguished by such noble qualities, the town of Aire has charged me to be their interpreter in communicating their thanks, and to offer you the homage of their esteem.—I have, etc. (Signed) Codroy, Mayor.

To Lieut.-Colonel Cameron,
Commanding Ninety-second Highlanders.

In the morning the wounded of both sides were placed in an hospital, and the magistrates waited on General Hill, thanking him and his troops for their forbearance in not plundering the town, and for the general good conduct of the soldiers, which gave the general, says one of them, as much satisfaction as his victory. The battalion remained in the town for ten days, and the men were provided with shoes and other necessaries.

CHAPTER XXII

MEANWHILE Soult had retreated by Tarbes, having determined to make his next stand at the fortified position of Toulouse. Misfortune had caused discontent in the French army, who behaved with so much violence on the retreat that their countrymen contrasted their conduct with the discipline of the Anglo-Portuguese, who paid for everything on the spot, and Soult wrote to his Minister of War that the population "appear more disposed to favour the invaders than to second the army." Many of the inhabitants, indeed, were in favour of the Bourbons and against the Government of Napoleon. In this state of affairs, Wellington had been able to detach 12,000 men to take possession of Bordeaux and retain it as a port for the Allies.

Information having been received that Soult was making a movement to his left and menaced the British right, Wellington made a corresponding move, and the 92nd left their quarters at Aire on the 13th of March and advanced on the road to Pau to a position near Garlin. On the 18th the battalion marched to Conchez, the enemy retiring on the approach of the Allies; on the 19th a skirmish took place at Vic Bigorre, and the enemy retired across the Adour. The battalion halted at night in front of Vic Bigorre. On the 20th, the French made a stand at Tarbes, and the Gordons hoped to celebrate the anniversary of the battle of Alexandria by a victory on the 21st; but after a combat in which Hill forced the passage of the Adour, the enemy retreated under a cannonade from that general's artillery. The nature of the country prevented a pursuit by cavalry, but the 92nd, with the other troops, continued to advance. At St Gaudens some French cavalry which were drawn up across the road were overthrown by the 13th Light Dragoons. The 92nd continued in pursuit by St Julien, constant rain impeding the operations till the 26th, when they halted at Muret. Next day Hill's troops were withdrawn to St Roque, and saw no more of the French till they got within sight of Toulouse. The division crossed the Garonne by a pontoon bridge, after considerable delay in its construction, and proceeded to Miremont; and on the 1st of April they marched fifteen miles to Cintagabelle, in order to seize the bridge over the Arriège, the intention being to attack Toulouse on the south side, while Wellington assailed St Cyprien; but the state of the roads preventing the rapid movement of artillery, Hill wisely renounced the project and recrossed the Garonne at St Roque. Here the battalion stayed till the 5th of April, when it was put into a large and handsomely furnished mansion at St Simon, the inmates of which had fled. They stayed in this comfortable situation till the floods which had obliged the British general to remain inactive had subsided, when, leaving Hill with two divisions to distract the enemy's attention in that quarter by menacing the suburb of St Cyprien on the left bank of the river, Wellington, with the main body of his army, crossed to the right bank on the 9th. Having carefully examined the enemy's position, which was admirably chosen and strengthened by field works, he gave the signal for battle at seven o'clock on the morning of the 10th of April. The 92nd advanced with Hill's troops, who drove in the French outposts, and by a vigorous attack forced the first line of entrenchments covering St Cyprien, and menaced the second line, which was so very strongly fortified that it could not be stormed. For the rest of the day Hill's troops could only remain spectators of the desperate battle which was being fought out on the opposite side of the river. They could see the dangerous flank march of Beresford's Corps, the disastrous flight of Freyre's Spaniards ;[1] how Picton, notwithstanding his instructions only to engage the enemy's attention by a feigned attack, impetuously turned it into a real one, and was beaten back from an impracticable redoubt with the loss of 500 killed and wounded, among the former being Colonel Forbes of the 45th, one of the original officers of the Gordon Highlanders. They could hear the roar of the guns and the shouts of the combatants when Beresford repelled the attack of Taupin's Division, and observed the havoc made by the Congreve rockets,[2] seldom, if ever, used against infantry before. They remained an inactive but necessary part of the combination while their countrymen of the 42nd led the assault on Columbette, and gained it only to be overwhelmed, and driven out with enormous loss ; re-forming, however, with other troops and declining to retire, Cole and Clinton's Brigades came to their assistance, and the French were again forced back, fighting hard. The Columbette redoubt was taken by the 79th, and at 4 p.m. the Allies, at last successful at all points, had gained the bloodstained heights. Soult withdrew his troops within the second line of defence, and the allied forces fell on the retiring columns till stopped by the fire from the *têtes-du-pont*

[1] Wellington had granted Freyre's request to lead the battle at Calvinet. At first the Spaniards advanced resolutely, but the fire of the French cannon and musketry thinning their ranks at every step was more than they could endure ; they wavered, the French charged, and they at last broke in headlong flight. Wellington, who was at hand, covered the panic-stricken troops with Ponsonby's cavalry, the Portuguese guns, and a brigade of British infantry ; the victorious French then retreated to their entrenchments, having killed and wounded 1500 Spaniards. Wellington is said to have remarked, " Well, d—— me, if ever I saw ten thousand men run a race before."

[2] " The roaring of the cannon and the horrific blazing of the rockets was awful and grand beyond description."—Letter from Lieutenant H. Innes, 92nd.

NOTE.—War rockets were invented in 1805 by General Congreve. Sir H. Shrapnell invented the spherical case which took his name, and which was first used at Vimiera.—" History of British Army," by Colonel Cooper King.

on the canal, behind which the whole French army was ranged. At the same time, Hill on the other side of the Garonne drove the enemy from their second line of defence within the old city wall, so that they were now entirely cooped up within the town of Toulouse.

In this terrible battle the Allies lost four generals and 4659 officers and soldiers killed and wounded, of whom 2000 were Spaniards. The French lost five generals and 3000 officers and men killed and wounded, and one piece of artillery.

The Gordon Highlanders, as we have seen, took but a small part in the fighting, though they assisted to seize and to hold a most important post, thereby having a material effect on the fortune of the day. Cannon's Record does not mention any loss to the battalion, and the regimental returns of the time are lost. Sergeant Robertson says, " We had only a few wounded, of whom one died after his leg had been amputated." When medals were given to the survivors of the Peninsular War,[1] the men of the 92nd got the clasp for Toulouse, but it is not on the colours of the regiment.

After the battle the battalion went into cantonments. There was only a slight skirmish on the 11th, which was passed in getting up ammunition for the guns. The attack was arranged for daylight next morning, the 92nd being in orders for storming,[2] but Soult, feeling that Toulouse was no longer tenable, silently evacuated the city at night, and conducted the retreat with such ability that by daylight his troops were at Ville-Franche, twenty-two miles distant ; he abandoned his magazines, and left his wounded to the humanity of the British. On the morning of the 12th, the 92nd, with the Second Division, was sent in pursuit ; the troops were ordered to put bay-leaves in their bonnets and "chacos." At the gate of the city the Mayor presented the keys to Sir Rowland Hill, and many of the people mounted the white cockade, the royal colour of France. The battalion passed rapidly through the streets and along the road to Ville-Franche ; hundreds of French soldiers, who were unable or unwilling to proceed, were taken prisoners, and after a long march the battalion was cantoned in a village near Ville-Franche. In the afternoon intelligence was received of the abdication of Napoleon.

At first Soult refused to consider the news authentic, and the outposts were on the point of engaging on the 17th, when, having received official information, he notified his adhesion to the new state of affairs. Had the express not been delayed on his journey by the French police, the battle of Toulouse would have been prevented. A disbelief of this intelligence also caused much

[1] In 1848. [2] Letter from Lieutenant Innes.

unnecessary bloodshed at Bayonne, where the garrison made a desperate sortie on the night of the 14th of April. In repelling it Lieut.-General Sir John Hope (colonel of the 92nd), who commanded the British, was wounded and taken prisoner.

This was the last action of the Peninsular War. In memory of the part taken by the First Battalion 92nd, which was as distinguished as that of any battalion of the army,[1] the word "Peninsula" was, in 1816, added to the distinctions on the regimental colour.

Since the last campaign opened, about six weeks before, Wellington had driven the French, except the garrison of Bayonne, from the neighbourhood of that city to Toulouse, a distance of 200 miles, had conquered the whole country from the Pyrenees to the Garonne, had forced the passage of many rivers, defeated Napoleon's best general in two pitched battles and several combats, and brought about a revolution in favour of the Bourbon dynasty at Bordeaux and Toulouse. Moreover, he had established such admirable discipline in his army, composed though it was of the troops of different nations, that but few instances of individual plunder had disgraced their steps. There was, consequently, an almost total discontinuance of punishment;[2] and so great was the confidence of the French peasantry that they even sought refuge in the British lines from the pillage of their own troops.[3]

And if Wellington compelled the admiration of the world as a commander, and his troops the respect of the French population by their moderation, "How," says Scherer in his "Recollections of the Peninsular War," "shall I picture the British soldier going into action? He is neither heated by brandy, stimulated by hopes of plunder, nor inflamed by feelings of revenge; he does not even indulge in feelings of animosity against his foes; he moves forward confident of victory, never dreams of the possibility of defeat, and braves death with cheerful intrepidity." At that time there was no Victoria Cross, no medals were given to any below the officer commanding a battalion; subaltern and n.c. officers or soldiers were seldom, if ever, mentioned in dispatches—their reward for deeds of gallantry was found in the approval of their superiors and the cheers of their comrades. Many were, however, promoted from the ranks to commissions.[4] In the Army List of 1814 no officer of the Gordon Highlanders is mentioned as having a medal except the colonel (Sir John Hope) and Lieut.-Colonel Cameron,[5] and the latter, though he had commanded his battalion throughout

[1] Some regiments had two and the 95th Rifles three battalions engaged.

[2] Wellington to Royal Commission on military punishments.

[3] Alison. The inhabitants of the country spoke of them to the author in the same strain.

[4] Some twelve or thirteen sergeants of the battalion and nine or ten Volunteer privates had received commissions.

[5] The medals given to officers by the Sultan of Egypt seem not to have been noticed.

the war and a brigade in several actions, and been four times wounded, was only promoted brevet-colonel in the following June.

On the 20th of April the 92nd entered Ville-Franche, on the 24th marched to Besiège, and on the 25th occupied quarters in Toulouse.

A treaty of peace was made between Great Britain and France. Louis XVIII was placed on the throne of his ancestors, and Napoleon was given the Island of Elba for his residence.

At Toulouse all the inhabitants, except those who had been in Napoleon's army, were very friendly to officers and men. As one of the latter remarked to me, "The Frenchmen were very kind to us and the French lassies forby!" Provisions and wine were plentiful and cheap, so they lived very comfortably. Large hospitals were established by both nations for their wounded, and the Highlanders remarked that the British were the best cared for, till on the arrival of the Duc d'Angoulême, nephew to Louis XVIII, he ordered that the French should be carefully attended to, and not be seen sitting in public places asking alms of the passers-by.

The 92nd was appointed to furnish the Duke's guard of honour while he remained; afterwards they were visited by the French Generals Soult and Suchet. Altogether their time in Toulouse was most agreeable, and the hearts of some were with "the girl I left behind me" when the battalion marched for Bordeaux on the 5th of June. Indeed, a good many men of the army were induced by liberal offers of employment to take their discharge and settle in the country. One handsome private of the Gordons named M'Intosh, from Daviot, won the affections of a lady possessing a considerable estate, and his nephew, a forester in Forfarshire, remarked in relating his uncle's good fortune, "An' if they had bairns, I 'm no a far awa' cousin to a French laird."

The battalion marched by Auch, where they were put in a barrack, the roof of which fell in, but no lives were lost; though some men were bruised. Here the Portuguese troops were ordered to return home. For nearly six years they had gallantly fought side by side with their British comrades, and the parting scene was very affecting, "more like the farewell between near and dear relations." The grief was heightened by an order that all the women who had followed the army from Portugal and Spain ("whether married or not," says Sergeant Robertson) were to return with the troops belonging to their respective countries and to be rationed by the way. None of them were to be allowed to embark for England.

The battalion continued its march, suffering from the great heat, but enjoying the hospitality of the villages where they were billeted[1] till the 19th of June, when they encamped at Blanchfort, six miles from Bordeaux. They were allowed to visit the theatres and sights of this beautiful city when they liked. The 21st being the anniversary of Vittoria, every regiment in camp had a laurel tree planted beside its colours, and every soldier a sprig of laurel in his head-dress. The French authorities, though glad to be freed from Napoleon's rule, objected to the celebrations of a victory over their countrymen ; but the regiments sent a party from each, headed by its band, to the Commander-in-Chief, telling him that the men wished liberty to celebrate the victory they had gained, in their own way without hindrance. Permission to do so was granted, with particular orders, however, that they were to give no offence to the people by going near villages or farmhouses. There was no parade that day ; dancing and other amusements were kept up with the greatest glee, and all forgot the hardships of the past in the enjoyment of the present, and the anticipation of being soon restored to home and kindred. In the case of several regiments, their hopes of home were disappointed ; they were sent off to America, where war had broken out between Great Britain and the United States. There was some talk of the 92nd being of the number, but fortunately they were reserved for events of greater importance.

On the 9th of July the battalion marched, and encamped at Pouillac, and on the 17th it was conveyed down the Garonne in small craft and embarked in H.M.S. *Norge*, 74 guns, which sailed at once and entered the Cove of Cork on the 26th of July. They disembarked at Monkstown on the 29th, and marched to Fermoy Barracks, where the thanks of both Houses of Parliament were communicated to the officers and soldiers for "the meritorious and eminent services which they had rendered to their king and country during the course of the war." Richly did the army deserve the thanks of their countrymen, who had long felt that their own and their children's freedom depended on the result of the war, so that its glorious termination excited feelings of relief and joy among all classes which it is now hardly possible to realise. Wellington was, with the approval of the whole nation, raised to the rank of a duke ; and among rewards to his generals, Sir Rowland Hill, under whom the 92nd had served so long, was created Baron Hill of Almaraz.

On the 6th of August the battalion was inspected by Sir W. Aylett, who expressed in Orders his great satisfaction at its appearance under arms, as well as its interior economy. The following

[1] "The people are kind and civil to a degree." . . . "Commonly billeted on good houses, and dine with the patron."—Letter from Lieutenant Innes ; Quartermaster-sergeant M'Combie's Journal to same effect.

morning state, dated Fermoy, August 7th, 1814, shows that the prisoners of war had not yet rejoined.

		Sergeants.	Corporals.	Drummers.	Privates.
Total in country (*i.e.* Ireland)		40	31	14	545
On foreign stations	Sick	..	4	..	122
	Command	6	2	..	16
In Great Britain		4	10	2	92
Prisoners of War		1	1	..	38
Total effective		51	48	16	813

The Order Book throws some light on regimental life of the time. A party of men from the different companies, in fatigue dress, went to market every morning at six o'clock under a subaltern, a piper attending "to play for them going and coming." The officer to report his return to the captain of the day, "who will inspect the bread and meat of the different companies and report as to its quality." The commanding officer is displeased with the messing of certain companies. "They must always have potatoes and beer at dinner, as well as bread and other vegetables." "He considers it of the utmost consequence for the well-being of the regiment that no trucking or sutling by the n.c. officers among their companies is allowed." He requests "officers to pay as much attention to orders on dress as to those of higher importance." He observes that some wear the "heckle" feather turned over the bonnet, which is contrary to orders. Gaiters are never to be worn except when actually on the march. Officers may wear "stocking hose" (instead of hose tartan). They may ride out in the old regimental cap till they get the new pattern cap and feather. No officer when walking out is ever to appear otherwise than regimentally dressed, and is not to absent himself from barracks for more than two hours "without the commanding officer's concurrence." Officers to wear their jackets buttoned across at morning parade, but buttoned back "to show two or three button-holes of the facing and lace of the lappell" at mess and at evening parade. In summer the pibroch to sound at 5.30 for evening parade at six o'clock.

All letters of inquiry as to deceased soldiers, prisoners of war, etc., to be immediately answered by the officer commanding the company to which the person inquired for belonged, and all possible information given.

The change from French wine to Irish whisky was a bad one for some of the n.c. officers and soldiers, who probably agreed with the song that the taste of whisky improves with each glass—

Ol a h'aon, cha n'eil e math
Ol a dha, cha'n fhiach e,
Ol a tri, cha'n'eil e cli
'Se'n ceathramh righ na riaghailte.[1]

The general " is sorry to remark the number of Courts-martial," which he attributes to the large sums of money in possession of the men. It seems a pity that they had not been sent to Scotland, where their arrears of pay and prize-money might have been better spent among their friends at home, and a good impression of the service would have been given. " It is with extreme regret that Colonel Cameron finds himself under the necessity of putting the sentence on Sergeant Wylie into execution. He hopes that this will prove a warning to others, and that he will be saved the mortification of inflicting so humiliating a degradation on men who have served their king and country so long and so faithfully." Sergeant Wylie was reduced to the rank and pay of a private sentinel for being drunk on duty. Some time afterwards, however, the colonel " is induced to give Private W. Wylie another chance as a n.c. officer in consequence of his never having been absent in the rear during the war " ; he is therefore appointed corporal.

While at Fermoy a large number of long-service n.c. officers and men were discharged to pension, also many seven years' men who had served their time. Convalescents and prisoners of war rejoined from time to time. In September a draft arrived from the 2nd Battalion ; and in October two captains and three lieutenants proceeded to Scotland to take charge of 12 sergeants, 13 drummers, and 160 rank and file transferred from the 2nd Battalion, which was disbanded at Edinburgh on the 24th of October. They took the colours of the 2nd Battalion with them to Fermoy.

I will now give a short account, from the few records available, of the 2nd Battalion since it left Weely on the 29th of October 1804, and embarked at Liverpool for Ireland, having marched from Weely to Blisworth, one day doing 28 miles, and from Blisworth to near Liverpool, going by canal-boat. The first authentic information is a letter from Major Cameron, commanding at Birr, in the spring of 1806. From there they marched about the end of April to Dublin, where the Duke of Bedford was Lord-Lieutenant, and the Duchess, who was a daughter of the Duke of Gordon, took a lively interest in the regiment and paid great attention to the officers. There are letters from Her Grace asking that the band may play at the

[1] " One drink is worth nothing,
And two are not worth much.
If you take three you are not far wrong,
But your fourth is king of the whole."

vice-regal parties, and also from the Marquis of Huntly, saying that his sister's letters about the battalion give him very great pleasure.

In January 1807, the battalion got the route for Scotland, but to their great disappointment it was countermanded, and they remained in Dublin. The major writes, "Without going to Scotland the battalion can never hope to recruit 400 more men, as we are expected by Government to do."

In June 1807, a draft of 100 picked men was sent to the 1st Battalion. In Dublin the major regrets the hard duty required of his men and the expense to his officers. A Highland gentleman is warned of the "shameful extravagance" of his son, whose mess-bill for drink alone amounted in five weeks to £3, which would not, I fancy, be considered a very large wine bill in the present day. In the end of June the battalion marched to Drogheda, "but we are to be stationed in seventeen different quarters. Weak as we are, this distribution is dismal." In August they received volunteers from three Scottish militia regiments quartered in Ireland, and at a later date the major mentions that he had received that day from one Scottish militia regiment "43 as fine fellows as ever stept."

Owing to a change of Ministry, the Duke of Richmond had succeeded the Duke of Bedford as Lord-Lieutenant. His Duchess was also a daughter of the Duke of Gordon, and was as proud of the Gordon Regiment as her sister of Bedford. Being on a visit at Slane Castle near Drogheda, the Lord-Lieutenant took the opportunity of reviewing the battalion, "the Duchess making a great work about them"; and she applied for two regimental bonnets, probably for her boys to wear. While at Drogheda, the battalion received much hospitality; among others from Allan Cameron (Lundavra), a Lochaber gentleman who held an important office in the Civil Service there.

It seems that 1806-7 was one of the often recurrent years of scarcity in the Highlands.[1] "What a melancholy state the Highlands is in; it makes my heart sick to think of it," says Major Cameron, who was engaged in buying a cargo of meal to send to Lochaber.

November 1807, the battalion moved to Ballyshannon. On March 2nd, 1808, the commanding officer writes from that place, "I have again completed this battalion within the twelvemonth to 600 and a few more, that is a few more than its establishment, from being a year ago scarcely half that number, but not without great anxiety, much trouble, and some expense." The men were nearly all from the Highlands; a great number appear to have been from the counties of Inverness and Ross.

[1] Dr Garnett mentions that the innkeeper at Achnacraig in Mull rebuked him for wishing to give oats to his horse when meal was so scarce for Christians.—"Tour in the Highlands," 1811, by Dr Garnett.

May 3rd, 1808, the major writes from Ballyshannon—"We are now 611 rank and file, besides sergeants and drummers, and we have about twenty recruits in Scotland still. The regiment are almost all Scotch, and two-thirds of them can scarcely speak a word of English"; "but as the 1st Battalion are going on service, we shall lose a hundred or two soon."[1] Letter-writing was a costly business in those days. Major Cameron mentions that his postage for the year amounted to nearly £50.

In November 1808 the battalion had moved to Tuam, and in January 1809 it was ordered to Athlone. Here Sergeant Neil Murray was promoted ensign in a garrison battalion. Sergeant Nicol writes that few could speak English at Tuam, only Irish-Gaelic.

Lieut.-Colonel Napier having been killed at Corunna, and Major Cameron having been promoted lieut.-colonel, he took command of the 1st Battalion, and the command of the 2nd devolved on Lieut.-Colonel Lamont (of Lamont), his adjutant being Lieutenant Ewen Campbell (Sonachan).

On the banks of the Shannon they were employed in duties now performed by the Royal Irish Constabulary, such as quelling faction fights; and detachments were sent to various places to put a stop to the treasonable proceedings of the "Carders," a body of men who administered an unlawful oath to numbers of people in Roscommon and Galway. A subaltern describes his quarters at Ballymoe Bagot, the only furniture between the clay under his feet and the thatch over his head being "large quantities of soot suspended from the roof in long beautifully curled rows as black as the feathers on a Highlander's bonnet"; but in these places they were received with genuine hospitality by the respectable inhabitants. At Roscommon, however, a large assemblage of town and country people attacked the barracks, breaking the windows and threatening to break the heads of the detachment. Ensign Little, who commanded, moved out his twenty men, fixed bayonets and charged along the market-place, clearing it of the rioters and wounding one of them. Mr Little, being called upon by the general for a detailed account, wrote a whimsical dispatch, in which, after giving the names of those who had distinguished themselves in the engagement, he concluded, "It now remains for me to express my sincere regret that the nature of the service on which I have been engaged will not permit me to recommend any of my brave followers for promotion." The ensign, poor fellow, fell in a more memorable battle, that of Quatre-Bras.

[1] In one of Major Cameron's letters from Tuam, he mentions that Lieutenant Donald Stewart (formerly sergeant-major) had been drowned on a boating excursion while at Lisbon on the way to join the 1st Battalion.

It appears from the Order Book of this time, a few sheets of which exist, that small parties under an officer were often employed in still-hunting, as—

Athlone, 16th November 1810.—One subaltern, one sergeant, one corporal, twelve rank and file will parade for Excise duty at three o'clock in the morning in the barrack square, by order of General Montresor. For the above duty Ensign M'Donald.

For every still captured, sergeants received 18s., corporals 10s. 6d., privates 7s. In the spring of 1811, a party of the 92nd with a few dragoons captured in thirty hours twenty-two stills, for each of which they received the above payments.

The men were taught a regular system of fixing flints. Agreeable to general order, twelve men per company were selected as marksmen, size and appearance not to be considered in choosing them, but activity, intelligence, and quickness of sight. A subaltern and two sergeants from each wing to be selected to command them when ordered out. Only eight of each company to be ordered out at a time, the rest to remain as supernumeraries to supply casualties. The marksmen of a battalion to be employed for its protection when in line or column from the annoyance of the enemy's skirmishers. They are to have half the practice ammunition of the battalion allotted to their use. Targets to be five feet in diameter, each shot to be pointed out so that the man may correct his fault. First practice at 100 yards, to be increased by degrees to 200 yards. Men to fire standing and kneeling, and may use a rest ; always to bring the piece up to the object. Men to learn to load without halting, and lying on the ground. Marksmen's station in battalion to be on flanks of sub-divisions. When battalion is three deep, they are to be on flank of front and rear ranks only.

5th December 1810.—A committee consisting of one sergeant, one corporal, and one private will assemble to-morrow at twelve o'clock to report on the quality and goodness of beer delivered to the troops.

24th December 1810.—At a meeting of pay-sergeants held this day by order of the commanding officer, they appear unanimously of opinion that the battalion can be served with beef and mutton cheaper than was offered by contract, viz. 4½d. per lb. The commanding officer is, therefore, induced to give it a trial this month.

Officers commanding companies are most minutely to inspect the meat purchased for their respective companies.

The troops regularly attended divine service.

9th March 1811.—The men to wear their humble bonnets till the new bonnets are put up.

28th March 1811.—At morning parades, garrison parades excepted, officers and men will appear in their undressed bonnets, and at evening parades in their feathered bonnets.

3rd April 1811.—Sergeant-Major Bryce is appointed quarter-master-sergeant *vice* Gordon, who has given in his resignation. Sergeant M'Kenzie of the Light Company is appointed sergeant-major.

Recruiting was carried on principally in the north of Scotland, and always with care as to the class of men taken. The following is a recruiting placard, date about 1811-12—

THE GALLANT
NINETY-SECOND
OR GORDON HIGHLANDERS

who have so often distinguished themselves at Copenhagen, Spain, on the plains of Holland and sands of Egypt, and who are now with Lord Wellington in Portugal, want to get a few spirited young men, lads, and boys, to whom the greatest encouragement and

Highest Bounty will be given.

From the character of the officers of the regiment who are from this part of the Highlands, they can depend that the interest and advantage of high-spirited and well-conducted soldiers from this part of the country will be particularly considered.

Printed at the *Journal* Office, Inverness.

It was a long tramp to bring recruits to the west of Ireland from districts north of Inverness, from which at this period the regiment got a great many men of the best class, who filled the vacancies in the 1st Battalion.

In 1811, the 2nd Battalion was moved to Scotland, and landed at Irvine on the 11th October. There are no records to show where it was quartered on arrival in its native land, but in 1813 it was in Glasgow,[1] from which city it marched on the 1st of August 1814, and occupied Edinburgh Castle on the 3rd. Here it remained till the peace of 1814, when it was reduced on the 24th October, having existed exactly eleven years, for though placed on the establishment from the 9th of July 1803, it was only formed at Weely on the 24th of October of that year. The officers were placed on half-pay; many of the n.c. officers and men were discharged to pension, the remainder were transferred to the 1st Battalion, taking with them the colours. The n.c. officers were borne as supernumeraries on the strength of 1st Battalion till vacancies occurred.

[1] The battalion seems to have been at Ayr before Glasgow. A letter from Lieutenant Hector Innes, dated Ayr, 20th September 1813, says that he with eight other officers are under orders "to join our brave friends, I may say *the remains of the Ninety-twa.*" They were twenty days from England to Santander!

Since the formation of the 2nd Battalion, it had never been necessary to receive volunteers from other regiments of the line as had been the case on several previous occasions, and notwithstanding the great drain of constant campaigning, the 1st Battalion had been kept complete with men in whose keeping its character was safe, both in quarters and in the field.

On the 26th and 27th of March 1815, the regiment, now consisting of one battalion, marched to Cork. "The colours of the 2nd Battalion will march with the left wing."

At Cork great attention was paid to putting the barracks into a thorough state of cleanliness. The commanding officer warns the officers against the "shocking practice" of signing without examining papers put before them by their n.c. officers. All copies of muster rolls, etc., to be in the officer's own writing.

There were still one drummer and twenty-four privates sick in the Peninsula; six sergeants, six corporals, twenty-six privates were recruiting at Inverness, Elgin, etc.[1]

It appears that a quarrel in a ball-room, resulting in a duel, had taken place between two officers. "It is with a feeling of deeper regret than he can express that Colonel Cameron has heard of an occurrence which, as it is the first of its kind in the 92nd Regiment he has heard of since he has been in command, he sincerely hopes it may be the last." . . . "He has been accustomed to feel proudly conscious of the high sense of propriety of conduct, together with the unanimity and harmony which prevailed among the officers of the 92nd." . . . "It is some palliation that this interruption had proceeded from very young men, of but very short service in the regiment, though by no means a sufficient excuse"; and the Order continues "that while he has the honour to command the 92nd, he will use his power to rid it of parties concerned in such transactions either as principals or accessories, the latter being frequently more to blame than the former."

In April the 92nd was inspected, and made "that soldier-like appearance under arms which has ever marked the regiment."

The Orders show that the women were employed to cook the men's dinners.

[1] On the 5th February the establishment was:—Sergeants, 65; corporals, 60; drummers, 22; privates, 1140; wanting to complete, 243 privates. Reduced 20th February to 55 sergeants, 50 corporals, 22 drummers, 950 privates; wanting to complete, 60 privates; but in the total effective were more n.c. officers and drummers than the establishment, the surplus being supernumeraries from 2nd Battalion.

CHAPTER XXIII

MEANWHILE a thunderbolt had fallen, causing consternation throughout Europe. Napoleon had secretly left Elba on the 26th of February. He was accompanied by about 1100 of his guard, whom he had been allowed to retain in his island kingdom, and with these, under Drouet (Count d'Erlon), he landed in France near Cannes on the 1st of March. On the 20th he entered Paris at the head of an army which had joined him as he advanced. Louis XVIII was compelled by the defection of his troops to withdraw from Paris to Ghent; and Napoleon, supported by the army, though not generally by the people of France, assumed his former title of Emperor of the French. The allied Powers refused, however, to acknowledge his sovereignty; Great Britain agreed to replenish the exhausted treasuries of the Continental nations; preparations for war were at once begun, and the Gordon Highlanders were ordered on active service.

The regiment embarked at the Cove of Cork, Headquarters on the *Atlas* transport. The embarkation strength was :—Field-officers, 3; captains, 6; lieutenants, 19; ensigns, 8; staff, 4; sergeants, 47; drummers, 16; rank and file, 621. 1 captain, 2 lieutenants, 8 sergeants, and 8 corporals, with the men left, were shortly after formed into a recruiting company, and sent to Scotland.

The regiment landed in boats near Ostend on the 10th of May, proceeded by canal to Bruges and next day to Ghent, where it was quartered. The troops disembarked in the market-place on the market-day, and the novelty of the scene made as great an impression on the Highlanders as their garb did on the country people, who treated them very civilly. At Ghent they were joined by the Royal Scots, 42nd, and 79th, and, says the sergeant, "a happier meeting could not have taken place, so many Scotchmen who had fought side by side in Egypt, Denmark, Spain, and France." The British regiments furnished a guard at the residence of King Louis; the officers were admitted to the royal presence, and dined at one of the tables.

In Orders the commanding officer compliments the men on their good conduct, and lessens the patrol duty on that account. The Peninsular Regimental Orders as to the line of march are repeated. The two field-officers are to be alternately riding along the flanks of the column to see that the proper order of march is attended to; no man on any pretence to be allowed to fall out without taking off his knapsack, which is to be carried by two of his comrades till he rejoins his company; officers paying companies to ride; officers to provide themselves with baggage animals; batmen with baggage to be regimentally dressed, with arms and accoutrements; the women at all times to move along with the baggage. "From the present weak state of the regiment, the commanding officer hopes they (the officers) will see the propriety of his requesting them not to take their servants to the rear when they go on duty or otherwise." [1]

The regiment marched at four in the morning of the 27th of May to Alost, along with the 28th, 32nd, 42nd, 79th, and 3rd Battalion 95th, all under the command of Colonel Cameron.[2] On the 28th they entered Brussels, where the 92nd was placed in the Ninth Brigade commanded by Sir Denis Pack, K.C.B., along with the 3rd Battalion 1st Royals, the 42nd, and the 2nd Battalion 44th. The Ninth Brigade belonged to the Fifth Division, which was under Lieut.-General Sir Thomas Picton, K.C.B.

The regiment was inspected by Sir Denis Pack on the 1st of June, and with the Fifth Division on the 3rd by Field-Marshal the Duke of Wellington, accompanied by Field-Marshal Prince Blücher. The Duke expressed his approbation of their appearance—" He was happy again to see some regiments that had served with great reputation in the Peninsula." [3]

The orders for inspection desire particular attention to be paid to the gartering of the hose ; best black gaiters to be worn ; camp kettles, bill-hooks, blankets to be carried.[4] Officers in blue web pantaloons without lace on the seams, and half-boots, jackets buttoned across, and gorgets. Field-officers with the Highland scarf. The above is no doubt the dress in which the officers of Highland regiments fought at Waterloo, except that the half-boots were replaced by shoes and gaiters (according to Regimental Orders for the actual line of march). The n.c. officers and men wore the Highland dress without purses.

While at Brussels each company was divided into three classes or squads, and were exercised in drill according to their proficiency. All were practised in firing ball and in turning out quickly with tents and baggage packed.

B.O., Bruxelles, 11th June 1815.—Officers will find it through life a useful maxim to be rather an hour too soon than one minute too late.

[1] A letter from Colonel Cameron shows that the regiment was not only weak in numbers, but that some of the young soldiers were not such strong men as he was accustomed to.

[2] General Orders, 27th May.

[3] Many of the Peninsular regiments were in America, and some battalions employed in the present campaign had no experience of war, and were composed to a great extent of very young soldiers and volunteers just received from militia ; but it should be remembered that the militia had been many years embodied and doing the duty of regulars, and that in all regiments the n.c. officers and many of the men were old soldiers, who steadied their younger comrades.

[4] *General Orders, Brussels, 31st May.*—Soldiers' greatcoats to be taken into store in order to lighten the weight ; only blankets to be carried on the knapsacks. Men to be paid daily.

At this time the Headquarters of the Prussian army under Marshal Blücher were at Namur ; Wellington was at Brussels, having over 25,000 men in and about that city, but the cantonments of both the allied armies extended to great distances. Wellington appears not to have expected an attack before the end of June, and Blücher wrote, "We will soon enter France ; we might stay here a year, for Napoleon will not attack us." But the Emperor had determined otherwise. Taking every precaution to prevent his movements being reported, he left Paris on the morning of the 12th of June and arrived on the 13th at Avesnes, where he found 122,000 men present under arms, while large numbers were on the way to join. They were seasoned soldiers who had been serving under the king, or who had rejoined the colours at the call of Napoleon, whose arrival raised their enthusiasm to the highest pitch. He hoped first to defeat the Prussians, that Wellington would then retreat, and that he would enter Brussels in triumph.

That city was filled with fashionable non-combatants of all nations ; numbers of British who had long been cut off from the Continent were drawn thither by curiosity, or from having relatives in the army. There was no sense of immediate danger, "and all went merry as a marriage bell." Among the social entertainments none was so much talked of as the now historic ball given on the 15th of June by the Duchess of Richmond, the same daughter of the Duke of Gordon who, when the 2nd Battalion was in Ireland, "made a great work with them." Colonel Cameron was among the limited number of guests, and a party of n.c. officers were also invited to show Her Grace's British and foreign friends a specimen of Highland dancing.[1]

Wellington was aware of the concentration of the French army, and his troops were all warned to be ready to march at a moment's notice ; but he did not hear till the evening of the 15th that Napoleon had crossed the Belgian frontier near Beaumont at daybreak that morning, had driven the Prussians out of Charleroi, and sent Marshal Ney towards Quatre-Bras, while he himself

[1] Having heard, when I joined the regiment in 1851, the tradition that four sergeants of the 92nd danced a reel and the sword dance at the famous Waterloo ball, I had the curiosity, in 1889, to ask Lady de Ros, a daughter of the Duke of Richmond, who was present, if she remembered this interesting incident. Her soldier brother had got orders to march ; she was at the moment helping his preparations and had only a faint recollection of it. She, however, wrote to her younger sister, Lady Louisa Tighe, who said in a letter dated January 13th, 1889 :—

"I well remember the Gordon Highlanders dancing reels at the ball ; my mother thought it would interest the foreigners to see them, which it did. I remember hearing that some of the poor men who danced in our house were killed at Waterloo. There was quite a crowd to look at the Scotch dancers."

The proud march of this little party into the hall, preceded by their pipers, formed a fitting prelude to the deeds that were about to distinguish their garb in the red field of battle. —Circumstantial account of Waterloo, 1816.

followed the Prussians, who retired to Ligny. Historians differ as to the time when Wellington heard of these movements. The 92nd officer whose journal I have so often quoted mentions that while walking in the park at seven o'clock on the evening of the 15th, he met an acquaintance who had dined with the Duke, who advised him to go home at once and pack up, telling him that during dinner Wellington had received a dispatch from Blücher that he had been attacked. On the cloth being removed, His Grace filled his glass and gave the toast, "Prince Blücher and the Prussian army ; success to them." He then rose from table and at once dispatched the necessary orders to every division of the army.

Wellington attended the ball, thinking that his presence there would have a reassuring effect on the people in Brussels ; he remained till past midnight, constantly, however, receiving messages and giving orders privately to the staff ; he then quietly withdrew. The officers gradually left the room and joined their regiments, many of them in their dancing-pumps.

Meanwhile bugles and bagpipes resounded through the streets, and the soldiers came swarming out like bees. Four days' bread and biscuit, which had been in possession of the quartermasters, was issued on the private parades, but this was so bulky that many of the men left part with the people on whom they were billeted. A gentleman who witnessed their prompt turn-out wrote, "The Highlanders have considerable tact in domesticating themselves in quarters, and are in general so quiet, sober and orderly, that they become great favourites everywhere ; they nursed the children and generally assisted in the household labour." The Belgians called them "les bons petits Ecossais," and bid them a sorrowful adieu, thinking the kindly Scots, who rocked the cradle and helped to make the soup, would be no match for the fierce-looking French grenadiers.

The whole army had been directed to concentrate on Quatre-Bras, but the Fifth Division, which had assembled in the park, did not march off till sunrise on the 16th, when they left the city by the Namur Gate. A lady mentions seeing tearful partings between soldiers and their wives and children, but that many of the wives marched with the regiments. "The 42nd and 92nd Highland Regiments marched with their bagpipes playing before them, while the bright beams of the sun shone on their polished muskets and on the dark waving plumes of their bonnets. We admired their fine athletic forms, their firm, erect, military demeanour and undaunted mien. We felt proud that they were our countrymen."

They halted on the verge of the Forest of Soignies, close to the then obscure village of Waterloo. Here, while they cooked and rested, Wellington rode past towards Quatre-Bras, a hamlet situated where the road from Brussels to Charleroi crosses that from Nivelles to Namur, eighteen miles from Brussels.

The position was important, as it commanded the communication from Nivelles and Brussels with Blücher's position at Ligny. It was held by the Prince of Orange, with only 7800 bayonets and 14 guns,[1] and Napoleon had ordered Marshal Ney to take possession of it. The French General had a corps of 19,000 infantry, 3500 cavalry, and 64 guns, and a reserve of 20,000 men,[2] who did not, however, come into action ; but he was uncertain what force was opposed to him. " It might be like a battle in Spain," said Reille, " where the British only showed themselves at the right moment," and Ney waited till more of his troops had come up. Wellington arrived at ten o'clock, took in the situation, and sent orders to Picton and the Brunswick Corps to continue their march on Quatre-Bras. Accordingly, the 92nd advanced with the division. They met some waggons with Prussian soldiers wounded on the previous day. They moved slowly, for the heat was excessive and the dust choking ; at Genappe the inhabitants had tubs of water, and some of them milk, ready to hand to the thirsty soldiers as they passed along the street.

Hardly were they out of the town when the stimulating sound of cannon freshened them up, so that the men quickened their pace and " Forward " ran through the ranks ; but Colonel Cameron knew better than to allow his men to exhaust their energy, and kept them to the ordinary rate. The sounds of strife increased as they neared Quatre-Bras, where they arrived about 2.30, in the nick of time. Wellington had just returned to Quatre-Bras after a meeting with Blücher, who, from the top of the mill of Bussy, had shown him Napoleon taking up his ground, and the dispositions of the Prussian army. Wellington did not approve of these. " If they fight here," he remarked, " they will be damnably mauled ";[3] and they were.

The hamlet of Quatre-Bras consisted of three farms and two other houses overlooking the undulating corn-fields round it. To the east, the road to Namur was fenced by a bank and ditch ; less than a mile and a half along this road, and five hundred yards south of it, stood the farm of Piraumont ; on the south-west of the road to Charleroi, about two miles from Quatre-Bras, was the farm of Pierrepont, and half a mile behind Pierrepont was the coppice wood of Bossu, extending back to the hamlet. The large farm of Gémioncourt, three-quarters of a mile along the road to Charleroi, and close to it, formed, with the other farms and the wood, natural positions of defence.

[1] Houssaye. [2] *Ibid.* [3] Sir H. Maxwell's " Life of Wellington."

Ney had at last begun the action by a cannonade on the farm of Piraumont. He then sent a brigade of infantry and Piré's Lancers against it. The Dutch troops who held it were too few to sustain the attack ; they were obliged by the infantry to retreat and were routed by the cavalry, so that the French were masters of the position except on the right, which was held by four battalions under Prince Bernard of Saxe-Weimar ; and a strong body of French cavalry and infantry were marching to attack these battalions.

It was at this critical moment that the Fifth Division, followed by the Brunswickers, arrived, and without halting, deployed on the Namur road, the Ninth Brigade next the hamlet, the Eighth farther down the road. The 79th formed the extreme left, and the 92nd the right, of the division, which instantly advanced and became engaged with the enemy in the wheat-fields towards Gémioncourt. Wellington, however, desired Colonel Cameron to form the 92nd in line on the road, with his right resting on the houses of Quatre-Bras. His Grace took his station on foot with his staff at the left of the regiment.[1] The enemy opened a cannonade on them from a battery on a height near the road to Charleroi ;[2] the Duke ordered the regiment to lie down under cover of the ditch and bank of the road, and he and his staff did the same. Soon the terrible French cuirassiers, under cover of their artillery, with loud cries came charging up the fields in front of the regiment, which still remained *in line*, and was not afraid. Wellington moved in rear of the centre and said, " 92nd, don't fire until I tell you "; and when they came within twenty or thirty paces he gave the word. A crashing volley —and the ground was covered with fallen horses and their steel-clad riders, while the rest were galloping away. Many, dismounted but unhurt, tried to escape, but in their jack-boots were no match for the Highlanders ; some who resisted with their swords felt the bayonets of the angry Scots ; those who surrendered were made prisoners.

Meanwhile the rest of the division were sustaining with equal confidence the repeated charges of the French cavalry, who rode

[1] " Waterloo Letters," 165.

[2] " Did you see that, sir ? " said a soldier to his officer during this cannonade, as a cannon ball sent his bonnet flying. He was not hurt, though he had a wild look, and the concussion addled his brains for a day or two. During the day shells frequently came dancing to their own music across the road which the Highlanders were lining. " When one of these dangerous characters intruded himself into our society, a ludicrous scramble took place for the honour of being undermost in the ditch." Another 92nd officer says, " This heavy fire was maintained against us in consequence of the Duke and his staff being only two or three yards in front of the 92nd, perfectly seen by the French, and because all the reinforcements passed along the road in which we were. Here I had a remarkable opportunity of witnessing the *sang-froid* of the Duke, who, unconcerned at the shot, stood watching the enemy and giving orders with as much composed calmness as if he were at a review."

boldly up to the bayonets, firing carbine or pistol in the hope of breaking the squares. Their indefatigable lancers got near the 42nd, who at first mistook them for Allies; two companies were cut off before they could form square, and suffered severely, but a number of the lancers got hemmed in between the square and the remainder of the two companies, and were instantly bayoneted. The French infantry, assisted by a powerful artillery and magnificent cavalry, behaved with the greatest gallantry in their endeavours to force their opponents to retire. The British had only a few guns and the mounted corps of the Duke of Brunswick, but they withstood every assault with astonishing firmness; the tide of battle ebbed and flowed, fringed by a foam of French skirmishers, and of the Light Companies and marksmen of the British Brigades.

As the day wore on allied reinforcements kept arriving, but some had no stomach for the fight, and spread dismay as they fled by false statements of French success. Mercer, marching with his troop of Horse Artillery on the Nivelles road towards the scene of action, met numbers who cried that all was lost and the British defeated. At last a Gordon Highlander came limping painfully along, with his musket under his arm and a bullet in his bandaged knee. Mercer halted to ask for information, telling him what the Belgians had said. "It 's a damned lee. They were fechtin' when I cam' awa', an' they 're fechtin' yet." He then sat down on a wall and lit his pipe, while the artillery surgeon extracted the bullet.

The Brunswickers had been stationed near the 92nd, and about four o'clock the Duke of Brunswick led his hussars (who wore black uniforms with a silver death's head in memory of his father, who had been killed in fighting the French) past the 92nd, to repel some French cavalry who had broken the Brunswick Infantry; the hussars charged gallantly, but were disorganised by musketry fire. The Duke was mortally wounded, and they fled before the Red Lancers and Light Cavalry, who were soon up with them, "cutting them down most horribly." Wellington, who was in front when the hussars charged, was carried away in their flight, and in danger of being taken; he galloped to the bank lined by the 92nd, and calling to them to lie still, rode at the fence and jumped it, men and all. He had his sword drawn, and as soon as the Highlanders were between him and his pursuers, he turned round with a confident smile, and ordered the regiment to be ready. The flying Brunswickers passed round the right flank close to the bayonets; the French mingled with them, cutting and slashing. As soon as the Brunswickers had cleared their right, the Grenadier Company wheeled back on the road; the rest were to fire obliquely on the mass of rapidly advancing cavalry, who, elated at their success, charged them. The Highlanders received them as they did the cuirassiers, with a volley at close quarters. "The volley was decisive. The front of the French charge was completely separated from the rear by the gap which we made, and nothing was seen but men and horses tumbling on each other; their rear retreated, and the front dashed through the village, cutting down all stragglers; our assistant-surgeon, dressing a man behind a house, had his bonnet cut in two and a lance run into his side."[1] The survivors, not expecting such a finale to their victorious career, fled in confusion. One of their officers, having got too far in front of his men, made a neck or nothing dash to escape along the rear of the 92nd. Some thought he was riding at the Duke, and one or two mounted officers tried to stop him, but he parried their cuts and galloped down the line past His Grace, who called out, "Damn it, 92nd, will you let that fellow escape?" Some men of the 6th and 7th Companies turned round, fired, killed his horse, and a ball at the same time passed through both feet of the gallant young officer. Lieutenant Winchester, 92nd, wounded at Waterloo, was afterwards billeted with him in the same house at Brussels for six months, and travelled with him to Paris, where he received much kindness from him and his family; his name was Burgoine. Three others came down the road at full speed on the Grenadiers, brandishing their swords; two were brought down; an officer with his sword cut the hamstring of the other's horse, and he was taken.

Ney had received an urgent dispatch from the Emperor, desiring him to act vigorously, to drive away the British and to second his own attack at Ligny on the Prussians, so as to destroy them wholly. "The fate of France is in your hands," he wrote. Ney saw that his enemy was being reinforced, and that if Napoleon's orders were to be carried out, it must be now or never. He renewed his efforts on the key of the position, and about five o'clock the regiment was again charged by two columns of cavalry, who were received in the same manner as their predecessors, and with the same result. Some of the defeated cuirassiers sought shelter in the farmyard, remained there for some time unobserved, and finally galloped out and escaped.[2]

At this time the Third Division arrived, and prolonged the left of the British line; the Guards also came up, and were thrown into the wood of Bossu. As each battalion hurried past, it was loudly cheered by the 92nd, for never was help more welcome. The regiment, however, had little respite; the most severe contest of the day was yet to come.

[1] Letter from a wounded officer, 92nd, dated Brussels, 21st July 1815.
[2] Letter from Wellington.

Two hundred yards from the hamlet was a two-storeyed house beside the road, and from its rear ran a thick hedge a short way across a field. On the opposite side of the road was a garden surrounded by a thick hedge. Under cover of a cannonade, two columns of French infantry were now seen advancing, one by the Charleroi road, the other by a hollow in front of the wood of Bossu. The house and hedge were occupied by these troops, while some advanced still nearer to the 92nd, and the main body, 1200 to 1500 strong, took post about a hundred yards in rear of the garden. The colonel's fiery temper chafed to be at them, as he paced impatiently up and down, and he asked leave to charge. "Take your time, Cameron, you 'll get your fill of it before night," said the Duke, who was again with the regiment, for their position commanded a general view. Soon he said, "Now, 92nd, you must charge these two columns of infantry!" Instantly the regiment, about 600 in number, leaped over the ditch, headed by Colonel Cameron and General Barnes, the latter calling out, "Come on, my old 92nd!"—for their old brigadier loved the regiment, and though it was not his duty to charge, "he could not resist the impulse," writes a 92nd officer.[1] The Grenadiers and First Company took the high road, the other companies to their right advanced upon the house and hedge, and some upon the garden, the enemy pouring a deadly fire on them from the windows and from the hedge. The officer with the regimental colour was shot through the heart, the staff of the colour was shattered in six pieces by three balls, and the staff of the king's colour by one.[2] Cameron was struck in the groin by a shot fired from an upper window; he lost command of his horse, the animal galloped back to where the colonel's groom was standing with his second horse. There it suddenly stopped, and its rider was pitched on his head on the road.

"It was hot work then," said an old Highland soldier. "They were in the hoose like as mony mice, an' we couldna get at them wi' oor shot when their fire was ca'in' doon mony a goot man among us; but we had seen Cameron fall, an' oot o' that they had to come, or dee where they were; so we ower the hedge an' through the garden till the hoose was fair surrounded, an' they couldna get a shot oot where we couldna get ane in. In the end they were driven oot, an' keepit oot. Ay, but the French were brave men, an' tried again an' again to take it from us, but they only got beaten back for their pains, and left their dead to fatten the garden ground."

At the house the bayonet did its deadly work, and Cameron

[1] "Military Memoirs." General Barnes was adjutant-general. He had commanded the First Brigade at St Pierre, etc., 1814.

[2] Letter from 92nd officer, dated 21st July 1815.

was amply avenged by his infuriated followers,[1] but the main body beyond showed no disposition to retire ; they were greatly superior in numbers, and kept up a shower of musketry upon the Highlanders at the hedge, sufficient to appal any but the most experienced soldiers ; and but for the officers' attention to their duty, and the encouragement of their veteran comrades, some of the younger men would not have been kept steady under it. Lieut.-Colonel Mitchell, who had succeeded Cameron, was wounded, and Major Donald MacDonald took command. To extricate themselves from this situation part of the battalion was moved round the right side of the house and garden, and part round the left, while a third passed through the garden and forced the gates under a fire described by an officer of the party as "truly dreadful," and which they could not answer effectively till the whole battalion could be brought to bear upon the enemy. Then the word was given—the thrilling British cheer, which none can hear unmoved, rang out as they levelled their bayonets to the charge. For a few seconds the French awaited the assault, but as they marked the steadfast bearing and stern countenances of the rapidly-approaching Highlanders, their courage failed them, they gave way, and sought to escape by the hollow up which their left column had advanced. As soon as they turned their backs, the 92nd plied them with musketry ; "And so well did our lads do their duty, that at every step we found a dead or wounded Frenchman." "Never was the fire of a body of men given with finer effect than that of the 92nd during the pursuit, which continued for fully half a mile."[2] Being completely separated from the rest of the line, and threatened by a corps of cavalry, the regiment entered the wood and maintained their position there against all comers till relieved by the Guards, and about eight o'clock they were ordered to retire, and form behind the houses at Quatre-Bras. A night under arms, an eighteen miles' march, and five hours' fighting is hungry work ; "Malgré la chute des empires le dîner vient tous les jours" (Dinner time comes round though empires fall) ; the Highlanders cooked their suppers in the cuirasses worn by the cuirassiers they had killed a few hours before.[3]

In other parts of the field Wellington was equally successful, but the French gave way only foot by foot. Everywhere the mass of dead and wounded—over 4000 French and over 5000 of the Allies—bore witness to the fury of the fight.

[1] "The History of the Wars," describing the battle, says—"This heroic regiment (92nd), led on by Colonel Cameron, performed prodigies of valour." . . . "His death roused the spirit of the Highlanders to fury."

Another account says—"The bagpipes screamed out the notes of the 'Camerons' Gathering' as they levelled their bayonets and charged with the elastic step learned on the hillside."

[2] "Military Memoirs." [3] "Waterloo Letters."

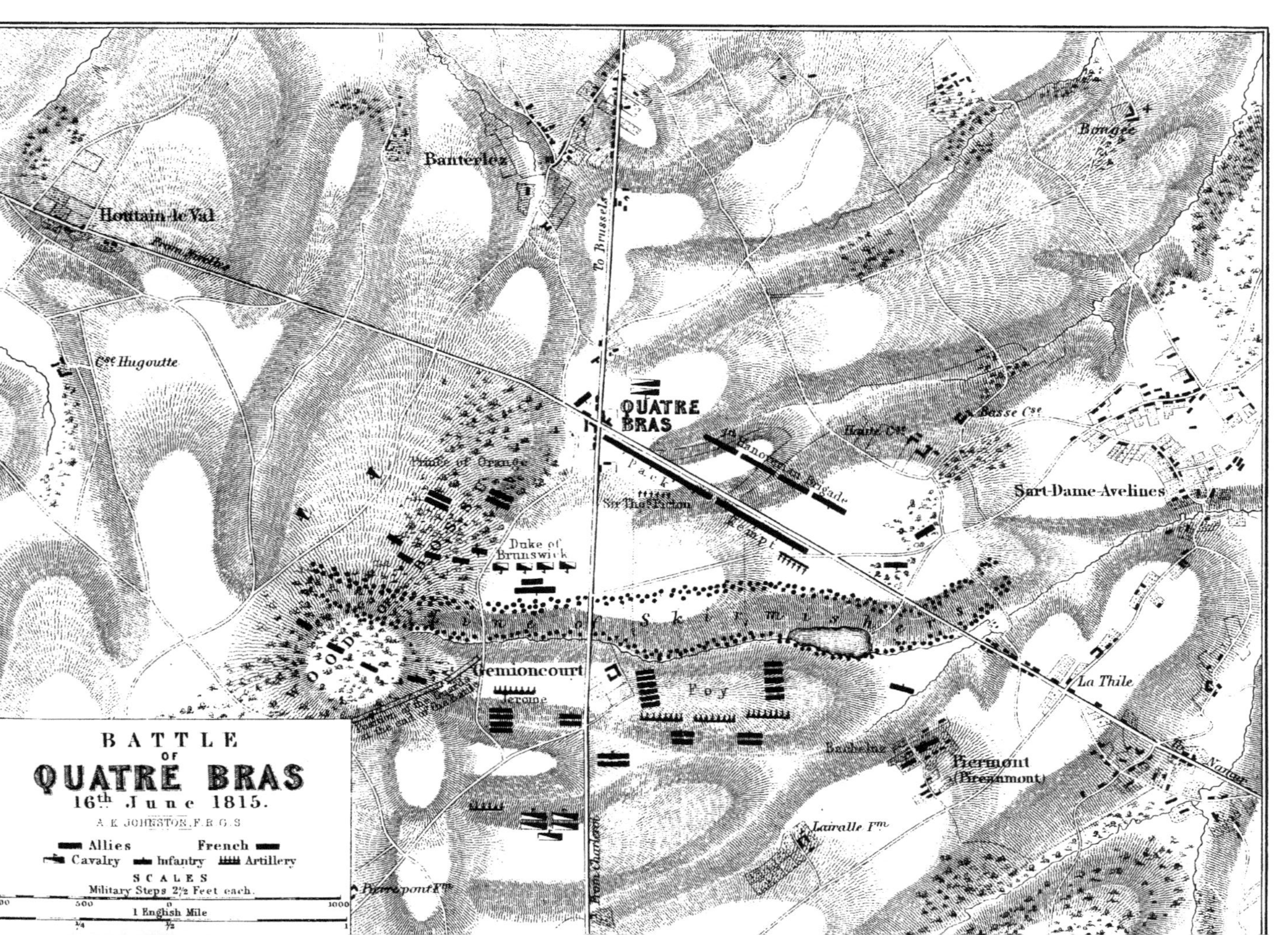

BATTLE
OF
QUATRE BRAS
16th June 1815.
A. K. JOHNSTON, F.R.G.S.
Allies
French
Cavalry
Infantry
Artillery
SCALES
Military Steps 2½ Feet each.
1000
500
0
1000
1 English Mile
0
¼
½
1
Houttain-le-Val
From Nivelles
Banterlez
Bongée
To Brussels
Cse Hugoutte
QUATRE
BRAS
Prince of Orange
Duke of
Brunswick
Pack
Sir Thos Picton
Kempt
Hanoverian Brigade
Haute Cse
Basse Cse
Sart-Dame-Avelines
Line of Skirmishers
WOOD
Gemioncourt
Jerome
Foy
La Thile
Bachelaz
Piermont
(Pireaumont)
Lairalle Fm
Pierrepont Fm
From Charleroi
To Namur

All firing had ceased by nine o'clock. The French reoccupied the ground they held in the morning ; the Allies had maintained their position, and completely defeated and repulsed the enemy's attempt to get possession of it. Wellington in his dispatch says—"The troops of the Fifth Division and those of the Brunswick Corps were long and severely engaged, and conducted themselves with the utmost gallantry. I must particularly mention the 28th, 42nd, 79th, and 92nd Regiments, and the battalion of Hanoverians."

"About ten o'clock," says an officer, "the pipe-major took post in front of the village." . . . "Long and loud blew Cameron, but his efforts could not gather above half of those whom his music had cheered on their march to the battlefield." Of thirty-six officers who went into action only eleven escaped unhurt. Colonel Cameron was mortally wounded ; Captain W. Little, Lieutenant James J. Chisholm, Ensigns Abel Becher and John Ross MacPherson were killed ; Lieut.-Colonel James Mitchell, Captains G. W. Holmes, Dugald Campbell, and William C. Grant (mortally); Lieutenants Robert Winchester, Thos. Hobbs, Thomas M'Intosh, James Ker Ross, Ronald MacDonald, Hector Munro Innes, George Logan, John M'Kinlay,[1] George Mackie, Alexander MacPherson, Ewen Ross ; Ensigns John Branwell, Robert Logan, Angus MacDonald, Robert Hewitt, and Assistant-Surgeon John Stewart were wounded ; 35 n.c. officers and men were killed[2] and 245 were wounded, of whom many died of their wounds. I have not the morning state of the 16th, but comparing that of the 18th with the number killed and wounded, it seems that about 650-660 sergeants, drummers, and rank and file went into action.

Sir Thomas Picton, commanding the Fifth Division, on returning to the rear after the action, not knowing how severely the regiment had been engaged till he saw the remnant of it, asked what this was. On being told it was the 92nd, he asked, "Where is the rest of the regiment ?"

The following is a copy of Major MacDonald's report of the action, written on the spot :—

Sir,—Colonel Cameron and Lt.-Col. Mitchell having been both severely wounded, I have the honour to report, for your information (not having been under your eye during the whole of the day), that the 92nd Regiment repulsed repeated attacks of cavalry, and by a rapid movement charged a column of the

---

[1] Lieutenant M'Kinlay belonged to the neighbourhood of Callander, and from him or his family was descended President M'Kinley of the United States of America.

[2] The names of the killed and those who died of their wounds are included with those of Waterloo. Cannon does not notice it, but Captain W. C. Grant died of his wounds at Brussels.—Stewart's "Highlanders," and "Scots Magazine" (Obituary).

enemy, and drove them to the extremity of the wood on our right. Our loss has been severe, as will be seen by the return of killed and wounded.

I have, etc.

(Signed) DONALD MACDONALD,
Major 92nd Commanding.

Major-General Sir Denis Pack, K.C.B.

After Colonel Cameron fell, he was taken on a cart by his faithful foster-brother and servant, Private Ewen M'Millan,[1] and one of the pipers, to the village of Waterloo, where he died during the night. Being told that the British had been victorious, he said, "Then I die happy"; and later, "I trust my dear country will think I have done enough; I hope she will think I have served her faithfully." His last hour was soothed by the plaintive music which awakes a Highlander's deepest feelings and most sacred memories of home and kindred, and his last supplications to the Throne of Grace were uttered in the mountain tongue.[2]

In Colonel Cameron the Gordon Highlanders lost an officer who had served with them in every campaign except Corunna; he had commanded a battalion since 1806, and had led them "always to honour and almost always to victory." His influence had a lasting effect on the character of the regiment. A strict disciplinarian, he understood that prompt obedience can be enforced without worrying officers or men, and that discipline and drill are but means to the end of making troops cool and handy in times of danger and difficulty. He was deservedly respected by his officers, to whom, however, his manner was rather reserved and distant. Thoroughly knowing his duty, he did not fear responsibility, and "never allowed the rights or comforts of his men to be disregarded or lost sight of by any one; they considered him their best and never-failing friend, and reposed the most implicit and unbounded confidence in him as a commander."[3] The effects of his fiery nature were never more severely felt than by the officers of the commissariat, if his men suffered from any neglect at the hands of that department.

A most respectable old soldier, John Downie by name, used to speak of him as being much liked by the good men, "though he made us do our duty"; "a splendid soldier"; "his only fault, his reckless bravery." The late Sir Duncan Cameron having heard it stated that his brother sometimes carried discipline to excess, sent for two Ballachulish men who had served in the 92nd. The first was a cunning calculating customer, whose desire was to please by speaking smooth things, and he made out the colonel's

[1] M'Millan was afterwards tenant of the farm of Carnas.—"Memoir of Colonel J. Cameron."

[2] "Memoir of Colonel Cameron."

[3] Letter from Captain Fyfe, 92nd.

COLONEL JOHN CAMERON OF FASSIEFERN

From a miniature now at Callart, painted about 1814

character to be so perfectly angelic that Sir Duncan, seeing through him, ordered him off without even showing him the hospitality for which his house was famous. The other was a manly old soldier, who spoke as he thought and felt. Sir Duncan questioned him, "Dē nis an scorsa duine a bha 'nam bhrathair?" "Innsidh mise dhuibh a Shir Donnachadh; se Duine gasda, foghainteach, àilidh, a bha 'n Colonel Ian, fhad 'sa bha gnothaichean a dol gu math, ach 'se fior dheamhan a bh'ann 'nuair a rachadh camadh na corraige air aimreidh." ["Now, what sort of a man was my brother?" "I will tell you, Sir Duncan. Colonel John was a fine, brave, splendid man when duty was well done (business went well), but the very devil when anything went wrong (*lit.* when the bending of the finger)."] Sir Duncan asked no more questions. "Ni sin an gnothach," and dismissed him to his dinner in the hall with a £1 note in his pocket. In short, Colonel Cameron may be described as one who used his authority "for the punishment of evil-doers, and the praise of them that do well."

He was a born leader of men, besides being a true Highland gentleman, devoted to the music and the poetry of his country.[1]

His remains, which had been interred near the place of his death by his friend Mr Gordon the paymaster, his servant, and a few wounded men, were removed the following year. They were brought to Lochaber by a man-of-war, and were committed to their final resting-place at Kilmallie, on the shore of Locheil, in Argyllshire. Three thousand people of the country attended the funeral, headed by Lochiel, MacNeil of Barra, MacDonald of Glencoe, Campbell of Barcaldine, and many other Highland gentlemen.

And Sunart rough and wild Ardgour,
And Morven long shall tell,
And proud Ben Nevis hear with awe,
How, upon bloody Quatre-Bras,
Brave Cameron heard the wild hurrah
Of conquest as he fell.[2]

A monument was erected to his memory at Kilmallie by an arrangement between his family and his regiment. The epitaph, written by Sir Walter Scott, is given in Appendix XI. A monument was also raised at Kinrara by his first commanding officer, the Marquis of Huntly, to the memory of Colonel Cameron, Lieut.-Colonel Sir R. Macara, K.B., of the 42nd, also killed at Quatre-Bras, and the other Highland officers and soldiers who fell in the war.

The Duke of Wellington thus expressed himself in his dispatch

[1] In his pocket-book when he died was found a list of the men who joined with him in 1794, with notes as to each. See Appendix X.

[2] Sir Walter Scott's "Dance of Death."

to the Secretary of State, transmitting the lists of killed and wounded at Quatre-Bras and Waterloo :—

"Your Lordship will see in the enclosed lists the names of some most valuable officers lost to His Majesty's service. Among them I cannot avoid to mention Colonel Cameron, of the Ninety-second, and Colonel Sir Henry Ellis, of the Twenty-third Regiments, to whose conduct I have frequently drawn your Lordship's attention, and who at last fell distinguishing themselves at the head of the brave troops which they commanded.

"Notwithstanding the glory of the occasion, it is impossible not to lament such men, both on account of the public and as friends."

In acknowledgment of his distinguished services, Colonel Cameron's father was created a baronet.

Ailean Dall (Blind Allan), the bard of Glengarry, composed a lament, which is given in Appendix IX.

"While I live," says an eye-witness, "I shall ever retain a vivid recollection of the farmyard at Quatre-Bras on the evening of the 16th and morning of the 17th."[1] The whole yard was dyed with blood and the walls stained with it. A surgeon stated that at one period of the battle nearly 1000 wounded soldiers were in that narrow space. In those days there were no anæsthetics to save the shock of amputation to the system ; no antiseptics to keep the wound clean. The only means available to the surgeons to prevent hemorrhage was an application of boiling oil or pitch. There were no organised ambulances or trained nurses, but the soldiers' wives were always helpful on such occasions. Those of the wounded from Quatre-Bras who could walk went to Genappe ; the others, in carts to Brussels and Antwerp, where the people vied with each other in their attentions.

About five o'clock in the morning Wellington arrived from Genappe, where he had slept. It was cold, and on dismounting he asked, "Ninety-second, will you favour me with a little fire ?" The men quickly made a fire opposite the door of a little hut made of boughs of trees, which they endeavoured to make suitable for the Commander-in-Chief, His Grace expressing his thanks for their attentions. In this airy residence he received definite information that Blücher, after his defeat at Ligny the previous evening, was retreating on Wavre. "Old Blücher," he said, "has got a d——d good drubbing. He has gone eighteen miles to the rear ; we must do likewise. I suppose they will say in England that we 've been thrashed. I can't help it."

[1] Private Robertson, one of the wounded, showed this officer a Life of Sir William Wallace which had belonged to a dead Frenchman ; it appeared, he says, that the histories of Bruce and Wallace were as favourite reading with the French soldiers as with the Scots.

In passing by Brussels the year before, he had observed the position of Mont St Jean ; he now sent word to Blücher that he would establish himself there, and expect support from at least one division of Prussians. Should he march at once, or let the troops breakfast ? "I know the French," he said. "They won't attack till they have made their soup." And he decided to let his men rest. Here he gave his orders, and received the Prince of Orange and many other distinguished officers, among them Lord Hill, under whom the 92nd had served so long in the Peninsula, and whom they "dearly loved for his kind and fatherly conduct towards them." Officers and men joined in testifying their regard. Hill approached them hat in hand, much affected, and going among them spoke to officers and men "in a very kindly manner." The cheers brought Wellington hurriedly out of the hut, but on perceiving the cause, he laughed and seemed "quite delighted with the mark of respect which 'her nainsel' had paid to his friend and favourite general." Hill gave them a high character, to which the Duke replied that he himself knew well what they could do, and that by and by he would give them something to keep their hands in use.

All felt depressed when it became known that Blücher had retreated, but Wellington appeared unmoved and confident. A 92nd officer describes him walking quickly up and down for a time, "dressed in white pantaloons, half-boots, a military vest, white neckcloth, blue surtout and cocked hat," his left hand behind his back, and in his right a switch, one end of which he frequently put to his mouth ; interrupted often by a courier with dispatches, or stopping to talk with animation to the Prussian Envoy (Müffling). About ten o'clock the infantry were put in motion, but, to conceal his design from the enemy, the Duke kept the Third and Fifth Divisions in front of Quatre-Bras till the other troops were well on their way, when they also began their retreat. The 92nd formed the rear guard of the infantry, whose retreat was by no means hurried, for the Fifth Division halted half an hour near Genappe ; during which time some men of the division were tried for wantonly firing away their ammunition instead of drawing the charge at the end of the day, a practice dangerous to comrades and apt to give rise to false alarms. While the regiment was ascending the heights in rear of Genappe, the body of the gallant Duke of Brunswick was carried past on a waggon. The Black Hussars who guarded it pointed out to the Highlanders the fatal wound in his breast, and swore to avenge the death of their prince.

The cavalry, who had arrived from their various cantonments after the action of the 16th, covered the movement of the whole. They remained with Wellington, who still watched his adversary from Quatre-Bras, till the afternoon, when Napoleon, who now commanded in person, began to advance in force. There was a sharp affair of cavalry [1] at Genappe ; the Emperor himself could be seen by the British rear guard encouraging the pursuit. "Tirez! Tirez ! Ce sont des Anglais !" (Fire ! Fire ! They are English!) he cried, with an accent of anger and hate, to the gunners of his advanced guard.[2]

The day had become sultry and black clouds darkened the sky ; the first cannon-shot seemed to break them, peals of thunder and flashes of lightning mingled with the roar of the guns, a perfect deluge of rain turned the roads into water-courses and the fields into a morass ; it was with difficulty the horses could drag the guns, or the infantry their feet, from the moistened clay. At La Belle Alliance the 92nd were relieved by a corps of foreign troops, when they crossed the narrow valley, but before they took their place in position, the enemy were cannonading them from the heights of La Belle Alliance. This was between six and seven p.m. Shortly after, firing ceased on both sides.

[1] In this affair an officer of the 7th Hussars was wounded and taken before Napoleon, who questioned him as to the strength of the British cavalry. Getting an unsatisfactory answer, the Emperor became angry and treated the officer with discourtesy. An aide-de-camp interfered on behalf of the captive, with whom he withdrew, and whom he treated with the greatest kindness. The next time they met was at the wedding of the French aide-de-camp, Count de Flahault, to Baroness Keith of Tulliallan, cousin of the hussar, who was the father of Captain the Hon. E. Elphinstone of the 92nd.

[2] Houssaye.

THE HAMLET OF QUATRE-BRAS. (*From a sketch done three weeks after the battle, showing the graves of those who fell on the 16th June.*)

CHAPTER XXIV

ON arrival on the position in front of Waterloo on Saturday evening the 17th June, the regiment bivouacked in a ploughed field, a picket being thrown out to its front. Every one was drenched to the skin, the baggage could not be found, the peasants had removed the ropes from the neighbouring draw-wells, so that a drink of clean water could not be got; the men cut clover in a field to keep them off the wet ground, and unless one had something in his haversack, there was nothing to eat. Imagine the situation of a man, cold, wet and hungry, without fire, meat or drink, sitting on a bunch of damp grass with his feet deep in a puddle; a powerful army opposite led by the greatest general of the age—not a situation in which to feel heroic! "But," says Sergeant Robertson, "we tried to cheer our drooping spirits by the thought that we had never run out of the field; and the call began to pass from one to the other what we should do when we had beaten the enemy." A false alarm, owing to Belgian cavalry answering the sentries' challenge in French, caused them to stand to their arms at midnight; it still rained hard, but the men lay down again in fours, covered by their united blankets. At daylight on the morning of Sunday the 18th June the troops stood to their arms, shivering with cold; an allowance of gin was served out, which put some warmth into them, and appears to have been highly appreciated;[1] beef was also brought. Soon the day cleared and the sun shone out; the regiment was withdrawn from its position in the puddle to a drier one further to the rear, where they were able to light fires and cook, clean their arms and dry their clothes. Many were fast asleep when the order was given to fall in, prime and load, and take up their position; then Wellington, looking calm and confident, rode along the line on his favourite horse, named after his first European victory, "Copenhagen."

The allied army in position in front of Waterloo amounted to about 67,600 of all arms,[2] of whom 12,500 were cavalry, with 156 guns; about 24,000 were British, and 5800 were of the gallant King's German Legion in British pay. The rest were Hanoverians, Brunswickers, Nassau contingent, Dutch and Belgians; Napoleon, in estimating the chances of success, which he thought greatly in

[1] "Am bheil sin math, a Dhonnachaid?" asked a Highland captain of his piper as he quaffed the reviving spirit. "Nan robh bainne mathar cho maith, bhithinn a' deoghal fathast!" ("Is that good, Duncan?" "If mother's milk was so good, I would be sucking yet!")

[2] Alison. Wellington had in addition troops stationed at Hal, Antwerp, Ostend, etc., but these were not engaged.

On the 18th June the 92nd had present:—Field-officer, 1; captains, 2; subalterns, 15; staff, 4; sergeants, 27; drummers, 12; rank and file, 361.—Wellington's Dispatches.

N.B.—Several officers and men were able to take their places, though smarting from wounds at Quatre-Bras. Batmen, baggage guard, surgeons' orderlies are included, though not actually in the ranks.

his favour, considered one British soldier equal to one Frenchman, and one French soldier equal to two of Wellington's other troops. The British were not in composition and discipline equal to the army which fought on the Pyrenees and at Toulouse, with which its leader said he "could go anywhere and do anything." A large part of that army was in America, and was replaced by battalions which had seen no service ; but several of the most distinguished Peninsular regiments were there. Many of the Continental troops had served under Napoleon, the country of others had been conquered by him, and all were inclined to believe him invincible ; Wellington, therefore, placed most reliance on the British and the German Legion, the other troops being posted alternately between them.

The position taken by Wellington was along the ridge, half a mile south of Mont St Jean, a hamlet two miles south of the village of Waterloo, which is nine miles south of Brussels. From the hamlet to the crest of the ridge the ground rises gently, and along the ridge runs a road or lane from Wavre and Ohain on the east, to Braine l'Alleud on the west. This lane crosses the high road from Brussels by La Belle Alliance to Quatre-Bras and Charleroi ; on the east of the crossing it was fenced by hedges, and on the west it was formed by a cutting with high banks on each side. These banks and hedges were pierced for the passage of cavalry and artillery. From this road the ground slopes down into a shallow valley. About three hundred yards down the west side of the road to Charleroi stood the farm of La Haye Sainte, and on the other side of the road was a gravel pit. Near La Haye Sainte an abattis was placed across the road, which then crosses the hollow and ascends gradually to La Belle Alliance on another ridge parallel to that of Mont St Jean. About twelve hundred yards east of La Haye Sainte, and five hundred yards south of the Ohain road, were the farms of Papelotte and La Haye, and a little further south was the hamlet of Smohain. At some distance west of the Charleroi road, and about five hundred yards south of the Ohain road, was the chateau of Hougoumont, with walled garden and woods. The position had its right thrown back to a ravine near Merbe Braine, which was occupied, as were the house and gardens of Hougoumont and the farm of La Haye Sainte, and also Papelotte and La Haye, while a picket of the 10th Hussars was posted at Smohain. By the left Wellington communicated with Blücher at Wavre through Ohain, and the Marshal had promised to support him with one or more corps as might be necessary.

The French army, 74,000 strong, of whom over 15,000 were cavalry, with 246 guns, was superior in numbers, in horsemen, and in artillery, and was composed of experienced soldiers of the same nation, animated by an enthusiastic confidence in their leader and in themselves. Napoleon also expected Grouchy's Corps of about 20,000 men, which had been sent to observe the Prussians.

The French position was along the ridge or plateau of La Belle Alliance, opposite and parallel to that of the Allies. The distance from right to left of the army was less than three miles ; at no point were the opponents a mile apart, and in some places they were much nearer.

Two men seldom agree in their account of a fox-hunt or a battle, and I am not going to try to tell exactly what took place on the whole field of Waterloo, but only on that part of it where the Gordon Highlanders were posted, so far as it can be gathered from dispatches, histories, and the letters or reminiscences of those who were present. Suffice it to say generally that the battle was a defensive one on the part of the Allies. Wellington's object was to keep Napoleon at bay till the Prussians could join him. Napoleon strove to break through the British before Blücher could come up, but he was prevented from attacking till the sun had rendered the wet clay land fit for the operations of cavalry and artillery. At about ten to eleven o'clock [1] he commenced a furious attack on Hougoumont, held by a few foreign troops and a detachment from Byng's Brigade of Guards, who maintained the post throughout the day with the greatest gallantry. This attack was accompanied by a heavy cannonade upon our whole line, followed by repeated attacks of cavalry and infantry, occasionally mixed, but sometimes separate. In one of these the enemy carried the farm of La Haye Sainte, after a desperate resistance by the detachment of the King's German Legion who held it. They had expended their ammunition, and the enemy occupied the only communication there was with them. The repeated attacks of cavalry were uniformly repulsed by our artillery and infantry, and they afforded opportunities to our cavalry to charge. These attacks were continued till about seven p.m., when the enemy made a desperate effort with cavalry and infantry supported by artillery, to force our left centre near La Haye Sainte, which, after a severe contest, was defeated. Wellington observed that the troops retired from this attack in great confusion, and that the march of Bülow's Prussians by Frischermont upon La Belle Alliance had begun to take effect ; and as Marshal Prince Blücher had joined in person with a corps of his army to the left of our line by Ohain, he determined to attack the enemy, and advanced the whole line of infantry, supported by cavalry and artillery. The attack succeeded at every point, the enemy was forced from his position, and fled in the utmost confusion.

[1] Historians differ as to the hour. Wellington's dispatch says "about ten o'clock." Houssaye says about 11.35.

Many of the Allies kept their ground during the day, as firmly as the British and Germans, but others gave way, and some could not be rallied even in second line ; some without firing a shot rode off to Brussels, filling the city with consternation and dismay.

The Fifth Division, commanded by Sir Thomas Picton, was stationed immediately on the left (*i.e.* east) of the Charleroi road, having Bylandt's Dutch-Belgian Brigade in front. The Eighth or Kemp's Brigade formed the right of the division ; the Ninth or Pack's Brigade was some little distance to the left, and in rear of the Ohain road on the reverse of the ridge, in the following order : the 1st Royals on the right, 42nd, 92nd, and 44th on the left, the Light Companies being extended in front. On the left of Pack were Hanoverian and Belgian infantry, some of them holding the farm of Papelotte, supported on the extreme left by Vandeleur's and Vivian's British Light Cavalry, while the Third and other Divisions were on Picton's right. The reserve, including most of the cavalry, was in the rear, and artillery was posted at intervals along the front line.

Napoleon's only fear had been that the Allies would retreat during the night to join the Prussians. " Now," he said, " Wellington has thrown the dice, and they are for us." Napoleon had never encountered British infantry. Soult, who knew their quality, advised him to hasten Grouchy's recall. " Because you have been beaten by Wellington," retorted the Emperor, " you think him a great general. I tell you Wellington is a bad general, and the British are bad troops." He then asked Reille's opinion, as he also had often fought against them in Spain, who answered, " Well posted as Wellington knows how to do, and attacked in front, I look on the British infantry as invulnerable by reason of its calm tenacity and the superiority of its fire, but they do not manœuvre so quickly ; if one could not vanquish them by a direct attack, one might do so by out-manœuvring them."[1] Napoleon broke off the conversation with an exclamation of incredulity, and proceeded to pass his columns in review as they took up their ground. It was little more than 1000 yards from where the Emperor made this his last inspection to the station of the Gordon Highlanders, who could see the formidable force of their antagonists, and with a telescope could distinguish the gorgeous columns ; the infantry in blue coats, red epaulettes, white breeches and long white gaiters ; light infantry in blue, green epaulettes and black gaiters. The Old Guard in campaigning dress, blue pantaloons, long greatcoat and bearskin cap (their full dress, white breeches and gaiters and red plumes, was carried in their haversacks for the expected triumphal entry into Brussels) ; they had red epaulettes and wore the queue,

[1] Houssaye.

and had earrings as big as a crown piece. The gunners of the Guard in bearskins marching beside their twelve-pounders, which Napoleon called his "belles filles";[1] cuirassiers with steel helmets and floating horse-hair, short-tailed blue and red coatees, white breeches and high jack-boots, armed with long straight sword and pistol; gigantic carabineers wearing Grecian helmets with red crests, white coats and breeches, and gilt cuirasses; horse grenadiers in high bearskins and red hackles, in blue turned up with scarlet, and epaulettes and aiguillettes of orange, armed like the cuirassiers. Green and white dragoons, chasseurs, and hussars, armed with sabre and carbine, whose pelisses and dolmans differed in each regiment; chestnut and blue, green and scarlet, red and sky-blue, grey and blue; green lancers, lancers with yellow epaulettes on red jackets, and red overalls, armed with lance, sabre and pistol, a brilliant kaleidoscope glittering with gold and silver.[2] Their music and their shouts of "Vive l'Empereur!" could be faintly heard by the Highlanders, who fully realised the power they had to contend with, and that the fate of Europe depended on the result of the approaching combat.

After this they were visited by the Duke of Richmond, who had ridden from Brussels. After congratulations to officers present on having come safe through the affair of the 16th, he told them he had just seen the General-in-Chief, and that Prince Blücher was on the march to their assistance. This news produced an extraordinary effect on the minds of the soldiers; "but it may fairly be doubted," says the officer to whose memoirs I am so much indebted, "whether the speedy prospect of being succoured by the Prussians, or the two following verses, produced the most powerful effect on the hearts of the Highlanders." The lines were chanted by the men, having been altered to suit the occasion by one of them.

> Now 's the day and now 's the hour,
> See the front of battle lour,
> See approach Napoleon's power,
> Chains and Slavery.
>
> Lay the proud usurper low,
> Tyrants fall in every foe,
> Liberty 's in every blow,
> Let us do or die.

The battle had begun by a furious attack on Hougoumont, Napoleon hoping that Wellington would weaken his centre to repel it; but being disappointed in this, and being aware by an intercepted dispatch that the Prussians were advancing, he ordered Marshal Ney to force the centre by an attack which was accompanied by a cannonade along the whole line. A battery of eighty

[1] Pretty girls. [2] Houssaye.

guns thundered from its position immediately opposite the 92nd, which suffered some loss. Wellington retired his infantry (except Bylandt's Brigade, and part of the 95th, who held the gravel pit) so as to shelter them behind the crest of the ridge, while his artillery remained in advance and replied to that of the enemy for half an hour. About 1.30 four divisions of infantry in echelon, supported by cavalry and led by Ney and d'Erlon, with drums beating and colours flying, descended into the valley (which was covered with crops of grass and standing corn, and without fences); the cannon-balls crossing each other above them till they neared the British, when the French guns ceased, but those of the Allies continued firing. The valley was filled with smoke by the musketry of the light troops who covered the assault on La Haye Sainte. The advanced companies of the 95th were driven from the gravel pit, and Bylandt's Dutchmen, who had suffered terribly from artillery fire in their exposed position, were broken, after some resistance, by Donzelot's troops ; they finally fled through the Fifth Division, and were no more seen. Picton was shot through the temple as he gave Kempt's Brigade his last order to charge. Donzelot's masses were driven back, but renewed the fight ; so close was it that the wadding of the cartridges remained smoking on the coats of the combatants. The post abandoned by Bylandt's Brigade was re-occupied by the 3rd Battalion Royal Scots and the 2nd 44th ; these two weak battalions poured a heavy fire of musketry on the assailants, and for some time maintained their ground with the unflinching courage of true British soldiers, but were at last compelled by superior numbers to give way for the moment.[1] Papelotte had been taken from the Nassau troops who held it. "The situation was very critical," Kempt says in his report ; and no doubt it was so, for had the French columns penetrated the allied line at this point and gained the heights, the victory would have been more than doubtful.

There had been changes of position by the battalions of Pack's Brigade, and the 42nd was posted at an important spot considerably to the left of the 92nd, who were now reduced to less than 300 men, and were lying down concealed from the French. Marcognet's column, 3000 strong, had passed by Donzelot's right ; his leading regiment had reached the hedge of the Ohain road, crying "Victory !" when the Gordons were ordered to stand to their arms. Sir Denis Pack said earnestly, "92nd, all the troops in your front have given way, you must charge this column."[2] He then ordered the line to form fours deep, and close on the centre.

[1] Cannon's "Record," and "Military Memoirs."

[2] "Waterloo Letters," No. 169, and Cannon's Record, probably alluding to Bylandt's Brigade. In one account it is put, "Everything has given way on your right and left."

The enemy on reaching the hedge, thinking themselves victorious, had ordered arms ; but becoming aware of the advancing Highlanders, were in the act of shouldering when they received a volley at twenty yards from the 92nd, which they at once returned. At this moment the Scots Greys, who were the left regiment of the Union Brigade (Royal Dragoons, Greys, and Inniskillings), came up, the pipers played, and the regiments mutually cheered each other, calling out, " Scotland for ever ! " The Greys doubled round the flanks, and through the openings made by the Highlanders, as best they could, and both regiments charged together, the wildest enthusiasm pervading the ranks of each ; many of the Highlanders caught hold of the Greys' legs and stirrups to support them as they ran, determined not to be last. " I must own it had a most thrilling effect on me," says one of the Greys. " 'I didna think ye wad hae saired me sae,' cried some of the Highlanders who were knocked down by the horses, but in our anxiety we could not help it."[1] " The Highlanders seemed half mad, and it was with the greatest difficulty the officers could preserve anything like order in the ranks."[2] " I never saw the soldiers of the 92nd Regiment so extremely savage as they were on this occasion."[3] An old piper, it was said, in his excited imagination cried out that he saw " Fassiefern " waving his bonnet in front as of yore.[4] Long after that day, individuals who witnessed the charge used to speak of the thrilling sensation which overcame them when they beheld the small body of bonnets and plumes lost, so to speak, amid a crowd of shakos.[5]

[1] J. Armour, Scots Greys. Alison (Appendix).

[2] " Waterloo Letters," No. 39.

[3] " Military Memoirs." This feeling was produced principally, says the author, by a report circulated the previous day that the enemy had put to death in cold blood British and Prussian prisoners who had fallen into their hands. Towards the close of the engagement, however, when the French were more to be pitied than feared, they assumed a very different air, and treated the prisoners with the kindness characteristic of the British soldier.

[4] This tradition is alluded to in a poem on Waterloo by David Home Buchan :—

> The spirit, too, of him is flown
> Who led the gallant Gordons on,
> Bold Cameron is no more ;
> Descendant of the great Lochiel,
> In honour's glorious field he fell,
> While all his loss deplore.
> Of vision keen and versed in spells,
> Strange tales the colonel's piper tells.
> How he with more of joy than fear
> Again beheld his chieftain dear,
> High riding in a misty cloud
> While war's artillery thundered loud
> And broke o'er Waterloo ;
> That though he heard not there his voice,
> He saw him wave his bonnet thrice.

[5] Gleig's " Battle of Waterloo."

Staggered by the Highlanders' volley, charged in front and flank by cavalry and infantry, paralysed by its own press, the heavy French column could make but a poor resistance. The men fell back on one another, grouping together so closely that they had hardly room to strike or fire at the Highlanders and horsemen who penetrated their confused ranks ; the dragoons traversed these splendid regiments, scattering them and cutting them off as the broken battalions rolled down the slope under their blows. "In three minutes the column was totally destroyed, and numbers of them were taken prisoners.[1] The grass field in which the enemy was formed, and which had been as green and smooth as the Fifteen Acres in Phœnix Park, was in a few minutes covered with killed and wounded, knapsacks and their contents, arms and accoutrements, etc., literally strewed all over, so that to avoid stepping on one or the other was quite impossible ; in fact, one could hardly believe, had he not witnessed it, that such complete destruction could have been effected in so short a time. Some of the French soldiers who were lying wounded were calling out 'Vive l'Empereur !' and others firing their muskets at our men who had advanced past them in pursuit of the flying enemy."[2] To those who cried "Quarter !" "Prisoner !" the answer was generally, "Well, go to the rear, d—— ye."

After this brilliant affair[3] the regiment was recalled ; Sir Denis Pack rode up and said, "You have saved the day, Highlanders, but you must return to your former position ; there is more work to be done."[4] As they re-formed, they saw the impetuous career of their countrymen as they encountered two batteries half-way up the opposite slope, sabred the gunners and drivers, upset the guns into a ravine, and then, intoxicated by success, assailed the great battery. The recall was sounded in vain, they did not or would not hear it. Napoleon also noticed and admired "ces magnifiques chevaux gris," and sent a regiment of cuirassiers and one of lancers against them. The Greys had lost their colonel and many officers ; their horses were blown ; every man did what he could, and faced his foes manfully. "Conquer or die" was the word. "At night," wrote Colonel Wyndham, then a lieutenant of the Greys, "Brevet-Major Cheney brought out of action four or five officers and

[1] 2000 prisoners were taken in the charge of the Union Brigade, also two eagles—one by the Royal Dragoons, the other by the Greys.

[2] "Waterloo Letters," No. 168.

[3] This famous charge is the origin of the firm friendship between the Greys and Gordons, which is still constantly expressed in mutual acts of good comradeship. Alison justly remarks that they could not have a finer motto on their crests than "Scotland for ever."

[4] Cannon's "Record."

under thirty men. The lancers did us as much mischief as the round shot and shell."[1]

Nor had the conduct of the troops on the right been less effective. The infantry repulsed their assailants, the cavalry routed them. Life Guards and dragoons rode across the valley, their swords ringing on the armour of the cuirassiers with the " noise of smiths at work." The fight at Hougoumont still raged, but elsewhere the action was arrested while both parties regained their positions. During the interval a cuirassier left his regiment and crossed the valley towards the British, who, thinking him a deserter, did not fire ; he rode close up to the hedge, stood up in his stirrups, waved his long sword, crying " Vive l'Empereur ! " to the British infantry on the other side, and galloped back unhurt !

At the close of the charge a 92nd officer captured a beautiful charger ; it was richly caparisoned, and its new owner expected to furnish himself with a handsome pair of pistols from the holsters, but was by no means disagreeably surprised to find a cold fowl in the one and a bottle of champagne in the other. Thirsty as they were, he and those with whom he shared voted it the best glass of wine they ever tasted.

It was now about 3.30. Napoleon determined again to try to drive away the British before the Prussians could arrive. " Heavy," says a 92nd officer, " as was the cannonade in the early part of the action, it was trifling in comparison to what followed " ; the balls rolling along the ground towards them, flying over their heads, or carrying off men in the ranks. " Never," said General Alten, " had the oldest soldier seen such a cannonade." Then Ney and d'Erlon led the second attack ; a brigade, deployed as skirmishers,

[1] The French General Durutte, speaking of his lancers at Waterloo, said, " Never did I so well see the superiority of the lance over the sabre " (*Houssaye*). It was only after Waterloo that the lance was introduced into the British army, nor did the Life Guards then wear the cuirass.

AUTHOR'S NOTE.—It is difficult to give the correct details of this celebrated charge of the Union Brigade, and to reconcile the different accounts of those who took part in it at five hundred or even fifty yards apart. The smoke, the excitement, prevented their appreciating events which succeeded each other so rapidly. The Royal Dragoons were on the right, the Inniskillings in the centre, and the Greys on the left rear. Sir De Lacy Evans, K.C.B., who was aide-de-camp to General Ponsonby and who was with the right of the brigade, says in " Waterloo Letters," No. 31, " The brigade (? division) you speak of under Sir T. Picton (and afterwards under Sir T. Kempt) was successful, as your letter states, but the infantry in our front had, I think, been obliged to give way, at all events it passed round our flanks." " The Dutch-Belgian infantry yielded with slight or no resistance to the advancing columns and got quickly to the rear, and not in the stubborn, deliberate way of our infantry." " Our artillerymen as well as our infantry kept firing into the column as long as possible. The 32nd and 79th had *regained their position* as we charged past their left, but I am certain their position was some fifty or a hundred yards in front of the hedge." The situation of the 92nd before the charge, and the part they took in it, as described above, is taken from Cannon's " Record," printed by authority, and from the writing of two officers and a sergeant of the regiment who were present, etc. There seems to have been no other infantry close to the regiment at the moment.

ascended the slope east of the Charleroi road, but were driven back. Wellington had again retired his infantry for shelter behind the crest, and Ney, seeing this and the escorts of prisoners and wounded going to the rear, thought the British were giving way and ordered a charge of 5000 horsemen. But the British General had no idea of retreat ; his infantry stood up and formed squares ; the artillery horses were sent to the rear, but the guns remained in action, their fire at close range causing such destruction to the cavalry coming up at a trot through the deep ground and standing corn, that the survivors of the leading squadrons hesitated. Their trumpets again rang out the "Charge." "Vive l'Empereur !" The gunners ran to the shelter of the squares, the guns were taken, but the enemy had not the means of spiking or carrying them off. "The bullets rattled on the cuirasses with the noise of hail on a slate roof," yet the musketry, though it broke their ranks, did not dismay them. Cuirassiers, lancers, chasseurs rode on, whirling round and between the squares, striking at the front rank with the sword, firing pistols in their faces, some even throwing their lances at the men's breasts ;[1] but as a breach was made, "Close up !" was heard, and the place of the fallen filled. Like wave succeeding wave, squadron succeeded squadron, but the squares stood firm.

The cavalry even charged squares in the second line. Then Uxbridge sent the Dutch and Brunswick cavalry, who had not been engaged, to charge the enemy in their disorder ; they fell back between the squares, escaping the sabres only to fall under the bullets. They abandoned the plateau, the gunners ran back to their guns, and again the British batteries belched forth their destructive fire. But these gallant horsemen rallied and re-formed in shelter of the ground, then once more advancing slowly, but with admirable courage and discipline, under a fire of grape-shot, they again crowned the height of Mont St Jean.

The French on La Belle Alliance, seeing them through the smoke, cried "Victory !" Fresh brigades were sent to smash the British by their weight, more than sixty squadrons were advancing ;[2] it seemed like a sea of steel. Some among the Allies thought all was over, but the squares stood firm.[3] The gunners felled entire ranks as a mower cuts a swathe of grass, ere they ran to the sheltering bayonets ; still the squadrons in rear came boldly on ; charge succeeded charge till the squares seemed submerged in the tide of cavalry, but as the smoke of their fire cleared away they reappeared,

[1] Sergeant Robertson and others.

[2] Houssaye.

[3] "During the greater part of the time the 92nd was formed in square."—"Waterloo Letters," No. 168. "Our regiment was charged by the cuirassiers of the Guard, and we gave them a noble peppering."—Letter from a wounded officer, 92nd, dated Brussels, 21st July 1815.

hedged with glittering bayonets, while the squadrons receded from them like waves broken on the rocks.[1] The squadrons crossed and interfered with each other, their charges became less vigorous ; the atmosphere was like a furnace of fire and smoke ; the horses were blown, men could hardly breathe. Thirteen French cavalry generals were wounded ; Ney had three horses killed under him [2] and stood brandishing his sword near an abandoned battery ; the whole field was encumbered with the dead, wounded men dragging themselves away, riderless horses galloping about frightened by the whistling bullets. Wellington ordered his cavalry forward, and for the third time the French left the plateau.

These attacks continued with little intermission ; in vain the horsemen, reduced in numbers, pressed their tired chargers against the squares, now protected by a rampart of dead men and horses. After these two long and hardly contested hours, the infantry became assured they were invincible ; [3] it was the hostile cavalry who were demoralised, and retired discouraged, though not dismayed, from each attack. Nevertheless, the Allies' powers of resistance were severely tried. Papelotte had been taken by the French, many guns were without gunners, General Ponsonby was killed and his brigade was reduced to two squadrons ; several other generals were wounded, most of Wellington's staff were *hors de combat*, brigades were reduced to handfuls of men. Some of the allied troops had fled to Brussels, some even of the British made the excuse of helping the wounded to the rear ; ammunition was short, officers arrived telling Wellington of desperate situations, and asking for orders. "There are no orders, except to hold out to the last man." The state of affairs was summed up by a private soldier as he bit another cartridge, "The one that kills longest wins."

Ney sent to Napoleon for fresh troops. "Troops !" exclaimed the Emperor ; "where does he expect me to take them from ? Does he think I can make them ?" He had, however, fourteen battalions of his Guard, and others of that chosen body

[1] Ossian's poem of "Fingal," translated from the Gaelic by the Rev. Dr Ross, is very descriptive of the scene.—"As roll a thousand waves to the shore, the troops of Suaran advanced. As meets the shore the waves, so the sons of Erin stood firm. There were groans of death ! the hard crash of contending arms. Shields and mails in shivers on the ground. Swords like lightning gleaming in the air ; the cry of battle from wing to wing ; the loud hot encounter, Chief mixing his strokes with Chief, and man with man."

[2] Houssaye. Ney had five horses killed before the end of the day.

[3] An Engineer officer, who had ridden for protection into a square, relates that the first time the cuirassiers approached, some of the young soldiers seemed to be alarmed and fired high, but they soon discovered they had the best of it, and welcomed a fresh charge as a relief from the artillery fire, from which they suffered in the intervals. "The jokes which the men cracked while loading and firing were very comical." The Scotch soldiers, comparing the armour of the cuirassiers to that of a crab, joked about "cracking the partan's shell" as they fired at the steel-clad horsemen.

had at four o'clock partially arrested the advance of the Prussians. The near approach of the First Prussian Corps now made Napoleon play his last card in the desperate game. Seeing the allied line disordered, he desired his batteries to redouble their fire ; he gave the command of the attack to Ney, and harangued his Guards, whose advance was seconded by the other divisions and by the cavalry.

Wellington passed along his line of battle giving his orders. The brigades were again advanced from their shelter, the reserve artillery was brought to the front, and ordered not to reply to that of the enemy, but to concentrate their fire on the columns of assault.[1] The centre was strengthened, and the 92nd, which at this time, about seven o'clock, was near the extreme left of the line, was ordered up to the left of the main road near the gravel pit at La Haye Sainte ; they were moving in column at quarter-distance when a shell fell in the midst of the battalion ; the companies in rear of it faced about, and doubled to the rear till it had burst, when they doubled back to their proper distance without any word of command, and arrived in time to take part with the Fifth Division in driving the enemy from the crest of the position above La Haye Sainte.

The French Guards advanced in squares, in admirable order, two guns of the Horse Artillery of the Guard with each, and were supported by the whole army. Those who attacked the British right were repulsed after a gallant struggle ; those above La Haye Sainte, who were fighting with the Fifth Division, saw the Guard on their left defeated, and the cries " La garde recule ! " " Nous sommes trahis ! " sounded the knell of the grand army of France. The charge of the British forced many pell-mell into the gravel pit from which they had before driven our troops. The enemy retired in confusion, but still protecting his rear with skirmishers along the other side of the gravel pit and of the farm of La Haye Sainte, and many, keeping their formation, still prepared to show a bold front. Then the Gordons could see an extraordinary movement on the right of the French, troops dressed like them in blue had turned their right, and were advancing. An aide-de-camp came galloping towards the Duke, calling out as he passed that the first troops of Blücher's army had come. All eyes were now turned on Wellington, who was on the edge of the plateau. Quick to seize the favourable opportunity, he held his cocked hat high in the air, and waved it forward. The cheers that answered this signal for a general advance were the heartiest of the day. The army moved

[1] " There 's Bonaparte, sir," said an artillery officer to the Duke, who had approached ; " I think I can reach him. May I try ? " " No, no," replied the Duke ; " generals commanding armies have something else to do than to shoot one another."

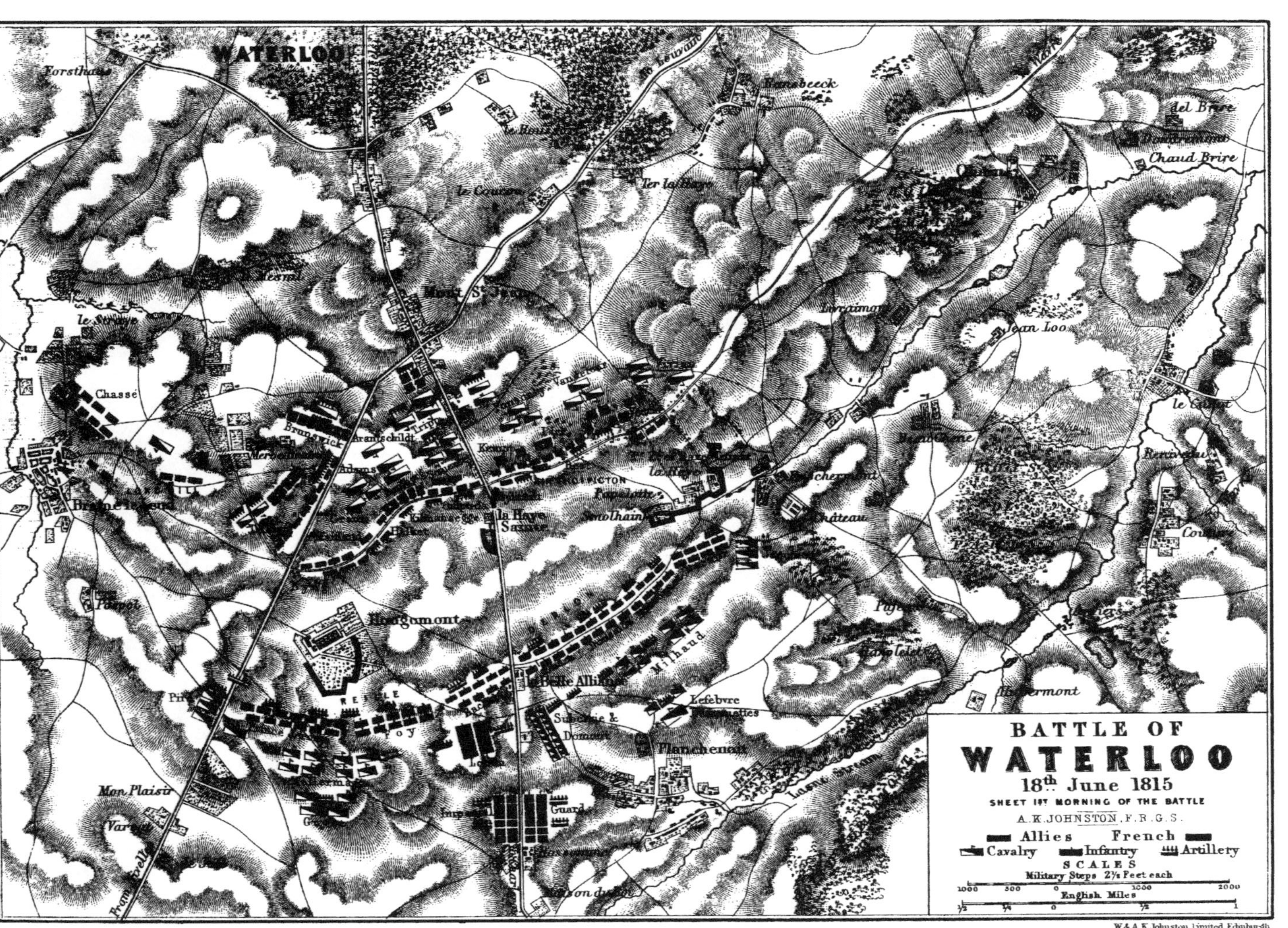
BATTLE OF
WATERLOO
18th. June 1815
SHEET 1st MORNING OF THE BATTLE
A. K. JOHNSTON, F.R.G.S.
Allies
French
Cavalry
Infantry
Artillery
SCALES
Military Steps 2½ Feet each
1000 500 0 1000 2000
English Miles
½ ¼ 0 ½ 1
WATERLOO
Forsthaus
Ransbeeck
del Brire
Chaud Brire
le Caureau
Ter la Haye
Mont St. Jean
le Strage
Chassé
Brunswick
Braine l'Alleud
La Haye Sainte
Papelotte
Lasalle
Chateau
Lavaimont
Jean Loo
le Canet
Couture
Hougomont
Milhaud
Lefebvre
Belle Alliance
Planchenoit
Frischermont
Mon Plaisir
Varzal
Reille
Foy
Guards
Maison du Roi
Paris
Ohain
Papelotte
Lebelet
Hubermont
W. & A. K. Johnston, Limited, Edinburgh

forward in the order it happened to be in, battalions, batteries, and squadrons side by side, to the sound of drums, bugles and bagpipes.

At this sight the last divisions of French infantry, and almost all the cavalry, precipitately regained the plateau of La Belle Alliance. When Napoleon saw the sudden crumbling of his line of battle, he felt that he was irretrievably beaten, but at first he tried to organise the retreat ; he formed the Old Guard in squares, and sent some squadrons against the British light cavalry, who, comparatively fresh, easily defeated them. The allied troops halted a minute in the valley, and then pressed on, sweeping all before them ; artillery drivers cut the traces and abandoned their guns, brigades were broken and dispersed. All was confusion, relieved, however, by many heroic instances of devotion and discipline. General Pelet, with the chasseurs of the Guard, was trying to cover the retreat of the French right ; they were surrounded by Prussian cavalry and infantry, their eagle was in danger of being taken. Pelet halted his men on a rising ground calling, " A moi, chasseurs ! Sauvons l'aigle ou mourons autour d'elle " (To me, chasseurs ! Let us save the eagle or die round it !). The band of heroes rallied, closed their ranks, and with levelled bayonets made a way through their foes, and reached the main line of retreat. Sergeant Robertson mentions one division at the house of La Belle Alliance who made an attempt to stand ; " such was our excited and infuriated state of mind at the time, and being flushed with the thought of victory, we speedily put an end to their resistance."

Wellington rode with his advanced troops, regardless of bullets from friend and foe which were falling round him ; when urged not to expose himself so much, he replied, " Never mind, let them fire away ; the battle 's gained, my life 's of no consequence now."

The rout was complete. None gave orders which none would have obeyed, each man saved himself as best he could. Napoleon fled, attended by a small escort. Grouchy's Corps remained intact, but when the Emperor quitted Laon for Paris on the 20th, 2000 soldiers assembled at Philippeville and about 6000 at Avesnes were all that remained together under arms of the French army engaged at Waterloo.[1]

Wellington and Blücher met about 9.15 near La Belle Alliance. The bands of the Prussian cavalry played "God save the King," their infantry soldiers sang Luther's hymn, " Now thank we all our God," and the troops mutually cheered each other. Blücher, it is said, wished the battle to be called after La Belle Alliance, in memory of this meeting ; but Wellington decided that his victory should

[1] Houssaye.

be named after the village where his headquarters were established the night before—Waterloo.

The British General represented that his men were exhausted and were hardly able to continue the pursuit. "Leave that to me," replied Blücher, "I will send every man and horse after the enemy"; and Ziethen continued the pursuit during the whole night without intermission, giving no rest to the wearied and dejected French, of whom only about 40,000 crossed the frontier out of the 74,000 who fought at Waterloo, and these escaped only to disperse; they lost 227 guns. The loss of the Allies (exclusive of the Prussians) on the 18th was 15,380 officers and men in killed, wounded, and missing, of whom 8481 were British and of the King's German Legion in British pay.[1] The honours of the day must be divided; the British gained the battle, the Prussians made the victory complete.

The Gordon Highlanders, who were commanded throughout the day by Major Donald MacDonald (Dalchosnie), bivouacked by the light of the young moon near the place where Bonaparte had stood most of the day.[2] They had to mourn the loss of Sergeant-major Taylor and thirteen other n.c. officers and soldiers killed in action. Captains Peter Wilkie and Archibald Ferrier, Lieutenants Robert Winchester, Donald MacDonald, James Ker Ross, and James Hope, and ninety-six n.c. officers and men were wounded, of whom many died of their wounds.

The following is a list of the names of those killed in action, and of those who died of their wounds at Quatre-Bras and Waterloo, with their place of enlistment or birth as far as possible:—[3]

Sergeant-major	George Taylor,	Fraserburgh, Aberdeen.
Sergeant	Alexr. Dunbar,	Duffus, Moray.
,,	Kenneth MacKenzie,	Ross.
,,	John Patterson,	New Machar, Aberdeen
Corporal	James Cameron,	Keith, Banff.
,,	Donald Ross,	Edderton, Ross.
,,	John Leadingham,	Banff.
,,	Donald Fraser,	Inverness.
,,	James Gibb.	
,,	James Grubb,	Monymusk, Aberdeen.
Private	William Appleton,	Edinburgh.
,,	Peter Bethune,	Kilmorack, Inverness.
,,	John Bremner,	Drumblade, Aberdeen.
,,	Charles Brown,	Forfar.
,,	Thos. Benson,	Durham.
,,	John Burnet,	Boyndie, Banff.
,,	William Burnet,	Fyvie, Aberdeen.

[1] Alison. [2] Sergeant Robertson. [3] 49 killed, 33 died of wounds—total, 82.

Private	Robert Breddon,	Monaghan, enlisted as boy in Glasgow.
,,	Donald Cameron,	Inveravon, Banff.
,,	Donald Cameron,	Kilmallie, Inverness.
,,	John Carr,	Dumbarton.
,,	Colin Campbell,	Reay, Caithness.
,,	Peter Dustan,	Ruthven, Banff.
,,	Lachlan Duncan,	Kent (soldier's son, enlisted as boy at Banff).
,,	Robert Duncan.	
,,	Robert Elliot,	Galway (? Galloway).
,,	John Fraser, 1st,	Beauly, Inverness.
,,	William Fraser, 2nd,	Fearn, Ross.
,,	Donald Fisher,	Enlisted at Birmingham.
,,	Donald Gow,	Dunkeld, Perth.
,,	Alexr. Grant,	Tulloch, Ross.
,,	John Grant, 1st,	Perthshire.
,,	Lewis Grant,	Kingussie, Inverness.
,,	John Glenny,	Edinburgh.
,,	David Graham,	Creich, Sutherland.
,,	Timothy Griffin,	Kilmorack, King's Co.
,,	Charles Higgin,	Antrim.
,,	James Hutton,	Brechin, Forfar.
,,	Alexr. Kenedy,	Inverness.
,,	Archibald Lamont,	Argyllshire.
,,	John MacCallum,	Cardross, Perth.
,,	William MacCrackin,	Inch, Galloway.
,,	Roderick MacLeod,	Snizort, Inverness.
,,	Angus MacLellan,	Inverness.
,,	Malcolm MacLeod,	Inverness.
,,	William MacLean,	Argyll.
,,	William Macgregor,	Urquhart, Inverness.
,,	Patrick MacGowan,	Antrim.
,,	William MacIntosh,	Abernethy, Inverness.
,,	William MacDonald,	Dunfermline, Fife.
,,	William MacDonald,	Pettie, Inverness.
,,	Donald MacDonald, 5th,	Kiltearn, Ross.
,,	James MacDonald, 2nd,	Enlisted at Paisley.
,,	John MacKenzie,	Inverness.
,,	Roderick MacKay, 2nd,	Sutherland.
,,	Thos. MacNeil,	Dumfries.
,,	John MacSwine,	Kilmuir, Inverness.
,,	John Mathers,	Rescobie, Forfar.
,,	Robert Millar,	Dunfermline.
,,	Charles Masterton,	Forfar.
,,	William Marshall,	Kirkmichael, Argyll.
,,	John Munro, 2nd,	Aberdeen.
,,	Peter Munro,	Old Machar, Aberdeen.
,,	Donald Mathieson, 1st,	Sutherland.

Private	Jas. Nellison (? Mollison),	Glenmuick, Aberdeen.
,,	Matthew Orr,	Ayr.
,,	Robert Pirie,	Culsalmond, Aberdeen.
,,	Alexr. Petrie.	
,,	William Reid,	Rogart, Sutherland.
,,	James Russell,	Rothes, Moray.
,,	Donald Ross,	Fearn, Ross.
,,	Donald Ross, 3rd,	Tain, Ross.
,,	Findlay Smith,	Strathmore, Inverness.
,,	Duncan Simson,	Inverness.
,,	John Stewart,	Inverness.
,,	William Stewart,	Bute.
,,	Donald Skinner,	Edderton, Ross.
,,	Alexr. Sutherland, 1st,	Moray.
,,	James Thomson,	Renfrew.
,,	Frederick White,	Glasgow.
,,	John Whitaker,	Essex.
,,	Frederick Zieger,	Germany (musician).

I have been able to trace the length of service of fifty-five of the above.

				Brought forward,	25	men.
Enlisted in 1794,	2	men.	Enlisted in 1808,		9	,,
,, 1798,	2	,,	,, 1809,		3	,,
,, 1800,	3	,,	,, 1810,		2	,,
,, 1802,	4	,,	,, 1811,		3	,,
,, 1803,	2	,,	,, 1812,		8	,,
,, 1805,	7	,,	,, 1813,		4	,,
,, 1806,	1	,,	,, 1815,		1[1]	,,
,, 1807,	4	,,				
Carry forward,	25				55	

It would appear from a comparison of the names on the prize-roll for Vittoria with that for Waterloo, that the proportion of Highlanders to Lowland Scots was greater at the former than at the latter period.

The wounded were taken to Brussels and Antwerp, but from want of carriage many had to get there as best they could, and suffered much. The people of Brussels showed great kindness and humanity. Every door was open to them, and the best rooms given up. One gentleman and his sisters nursed and fed thirty officers and men. Monsieur Troyans, a lacemaker, stopped his manufacture, converted his premises into a temporary hospital, furnishing the patients with linen, and giving the attendance of all his workwomen. A soldier's letter mentions that he and others were attended by ladies and gentlemen, who brought them deli-

[1] The German musician.

cacies and waited on them, though they could only understand each other by signs. The Sisters of Mercy also were constantly in attendance at the hospitals, the wounded French being treated with the same attention as the British. Mr P. L. Gordon, in his reminiscences, says the Highlanders were particularly favoured. Before, they were called the "Good little Scots," now they were "The good and brave Scots." Monsieur Hector, a brewer, visited the field at dawn next day with a dray laden with casks of beer to assuage the sufferers' thirst; he found a wounded Highlander, who had been billeted in his house, lying by the road, faint from loss of blood. The good brewer took him home, and had the satisfaction in a few weeks of seeing his guest able to sun himself in the park. Mr Gordon spoke to sixteen wounded Highlanders who filled a waggon passing his door. "We 're a' wantin' a leg or an arm," they said. He offered to bring them wine. "We wad rather hae a drink o' guid sma' ale," and they got it. There were not enough army surgeons, but a number of civilian surgeons from England arrived as volunteers. Mr Dorett, in charge of one of the hospitals in Brussels, extracted a bullet from a soldier's wound, which, however, showed no signs of healing; poultices and an incision produced a five-franc and a one-franc piece which had been in his pocket! The larger coin had been hit in the centre, and was in the form of a cup. "Donald, who belonged to the 92nd Regiment, recovered his health as well as his money." A Scottish lady at Antwerp describes the consternation there on the 17th and 18th, as reports of the retreat from Quatre-Bras, and still more gloomy and desponding accounts from Mont St Jean, were circulated—Complete defeat of the British! Wellington severely wounded! All the superior officers killed or prisoners! The French columns seen entering Brussels![1] All was panic and dismay. Early next morning, accompanied by a Belgian servant, she walked to the Malines Gate, hoping for news; she met waggons full of wounded soldiers; covered litters with serious cases borne silently along on men's shoulders; officers pale as death supported on their horses; but none knew the issue of the great battle. Then at a street corner she saw five Highland soldiers (probably from Quatre-Bras) with bandaged heads, arms, or legs, who, regardless of their wounds and fatigue, were throwing up their bonnets and vociferating, "Bony 's beat! Hurrah! hurrah! Bony 's beat! Hurrah!" They told the lady that a courier had just passed, bringing an account of the complete British victory. The townsfolk tried to learn the cause of their tumultuous joy, but could not understand the Highland English and the broad Scots in which

[1] It was true that a French column, over 2000 strong, did enter Brussels in the afternoon of the 18th, but as prisoners, not as conquerors.

they loudly told that " Bony was beat an' runnin' awa' till his ain country as fast as he can gang." One old lady caught a soldier by the coat, gesticulating and gabbling in her own tongue to induce him to attend to her, but was none the wiser for the reply—" Hoots, ye auld gowk, dinna ye hear? Bony 's beat, woman! Daursay the wife 's deaf—I say Bony 's beat, woman"; but when the news was translated, her ecstasy was as great as that of the Highlanders. The lady was greatly impressed by the " great fortitude and uncomplaining patience with which the soldiers in general bore their sufferings"; " not a murmur escaped their lips" as they lay in the long jolting waggons. Numbers on foot, almost sinking from fatigue and loss of blood, slowly and painfully made their way along the streets. " Your countrymen are made of iron," said a lady in Brussels, and told how she had met a wounded Highlander supporting himself by the rails as he made his way with difficulty. She said she feared he was very badly hurt and offered help, when he drew himself up, thanked her, and said, " I was born in Lochaber, and I do not care for a wound"; but the effort was too much, and he sank on the pavement.

The Scottish lady says she had a good deal of conversation with the wounded men of the 42nd and 92nd, some from the Highlands and some from the Lowlands; their account of the battles was simple and interesting; they could not be induced to speak of what they themselves had done, or boast of it in any way, but considered it quite as a matter of course, and rather made light of their sufferings, laying no claim to the admiration which their modesty increased.[1] But their conduct " was the theme of every tongue. The love and admiration of the whole Belgic people for the Highlanders are most remarkable." " In short, they are the best of soldiers and of men, according to the Belgians, and the idea they have of their valour is quite prodigious. They never speak of them without some epithet of affection or admiration, ' Ah, ces braves hommes! Ces bons Ecossais, ils sont si doux, si aimables! et dans la guerre! ah, mon Dieu! comme ils sont terribles!' " (Ah, these brave men! these good Scots, they are so mild, so amiable, and in war, my God! how terrible they are!).[2]

The following is an extract from a letter to Sir Walter Scott from Viscount Vanderfosse, first advocate of the Superior Court of

[1] A soldier of the 92nd who had been wounded at Quatre-Bras was observed by an English gentleman lying on the pavement in the shade of a house in Antwerp, patiently waiting his turn to be attended to. The gentleman entered into conversation, complimenting him highly on his conduct and that of his comrades. " Hoot man," replied the Scot, " what for need ye mak' sic a din aboot the like o' that; what did we gang there for but to fecht!"—Circumstantial details of Waterloo, London, 1816.

[2] " Narrative of a Residence in Belgium during the Campaign of 1815." Also " Paul's Letters to his Kinsfolk."

Justice of Brussels, dated January 5th, 1816—" Since the arrival of the British troops on the Continent, their discipline was remarked by all those who had any communication with them. Among these respectable warriors the Scotch deserve to be particularly commemorated, and this honourable mention is due to their discipline, their mildness, their patience, their humanity, and their bravery almost without example. Constant and unheard-of proofs were given, I do not say of courage, but of devotion to their country, quite extraordinary and sublime ; nor must we forget that these men, so terrible in the field of battle, were mild and tranquil out of it."

CHAPTER XXV

IT may be imagined with what anxiety news was awaited in England, and with what joy and relief the intelligence was received of a victory so unparalleled in importance to the country and to Europe. Tidings of passing events could not then be flashed from the other side of the world as they now are. Major the Hon. Henry Percy was almost the only one of the Duke's staff who remained unhurt. He posted to the coast, a sailing vessel landed him at Dover, from whence he drove post-haste to London, with the captured French eagles sticking out of the windows of the chaise, and presented them, with the dispatches, to the Prince Regent.[1] The mail coaches dressed with branches of laurel announced the great news to the expectant groups collected at every cross-road, distributing the few and scanty papers, which were read out to crowds collected in the towns and villages ; but it took days, and in the remoter districts even weeks might elapse, before it was known that the great peace-giving battle was gained.[2]

Early on the morning of the 19th June, all the men who could be spared were sent out to carry the wounded to the roadside, where waggons could be brought to convey them to hospital, friends and foes being treated alike. One of the 92nd described the " heart-melting spectacle," the bodies not merely scattered over the ground, but lying in heaps, men and horses together. There are bad men among soldiers as in other professions, and some of the wounded told how they had been robbed during the night by plundering scoundrels in both British and foreign uniforms.

The Gordons had not been long employed in their humane duty when they were recalled. The Fifth Division was ordered to cook, and after dinner the regiment marched in pursuit of the French by the Nivelles road.[3] At any other time they would have hailed the order with joy, but their minds were now pervaded by a very different feeling. They had to pass right through the field of battle, French and British calling on them to put an end to their sufferings, not to let them die among strangers. With hearts full of grief at having to leave so many of their brave comrades unburied, or dying with no friendly hand to close their eyes ; with no triumphal music, more like mourners than victors, the regiment

[1] The most graphic account of the battle is by the Duke himself in a letter to Lord Beresford, one of his Peninsular generals :—" You will have heard of our great battle of the 18th. Never did I see such a pounding match. Both were what the boxers call gluttons. Napoleon did not manœuvre at all. He just moved forward in the old style in column, and was driven off in the old type. The only difference was that he mixed cavalry with his infantry, and supported both with an immense quantity of artillery. I had the infantry for some time in squares, and we had the French cavalry walking about us as if they were our own. I never saw the British infantry behave so well."

[2] The news took a month to reach the Island of Coll, Argyllshire.

[3] Division Orders, 19th June.

moved on, awestruck, and as silent as the dead that lay around them. "I confess my feelings overcame me ; I wept bitterly and wished I had not been a witness of such a scene," says the honest sergeant whose journal I quote.[1]

After marching a few miles through a country whose inhabitants had fled to the woods, they took some prisoners and halted near Nivelles, where the Duke of Wellington came up. He thanked and praised them for their conduct during the engagement, but he had one fault to find with the 92nd, and that was for being too forward in crossing the hedge in the early part of the action. As it turned out, he said, all was well, but it might have happened otherwise, "and he urged upon us to pay attention to the words of command that might be given next day."

The first Regimental Orders after Waterloo are dated at Bavay, 23rd June, where they halted, but they were at Malplaquet and Mons (the birthplace so called of Mons Meg at Edinburgh Castle) on the 20th and the 21st. On the 20th a General Order was issued guarding the inhabitants of France as far as possible from loss, everything required to be paid for, but the inhabitants were cautioned to behave peaceably and to maintain no correspondence with the usurper.

At Englefontein, on the 24th, a Court of Inquiry is ordered to investigate the cause of absence of several men on the 16th and 18th inst. ;[2] also as to those men who have returned without their arms and accoutrements, the Court to judge what stoppages they should be put under. Such of the band as have lost their instruments will also appear. Subalterns are named to take charge of seven companies which have evidently lost their captains, and they are directed, if the companies are strange to them, to lose no time in making themselves acquainted with the men present, and in gaining all the information they can regarding those absent.

Cambray, Peronne, Avesnes and other fortresses were taken by the Allies, but the 92nd were not employed in any affair of consequence. Marching through a pleasant country, they reached St Denis, near Paris, on the 3rd of July, where the French guns fired on them from the fortifications, and a battle was expected, but on the 4th the following General Order was issued by the Duke of Wellington :—

"The Field-Marshal has great satisfaction in announcing to the troops under his command that he has, in concert with Field-Marshal Prince Blücher, concluded a military convention with the

[1] The tears were also seen on Wellington's cheeks when the lists of killed were brought him, and his words expressed deep sorrow rather than the pride of victory.

[2] Men who were left behind in Brussels, or who had gone to the rear with wounded officers and men on the 17th. Mr Innes in a letter mentions that he was left at Brussels to bring on those left there, and joined with them in time to take part at Quatre-Bras.

Commander-in-Chief of the French army near Paris, by which the enemy is to evacuate St Denis, St Ouen, Clichy, and Neuilly this day at noon, the heights of Mont Martre to-morrow at noon, and Paris next day. The Field-Marshal congratulates the army upon this result of their glorious victory. He desires that the troops may employ the leisure of this day and to-morrow to clean their arms, clothes, and appointments, as it is his intention that they should pass him in review."

The thanks of both Houses of Parliament were voted to the army with the greatest enthusiasm "for its distinguished valour at Waterloo," and the 92nd and other regiments engaged on the 16th and 18th, or on one of those days, were permitted to bear "Waterloo" on their colours.

The Highland Society of Scotland passed a unanimous vote of thanks "for the determined valour and exertions displayed by the regiment and for the credit it did its country in the memorable battles of the 16th and 18th June 1815."

In acknowledgment of the services of the army at Waterloo and the actions immediately preceding it, each subaltern officer, n.c. officer and soldier was allowed to count two years additional service. Silver medals were, for the first time in the history of the British army, conferred on all ranks.

Lieut.-Colonel James Mitchell, who succeeded to the command of the 92nd at Quatre-Bras, was appointed a Companion of the Most Honourable Order of the Bath, and had the Order of St Anne of Russia conferred on him by the Czar. Major Donald MacDonald, who commanded the regiment at Waterloo, was promoted brevet lieut.-colonel and appointed a Companion of the Bath, and received the Order of St Vladimir of Russia. The share of prize-money falling to the 92nd regiment amounted to about £5000. Subscriptions were set on foot and liberally supported in London and other places for the purpose of creating a Waterloo Fund, out of which immediate relief might be supplied to the families of the fallen, and pensions given to their widows ; sums of money were also given from this fund to wounded officers and soldiers. These honours and gifts were thoroughly appreciated by the troops. Lieutenant Hector Innes writes home—"Our country, I am proud to say, has more than compensated us for our dear-bought victory, but when so appreciated, who would not strive to serve and bleed for such a country !" All soldiers' letters were sent free, signed by the commanding officer.

The French troops began their march to the Loire on the 4th, and by the 6th the whole had proceeded to their destination. Guards were placed at the entrances of Paris, each with two field-pieces loaded in case of disturbance.

Wha keep guard at Vincennes and Marly ?
Wha but the lads wi' the bannocks o' barley ;
Bannocks o' bere meal, bannocks o' barley.

The allied armies entered Paris on the 7th July, and on the following day Louis XVIII was once more reinstated on the throne of France. The Gordons, after marching through Paris, encamped behind the neighbouring village of Clichy. Here the regiment was completed with knapsacks and other necessaries ; the tailors were set to work to mend the war-worn clothing ; and when all their appointments were clean and in good order, n.c. officers and soldiers had liberty to go into the city by turns, a hope being expressed that they would not quarrel with any French soldier that might happen to be there—half the officers remaining always in camp. Until a market was established in camp, parties were sent into Paris to buy vegetables for their companies. Sir Walter Scott, who was one of the many visitors to Paris, gives an entertaining description of the Highlanders making their bargains ; the soldier holding his piece of money between his finger and thumb with the grip of a smith's vice, and pointing out the quantity of the commodity which he expected for it ; while the Frenchman, with many shrugs and much chattering, diminished the equivalent as more than he could afford. Then Donald would begin to shrug and chatter in his turn, and to scrape back again what the other had abstracted ; and so they would stand for half an hour discussing the point, though neither understood a word of what the other said, till they could agree upon " le prix juste." " The soldiers," Sir Walter remarks, " without exception, conduct themselves in public with civility, and are very rarely to be seen intoxicated, though the means are so much within their reach " ; and he mentions meeting men of all regiments and conversing with them in the picture galleries and other public places.

Napoleon, who had left Paris after a two days' stay, concealed himself near the coast, intending to embark for America ; but, finding that the British cruisers made that impossible, he surrendered himself on the 15th of July to Captain Maitland of H.M.S. *Bellerophon*, which brought him to Plymouth. The island of St Helena was afterwards appointed for his residence.

R.O., Camp near Paris, 18th July 1815.—Opportunity now offering of paying the last tribute of respect to the memory of the late Colonel Cameron, officers will be pleased to wear a black crape round their left arm for one month.

On the 24th of July there was a grand review of the British and Hanoverian troops before the Emperor of Russia in the Place Louis Quinze. Officers were in blue pantaloons and half-boots ; the men in their best hose and new rosettes, without gaiters, but with knapsacks packed—blankets, etc.—every second man to carry a brush and towel to take the dust off on arrival at the ground. To spare the feelings of the Parisians, it was not intended that they should wear laurel, but almost the whole army did mount the emblem of victory. Immediately after the Gordons had passed the Emperor, their old general Howard, now Lord Howard of Effingham, who had commanded their brigade at Arroyo dos Molinos and other Peninsular actions, rode up in plain clothes, congratulated them on the additional honours the regiment had gained, and, glad to be again at their head, rode with the commanding officer as they returned through the streets of Paris to their camp. He was a great favourite with all ranks who had been with him in Spain. The Emperor of Russia was so much struck by the appearance of the Highland regiments that he afterwards requested that a sergeant, a piper, and a private from each of the three (42nd, 79th, and 92nd) should come to the Palace Élysée, where he resided. They were conducted by Lord Cathcart to the presence of the Emperor, who made a most minute inspection of their dress and appointments. He pinched Sergeant Campbell's (79th) skin to be sure he wore nothing under his kilt, examined his claymore, and asked if they learned any special exercise for the broadsword ; also as to their services and families at home. A private was put through the manual and platoon exercise. The pipers played the tune "Cogadh na Sith," the Emperor appearing delighted with the music. Then they were served with refreshments, and each received a present of money. The 92nd men were Sergeant Grant, Piper Cameron, and Private Logan.[1]

The life of the regiment was now the ordinary routine of the camp ; divine service on Sundays, with rather an extra dose of drill on week-days, and the Waterloo heroes grumbled a good deal at being turned out for "goose-step" at 4 a.m., with parades and field-days, regimental, brigade, or divisional, afterwards. Rolls were made out of all who had been present on the 16th, 17th, and 18th June, "No person to be included of whom the least doubt of his having been personally present on one or other of these days can be attached. The rolls will undergo a minute investigation at the War Office." Officers' servants and batmen having been on duty, though not actually in the ranks, were to be included.

On the 30th of August Lieut.-Colonel Mitchell resumed command.

R.O.—Hose and buckled shoes to be worn. The black gaiters only on marches, but one pair to be considered as necessaries. Greatcoats to be rolled on the top, and blankets on the back of the knapsacks as formerly (in Peninsula). Officers com-

[1] From the account of Sergeant Campbell, who commanded the party.

manding companies to send in returns of widows and orphans who were rendered such by the casualties in the actions of the 16th, 17th, and 18th June, and also of dependent relatives of men killed on those days, with a certificate for each family, also as to prize-tickets being filled up and signed.[1]

On the 22nd September there was a grand review of the British and Hanoverian army on the plain of St Denis. The Duke of Wellington received the Emperors of Austria and Russia and the King of Prussia. In the suite of the allied sovereigns were many celebrated men—Prince Blücher, Prince Schwartzenberg, Field-Marshal Barclay de Tolly, Platoff, Hetman of the Cossacks, the Archduke Constantine of Russia, two sons of the King of Prussia, etc., all the British and foreign visitors, but not the Parisians. The operations represented those of the allied army at Salamanca. Before moving from the original ground, Blücher, Platoff, and other foreign generals came to examine the dress of the three Highland regiments, and " cracked their jokes at the expense of the philabeg, but Donald, instead of getting sulky, revenged himself by his remarks on the foreigners, some of whom were in uniforms which he considered no less singular than his own." Lieutenant Innes wrote that the Russian Emperor " paid us kilted lads most marked attention, and conversed with several officers and men, and was pleased to pay us many very handsome and highly honourable compliments." Another officer mentions that the Archduke Constantine rode up to the 92nd, began to scrutinise the dress of the men, and told a young soldier to hand him his bonnet, but the Highlandman had no idea of being ordered by a queer-looking foreigner, and refused. The Emperor, understanding that British soldiers were treated with more consideration by their officers than the Russians, politely asked the lad to show his bonnet " to this gentleman, my brother," saying in French, " This is a brave regiment," when the man at once took it off and gave it to the Russian Prince, who examined it for some time before returning it to its owner. They then asked Captain Ferrier to let them examine his claymore.

The French revenged themselves on the allied troops by caricaturing them, and the Highlanders came in for their full share, generally with reference to the admiration of the fair sex ; there were also more respectful paintings of their picturesque costume as they were seen walking about in the streets of Paris. The drawings represent the feathered bonnet, kilt without purse, and buckled shoes.

[1] I find letters from officers to their friends in Scotland asking them to inform the families of deceased soldiers how to apply for the sums due to them, and giving particular accounts of the health and conduct of men belonging to their own districts, for the information of their parents.

There were a few regimental courts-martial during their stay at Paris, drunkenness being the usual crime, and there was some trouble with soldiers who robbed the market-gardens and orchards on the Seine. I find in Division or Brigade Orders fault found with some regiments for being slovenly on guard, etc., but not with either the 42nd or 92nd. A captain of a company describes with pride his success in reclaiming certain bad characters from the paths of vice " without the aid of the cat o' nine tails." He divided his men into squads, and put one of the desperate characters in charge of each. The men at first demurred to being made responsible for the conduct of others, but after a fortnight had passed without any thefts or plundering excursions being heard of in the company, they were satisfied that " a remedy had been found for the evils which had arisen since the battle of Waterloo." A number of n.c. officers and privates waited on their captain, apologised for their previous grumbling, and thanked him for what he had done ; and some time after, he had the satisfaction of enrolling four out of the six bad characters in the first class, and of seeing them draw in the lottery he had established for five prizes which he gave monthly to men of that class.

About this time each man received a present of a pair of shoes, called in Orders the " Waterloo donation shoes," but who the donor was does not appear.

On the 29th October the encampment on the banks of the Seine was broken up, and the regiment marched to St Germains, where they were billeted. Here they were able to visit the palace in which they were led to understand that Prince Charles Stuart had died. Lieutenant Innes, in a letter to his mother, says : " Many of our Highlanders were greatly affected on entering the chamber where their Prince died." It was not, however, really " Bonnie Prince Charlie " who died there, but his grandfather, James II (VII of Scotland). The incident, however, shows how fresh the memory of the '45 remained in the sentiment of many of the men, though they served King George III so loyally.[1]

The brigade under Sir Denis Pack left St Germains on the 31st, and were cantoned at Montainville and the surrounding villages. On the 2nd November the headquarters of the regiment

[1] This sentiment was aptly expressed to George III, who, wishing to see one of those who had been out in the '45, a grim old M'Donald from Knoydart, known as Raonull Mor a' Chrolen, was brought up and presented to His Majesty, who remarked that no doubt he regretted having taken part in the Rebellion—" Sire," promptly replied M'Donald, " I regret nothing of the kind " ; but the King, who had been taken aback by this bold answer, was completely softened when the old man added, " What I did for the Prince I would have done as heartily for your Majesty if you had been in the Prince's place."—" Among the Clan Ranalds," by the Rev. C. M'Donald, priest of Moidart.

"SCOTLAND FOR EVER!"

were at Neuf le Vieux, the companies being scattered at Crecy and other villages ; officers of companies being desired to take particular care that the men were comfortably housed, and that everything they required was paid for on the spot, and no damage done to property. The Prussians were not so particular ; but it is not to be wondered at if they meted out to the French people the measure their country had received when it was invaded by the French army. Two 92nd officers were billeted on a gentleman's house near Montainville, where they spent two pleasant days with the family. After doing justice to a substantial breakfast on the morning they marched, they went out to see after the men and baggage ; on returning to bid adieu, they found the servants busy packing up wine, cold fowls, ham, etc., with which the table was loaded, and to their surprise the butler asked respectfully where they wished him to put the parcels. On being told that British officers never carried things off from their billets, and receiving a handsome tip and thanks for his attentions, he first remained dumb with astonishment, and then exclaimed, "You English behave very differently to the Prussians !"

On the 10th November they marched to Montfort and the neighbourhood, where they remained till the 30th, when the brigade was broken up. The 4th Battalion Royal Scots, which had joined the brigade some time before, and the 42nd and 92nd, marched independently *en route* for embarkation at Calais, and the Gordons moved to Pontoise. Sir Denis Pack, in an Order of the 29th, taking leave of the regiments of his brigade, says : "The services rendered by the 92nd Regiment in the Duke of Wellington's campaigns in the Peninsula, and his Grace's late short and triumphant one in Belgium, are so generally and so highly appreciated, as to make praise from him almost idle ; nevertheless, he cannot help adding his tribute of applause."

The regiment marched by Chambly, Beauvais, Abbeville, Montreuil, and Boulogne. On arriving at the gate of Calais on the 17th December, they were surprised to find it closed against them ; and the commanding officer was informed that the troops (the 28th and 42nd had also arrived) would not be allowed to pass unless each company marched at a distance of a hundred yards from the one preceding it ; the men to reverse their arms, the colours to be kept cased, and the bands not to play through the town. Instant admittance was demanded and refused ; a second message was sent to the governor, who was probably one of Napoleon's old officers, giving him a quarter of an hour to open, or take the consequences. This brought him to his senses ; and the battalions passed in at the usual distance, with bayonets fixed, colours flying, and the bands playing the "Downfall of Paris" all the way to the place of embarkation.[1] The regiment immediately embarked, and sailed on the morning of the 18th for Dover, but a foul wind obliged their ship to anchor in the Downs. On the 19th the Highlanders landed at Ramsgate; and their reception on the shores of Britain would have been a cold one but for the attentions of a neighbouring gentleman named Sir Samuel Curtis, who showed them great kindness, especially to the soldiers' wives—helping them and giving each a present of money. He also invited all the officers to eat their first bit of the roast beef of Old England at his dinner table.

On the 20th they proceeded to Deal and Sandwich, and next day to Dover, where blankets and camp equipage were given into store; from thence they proceeded to Brabourne Lees, where they spent Christmas Day.

R.O.—Officers commanding companies will take care that the men are provided with a good dinner on Christmas and New Year's Day, for which purpose they will advance each man 1s. 6d.

The regiment marched on the 28th December, one wing by Feversham, the other by Ospringe, and on the 2nd of January 1816 they arrived at Colchester Barracks, where they were received by a guard of honour composed of wounded soldiers belonging to almost every regiment engaged at Waterloo. They were about 150 in number, and lined each side of the street, greeting the regiment with tremendous cheering. Most of them had lost an arm or a leg, some both. Sir John Byng, who had commanded a brigade of the Second Division in Spain and the south of France, and a brigade of the Guards at Waterloo, was the general at Colchester.

District Orders, January 2nd, 1816.—Major-General Sir John Byng is much gratified at having under his command the 92nd Regiment, with which he served so long a time in the Peninsula, a regiment which he highly respects for its invariable steadiness and gallantry in the field, and for its discipline and good conduct in quarters. This Order to be read at three successive parades.

In Regimental Orders mention is made of the men being credited with a daily payment called "the Waterloo gratuity penny." The men were "subsisted at the rate of 4d. a day for vegetables, salt, and washing, which, with 6d. for bread and meat, amounts to 10d. per day."

The Waterloo Roll was revised. No man to be entered who was not actually present with the regiment for some part of the period between the 15th of June and 7th July inclusive.

[1] Number embarked:—Field officers, 2; captains, 5; subalterns, 14; staff, 5; sergeants, 36; drummers, 16; rank and file, 449. The establishment was reduced February 1st, 1816, from 1000 to 800 rank and file.

In February they got new clothing, the men to make their kilts with as little delay as possible, and the white waistcoats to have a collar of the same stuff "at their option"; each man was provided with "a good pair of leather gloves."

On the 19th February 1816, the regiment marched for Hull, where it arrived on the 2nd March. On the march the men were ordered to wear their kilts inside out, new waistcoats and old jackets, the new jackets to be carried in the knapsacks. They were treated with great kindness on the road, and hospitably entertained in almost every place at which they halted.

The War Office seems to have given long credit in those days. A correspondence took place at Hull between the paymaster (Mr James Gordon) and the Commissary General, as to whether shoes issued to the regiment in Spain in 1808-9 should be paid for by the regiment. It was decided that they were to be paid for at the rate of 6s. 6d. a pair!

All the bonnets were set up afresh, at a cost of 1s. 8d. each; the colonel appears to have given four flat feathers yearly to each man, but as they were dear this year he only gave three. Heckles cost the sergeants and "music" 3s. 0½d. each, the men of the grenadiers and Light Company 1s. 6½d. each, and battalion companies 1s. 0½d.

Compensation was received for necessaries and musical instruments lost at Quatre-Bras and Waterloo.

In April 1816 the muskets, which hitherto had been kept brightly polished, were browned by instructions from the Board of Ordnance.

On the 23rd April the Waterloo medals were transmitted by order of "the Prince Regent in the name and on behalf of His Majesty"; ribbon was also received by which the medals were to be "suspended on the breast on the button-hole of the uniform." Thoroughly as the men of Quatre-Bras and Waterloo deserved their medals, it must have seemed rather hard to their comrades who had fought and conquered from Egmont-op-Zee to Toulouse that they should have no such distinction. (Medals were presented to the survivors of these campaigns by Queen Victoria in 1848.)

In May the "Recruiting Company" joined from Edinburgh.

There seems to have been some misbehaviour at Hull, absence and breaking out of barracks and the like, and the conviction of two soldiers of a more serious crime induced Colonel Mitchell to call attention to it by an order, in which he regrets that a court-martial for such an unsoldierlike and disgraceful offence should be entered on the records. "Since the return of the regiment from foreign service, the commanding officer has had every reason to be highly pleased with the general conduct of almost every individual under his command; but unfortunately a few men, regardless of their own character, have for a moment sullied the fair reputation of the regiment and brought upon themselves shame and disgrace." He "desires it to be clearly understood that drunkenness will never be received as an excuse for committing a crime, but, on the contrary, will be held as an addition to it."

On the 22nd August the regiment left Hull and marched by Berwick-on-Tweed to Edinburgh, being the second visit of the battalion to its native country since its formation.[1] On the road the church bells of the English villages rang peals of welcome as they passed, and they were entertained with profuse hospitality in most of the towns where they were billeted. Their reception in the Scottish capital was most gratifying; crowds were assembled on the road by which they approached the city, and the throng in the streets was so great that the leading company had difficulty in passing along. The 42nd, between which distinguished regiment and the Gordons there had always been a friendly rivalry—brothers in arms, but rivals in renown—had lately left the Castle, and a soldier of that regiment, standing looking on, cried, "This is nothing to the crowd when we came home; we could hardly get through them at all!" "You should have sent for us to clear the way for you, as we often did in Spain," was the ready retort of a Gordon Grenadier.[2]

In Edinburgh the regiment was the object of marked attention,

[1] "The 92nd Regiment or Gordon Highlanders, under command of Lieut.-Colonel MacDonald, and of which that most distinguished officer, the Earl of Hopetoun, is colonel, marched into Berwick on the 5th and 6th inst. on their route to Edinburgh Castle. They were received with every demonstration of joy, ringing of bells, ladies waving their handkerchiefs, and amidst the cheers of the inhabitants in general. The Mayor, in the name of the town, presented a handsome donation to regale the men, with an appropriate speech to Lieut.-Colonel MacDonald, who immediately distributed the money and returned thanks to the Mayor for this mark of attention on the part of the people of Berwick. The service of these brave fellows is not forgotten who so gloriously maintained the honour of Scotland in the ever memorable battle of Waterloo, as well as on many other hard-fought fields. It will be impossible ever to forget the gallant charge made by this regiment, who, reduced to about two hundred men, and when their commanding officer was told by Major-General Sir Denis Pack that everything in their front had given way to a column of 3000 men, and that the safety of that part of the position depended on the 92nd, this little band of heroes, led on by Lieut.-Colonel MacDonald and headed by Sir Denis Pack, forced their way with the bayonet through this solid column of Napoleon's Imperial Guards, who, panic-struck, began to throw down their arms, which was soon completed by the coming up of that fine regiment the Scots Greys to the assistance of their comrades, calling out 'Scotland for ever!' 1900 prisoners were actually taken by these two weak regiments."—*Edinburgh Courant*, September 7th, 1816.

[2] This story was told to me in 1856 by the then Cluny MacPherson, who served in the 42nd not many years after Waterloo. It is introduced in "Scottish Highlands, Highland Clans and Regiments," to the editor of which publication I gave it.

There was no *jealousy* between Highland regiments. The 92nd bard gave appreciative credit to the 42nd and 79th in his songs; as do Peninsular journals to the 71st.

but the ideas of the period on the subject of hospitality were rather excessive, and the gentlemen who gave banquets to the troops expressed their gratitude to the soldier for his glorious victory by doing their best to make him gloriously drunk ! Notwithstanding the unanimous admiration for the deeds of the British soldier of those days, it has often been repeated that his character for sobriety was indifferent. The Duke of Wellington himself had not a very high opinion of either the officers or men of his army, at any rate when he first commanded it, except as to their fighting qualities. The occasional jealousies or mistakes of his generals, the inattention to orders of some of the regimental officers, the love of plunder and want of sobriety on the part of many of the rank and file, had sometimes interfered with the success of his plans, and he seems perhaps to have been inclined not to distinguish between the good and the bad. His own pervading sense of duty, the energy of his iron constitution, and his abstemious disposition prevented his understanding or excusing either indolence or excess ; still, after having served with the troops of all nations, he said there would be nothing so intelligent as the British soldier if he would keep sober, and, in fact, he promoted many of them to commissioned rank. Napier, who had served chiefly as a regimental officer, and who had a far higher opinion of the soldier than Wellington, allowed that they drank, but drunkenness was the great fault of our countrymen, civil as well as military. To love liquor too well was a crime in the army ; in civil life it was pardonable conviviality.

The cock may craw,
The day may daw,
But aye we 'll taste the barley bree,

was the idea of good fellowship in Burns' days.

"That the British infantry soldier is more robust than the soldier of any other nation can scarcely be doubted by those who, in 1815, observed his powerful frame, distinguished among the united armies of Europe, and, notwithstanding his habitual excess in drinking, he sustains fatigue and wet and the extremes of cold and heat with incredible vigour. When completely disciplined, and three years are required to accomplish this, his port is lofty and his movements free ; the whole world cannot produce a nobler specimen of military bearing. Nor is the mind unworthy of the outward man. He does not, indeed, possess that presumptuous vivacity which would lead him to dictate to his commanders, or even to censure real errors, although he may perceive them ; but he is observant and quick to comprehend his orders, full of resources under difficulties, calm and resolute in danger, and more than usually obedient and careful of his officers in moments of imminent peril. It has been asserted that his undeniable firmness in battle is the result of a phlegmatic constitution uninspired by moral feeling. Never was a more stupid calumny uttered. . . . While no military qualification was wanting, the fount of honour was also full and fresh within him." [1]

Such was the opinion of British soldiers in general of one who had spent the best years of his life among them ; and the testimony borne to the character of the Gordon Highlanders by those under whom they served, and particularly by the inhabitants of the various countries among whom they lived, shows that theirs was a discipline "which, in the full sense of the word, is the fruit of moral and religious instruction." [2]

Before entering on the life of the regiment during the long succeeding years of peace which its services had helped to secure, I will now give a short notice of a few of the veterans from whose reminiscences I gleaned information as to the regimental habits and customs of early times.

Corporal John M'Innes, pensioned after Waterloo, 10s. 6d. a week, joined when the regiment was raised ; a fine-looking old Highlander when I saw him in 1852. He kept a shop at Tom 'a Mhulin, Glen Livat, and gave details of the first uniform and those who wore it ; he had served in all the campaigns, but said little of his own exploits.

John Cattanach, Badenoch, pensioner. He had no education and spoke little or no English ; he tried to enlist when the regiment was raised, but was too young, and was taken later. He spoke much of his captain, MacDonald (Dalchosnie) ; he and many other officers constantly spoke Gaelic to them, and would give news from home to those from their own districts, and Gaelic was commonly spoken in the regiment. He spoke often of "Fassiefern" as a hard man on soldiers who neglected their duty, but just, and a fine officer. On the retreat to Corunna he saw officers as well as men without shoes or hose ; some made "brogan gaelach" [3] out of hides of horses dead on the road. He liked the Spaniards, who often gave them wine and food ; they danced together ; but they were very cruel and barbarous to their French prisoners, hanging them to trees with weights to their feet, etc. John had been a fast runner as a boy at home, and was proud of

[1] Napier, Vol. III. p. 271. Sir John Moore, giving advice to a Highland officer in 1805, said that he considered the Highlanders under an officer who understands and values their character, and works on it, among the best of our military materials. Under such an officer they will conquer or die on the spot. "But it is the principles of integrity and moral correctness that I admire most in Highland soldiers, and this was the trait that first caught my attention." He also made observations on the character of Highland soldiers and duties of their officers with regard to their management of, and behaviour towards, their soldiers, and the necessity of paying attention to their feelings.—General D. Stewart.

[2] Baron Müffling.

[3] Highland brogues.

having won races in the regiment. He was taken prisoner in the Pyrenees, but knocked over one of his captors; another charged at him with his bayonet, when he caught the weapon, which ran through his hand; and jumping down a ledge of rock, he got away among the bushes and boulders, bullets hitting the rocks about him, but he escaped by his fleetness without knowing that a bullet had grazed his shoulder, till he saw the blood trickling down on his belts. This was the only time he was wounded. He never went home till the regiment came to Edinburgh after Waterloo, when he married, and was discharged three years later. He got a bit of land at Strone from the Duke of Gordon at a low rent, and had cows and sheep. Served in Holland, Egypt, Peninsula, Waterloo; had the Waterloo medal, but they never sent him the Peninsula medal. He lived to eighty-nine years of age. A cousin of his, whose father, an old soldier of the American War, had given him a good education, also enlisted in the Gordons, got a commission, and retired as Captain Cattanach. The captain was a frequent guest at Gordon Castle, where his wit made him a general favourite. He lived near Kingussie in the old cottage home, where he was often visited by the Duke of Gordon, when he would turn out the hens and ducks, saying, "There 's not room at the fire for both you and the Duke." He afterwards went to Canada.

John Ferguson, pensioner, a native of Dunean, Inverness, enlisted in Colonel Baillie of Dunean's Fencible Regiment, and volunteered to the 92nd. He served in Egypt, Corunna, Peninsula, Waterloo; was taken prisoner, and said the French were civil, but sometimes the prisoners got short commons, and never clothes or clean linen. Gaelic was ordinarily spoken in the regiment; officers used English on parade, but many spoke Gaelic to the men at other times. He mentioned drinking with the French soldiers, and being quite friendly when not fighting; liked the Spaniards, from whom he got lots of wine and other good things. Wine was sometimes easier to get than good water. He mentioned the Spaniards torturing French prisoners; spoke much of the sore eyes in Egypt, the dust storms and want of water, and of the hardships and severe discipline in the Peninsula. He could speak Spanish better than English; he was married when he enlisted, and came home after Waterloo; at first he had 6d. a day pension, increased to 1s. 3d. from the Kinloch Fund;[1] and lived to be a very old man.

[1] Founded in 1812 by William Kinloch, Esq., of Calcutta, a native of Arbuthnot, Kincardineshire, who bequeathed "the residue of his estate to the Governors and Managers of the Fund instituted in London for the relief of poor and indigent Scotsmen who have lost their limbs or eyesight, or have been otherwise maimed and wounded in the service of their country. About 300 disabled soldiers and sailors are now receiving pensions varying in value from £4 to £8 from that Fund."—*Royal Scottish Hospital Report*, 1900.

Alexander Achinachy, pensioner, a Lowlander from Banffshire, was still with the regiment as civilian messman in 1851, when I often talked to him. Enlisted in 1810, and had been through the Peninsula and Waterloo campaigns. He said that on these campaigns the officers wore nothing of the Highland dress but their bonnets and broadswords, but the n.c. officers and men always wore it except on fatigue, or when in cantonments, at night on guard. The fatigue trousers were grey, or anything they could get, and were not expected to be in good order. They never wore purses when marching and fighting, but left them with the heavy baggage, or carried them on the back of the knapsack. The men were generally decent and respectable, of the agricultural class, but some bad characters were among them. There were a few Irish and English, a good many Lowlanders, but the regiment generally was very Highland, far more so than in 1851. Grant's "Romance of War" had been lately published, and I remember telling him some of the incidents in that novel, and he recognised them, and said they were quite true. "Old Alexander" was greatly respected, and it was always said that, little and flat-footed as he was, he had never fallen out or been in hospital during the campaigns.

John Downie, native of Glenshee, Perthshire, pensioner with 9d. a day for wounds, increased to 1s. 3d. from the Kinloch Fund, joined 1810; a blacksmith by trade, but determined to enlist, and walked to Edinburgh for that purpose. An English regiment was quartered there. That did not suit him; but meeting a sergeant of the Gordon Highlanders, he took the shilling from him for seven years' service, and was soon sent to the 1st Battalion in Portugal. A good scholar, both in English and Gaelic, he kept a journal, which was unfortunately lost with his knapsack at Waterloo, where he was picked up insensible the day after the battle. His heroes were Wellington, "Fassiefern," Cameron the pipe-major, and Private Norman Stewart, and of these he had many tales to tell. There was no talk of teetotalism in these days, but moderation was expected. "Where is your fortitude, man?" the colonel would say to one who pleaded in excuse for being drunk that he seldom tasted liquor at all. Downie said that the night before Waterloo some took a desponding view of the situation—the retreat of the Prussians and the strength of Napoleon's army, and spoke of their great losses at Quatre-Bras. "Comadh co dhiubh," said a Lochaber man, "thug sinn buaidh, dh' aindheoin co' theireadh e" (No matter, we licked them, say what they may); and they began to speak cheerfully of what they would do in Paris, but the French prevented his seeing their capital by putting a bullet in him, which he carried with him to the grave. He had been wounded twice before by both bayonet and bullet. On one of these occasions, when he was making his way to the rear with a wounded comrade, they came on a man M'Intosh, who was one of the Duchess of Gordon's recruits ; he was sitting on a dead horse, and, pointing to his leg shattered by a cannon ball, said, in Gaelic, "What can I make of that?" "Mind, lad, ye got a kiss from the Duchess o' Gordon for that," was the rather unfeeling reply. Downie had a great contempt for the Spanish soldiers. "Clarty deevils," he called them, but thought very highly of the Portuguese troops. He had a great respect for the French army, and in 1870 was surprised at their being defeated by the Germans. "They can't be the same sort that fought against us in Spain," he said. Discharged after Waterloo, he first kept a small school in Glenshee ; when afterwards many people left the glen owing to the introduction of sheep-farming, he entered the post-office service at Alyth, where he was highly respected. He had always been a very religious man, and could read his well-thumbed Gaelic Bible without spectacles at the age of 93 ; he died at a still greater age.

Peter Stewart, pensioner. His parchment discharge mentions the actions of Vittoria, Maya, Donna Maria, Bayonne, Hillette, Garris, Aire, Orthes, Toulouse, Waterloo, his Peninsula medal having clasps for Vittoria, Pyrenees, Nive, Orthes, Toulouse. At some place in Spain the sentries had got into trouble from the spare ammunition being stolen. One night, Peter being sentry over it, a calf appeared, and kept getting nearer and nearer to the ammunition carts. It had rather odd action, and Peter challenged, when the creature answered with a prolonged "Bo," which had something of the human voice about it. He fired ; it fell ; "Bo noo, ye beggar," said Peter, and a Spaniard in a calf's head and hide was his bag. He said if he had his choice, he preferred close quarters and the bayonet to shooting and being shot at. He was in the Duke of Gordon's and afterwards in the Duke of Richmond and Gordon's service in the deer forest of Glen Fiddich ; and when the Duke built him a new house, he named it after what he said was by far the toughest fight he had ever seen—Maya—and by that designation he was afterwards generally known. When the Queen visited the Duke of Richmond at Glen Fiddich, his Grace presented "Maya" to Her Majesty, who, after speaking to him of his services, gave him a sovereign. Afterwards the Duke said : "Well, Peter, I suppose you will put the Queen's sovereign on your watch chain." "I 'll hae a dram to her health oot o't first." When old and infirm, the Duke took a house for him at Banff to be near the doctor, and there he died. His son is a gamekeeper on the Gordon estates.

Reminiscences of Private William MacKenzie, Onich, who enlisted at Fort-Augustus, and served with the Gordons from

Egmont-op-Zee to Waterloo; from notes by his grandson, the Rev. Dugald MacDonald, Episcopal Church, Oban :—

"In Holland William was up to the waist in water and got his powder wet. In the tussle there, they used their fists when grappling.

"At one fight, a piper named M'Lachlan had both legs broken, but continued to play 'Cogadh na Sith' till he fainted from loss of blood.

"He also mentions the incident of the piper's bag being spoiled at Fuentes d'Onor, as mentioned in the text from the account of another man.

"On the retreat to Corunna, some of them ground little bits of horse flesh between stones to squeeze out the blood, and ate it raw, as the old Highlanders used to do with a deer's liver when benighted in the chase. He mentioned the incident of the child on its dead mother's breast, as told in the text from other accounts. When the treasure was thrown away on the retreat, he said some of the men picked up gold doubloons as they passed and put them in their hose, which lamed them.

"At an engagement where his company had to stand long under fire of the enemy close in front, his Celtic impatience could not stand the strain, and he in his excitement shouted 'Hurrah, fhearaibh, bitheamaid aca' (Hurrah, men, let us be at them). His comrades, mistaking his voice for that of their officer, charged at once and routed their opponents. William was tried by court-martial for this breach of discipline, but in consideration of the circumstances, was pardoned. He mentioned that at one place the men used stones to hurl down on the enemy, their ammunition being exhausted (probably Maya). Though very strong, he was the shortest man in the battalion but one, being 5 feet 5 inches. In crossing the Nive he was carried off his legs by the stream, when Colonel Cameron rode to him, told him to catch his stirrup, brought him to land, and gave him a biscuit from his haversack and a drink of wine from his flask.

"He spoke of 'Fassiefern' as the finest soldier possible, but very strict. Men must turn out with arms and accoutrements clean, no matter how short the halt might be. He would allow no excuse for drunkenness or dirt. William thought him too hard on those who were 'heavy on the dram.' He lived to nearly a hundred, and would often say to the boys and their mothers, 'If I was young again it 's not sitting by the fire at home I would be, but with the lads with the yellow tartan' (gillean a' bhreacain bhuidhe), telling his young friends they should join the Gordons."

The veterans were generally very reticent as to their adventures, unless specially drawn out. One said, "If I was to tell people the hardships we endured and the sights we saw, they would not believe me."

I have often seen it stated both that the Highland regiments were not largely composed of Highlanders, and that they did not wear the Highland dress on active service. The three kilted regiments who fought throughout the great French war were recruited much in the same manner; more than once it appears from the remarks of 92nd men, as at Minorca, at Orthes, at Ghent, that they had brothers and neighbours in the 42nd and 79th, and that the regiments were delighted when they met to talk of mutual homes. Any one who knows the Highlands intimately must be aware that there is hardly a respectable family who cannot tell you of an ancestor who was in the army of that period; there were many from the Lowlands also in these regiments, but the Highland element must have been very strong. I myself met a Chelsea pensioner of the 42nd, about 1863, a Waterloo man, who, though himself a Lowlander from Renfrew, spoke Gaelic with ease, and when asked how he came to know it, replied that he learned it in the regiment.

With regard to the dress, the orders and the statements of those present show that the Gordons wore the kilt during every campaign they took part in, though after the Egyptian campaign of 1801, when their clothing was worn out, they came home in anything they could get, white, blue, and grey breeks! There is not much in the orders on active service as to officers' uniform, but from what there is, and incidentally, it appears that they wore the kilt in all the campaigns up to and including Sir John Moore's. In Wellington's campaigns they wore grey or blue pantaloons and shoes with gaiters; and the reason given for this by the late Sir John MacDonald, was that officers were encouraged to ride as much as possible, in order that they might be able at any moment to take a message quickly, but chiefly that they might be fresh at the end of a march to attend to the comforts of their men in the straggling villages and farms in which they were often billeted, for an officer's most important work began when the march was over; and though a great advocate of the Highland garb, as long as he commanded he would not allow an officer to wear it on the line of march.[1] The regiment seems first to have worn gaiters for marching when they went to Egypt, the low quartered buckled shoes having been found inconvenient in the sands of Holland.[2] The

[1] This difference in dress must have made officers liable to be picked off by sharpshooters. The officers of the 42nd and 79th dressed in like manner during these campaigns.

[2] The gaiters were at one time made of the same tartan as the hose, but generally of grey or black cloth.

AUTHOR'S NOTE.—In the Peninsula and Waterloo campaigns, the troops marched and fought carrying knapsacks with full kit, which, with arms and accoutrements, weighed between 50 and 60 lbs.; generally a blanket in addition, and occasionally two days' rations. Every sixth man carried a camp kettle.

sporran was never worn by the men except on occasions such as Sundays, guard, and inspections ; just as at home, when the kilt was commonly worn, a hillman never thought of going to the hill with his sporran on, but kept it for high days and holidays (sporrans seem not to have been taken at all on the Waterloo campaign). The feathered bonnet was the everyday wear of the men, and of officers on all duties, and often in walking out, though they had a sort of undress bonnet and feather also. Colonel Cameron's Waterloo epaulettes [1] are still at Callart ; they are of silver bullion with yellow and silver strap, having the crown in gold and " 92 " on a scarlet ground, surrounded by a silver belt inscribed " Gordon Highlanders."

Although the Highland uniform of the time was picturesque and showy, the brick-red colour of the coat was subdued to a great extent by lace, facings, and accoutrements, so that the various colours became blended, and the dress was perhaps less conspicuous at a distance than the dark figure of a rifleman. It is remarkable that among the various deaths from sickness and accident which are noticed in journals or regimental books, I have seen no notice of sunstroke, which speaks well for the feathered bonnet.

[1] Two pairs of epaulettes of different pattern belonging to Colonel Cameron were preserved at Callart. One epaulette of each pair was given in 1898 to the Officers' Mess of the 2nd Battalion by Mrs Cameron Lucy, great-granddaughter of Cameron, at the request of her nephew, 2nd Lieutenant I. A. Campbell, who was mortally wounded at Elandslaagte in 1899 while serving with the 2nd Battalion.

APPENDICES

I

It having been suggested that the account given of the retreat to Corunna did not do full justice to the hardihood and soldierly qualities of the 92nd, I have taken the number missing during the whole campaign from each of the thirty-three battalions of infantry in Sir John Moore's army, from the return published in Oman's "History of the Peninsular War," Vol. I., Appendix XIII. It appears from this that the total strength in October 1808 was 26,813, and on disembarkation in England, January 1809, 20,774, being a total deficiency of 6039, or an average number missing of 22·5 per cent.

Of the battalions which went through the whole retreat, the numbers missing were least in the following, viz. :—

	Strength, October 1808.	Strength on Disembarkation.	Deficiency.	Missing, per cent.
First Batt. 1st Guards,	1340	1266	74	5·5
[1] Second " "	1102	1036	66	6
First " 92nd Highlanders,	912	783	129	14·2
" " 38th Regiment,	900	757	143	15·8
" " 52nd L.I.,	862	719	143	16·6
" " 79th Highlanders,	932	777	155	16·6
" " 4th Regiment,	889	740	149	16·7
" " 42nd Highlanders,	918	757	161	17·6

The circumstances of different brigades may to some extent account for the difference in the numbers missing, but it may be taken generally as a criterion of discipline and physique, and this is borne out by the fact that a battalion which was specially named and rebuked by Moore at Salamanca for its lack of discipline before the retreat, had 44·3 per cent. of its men missing, and another, which he mentions as weakly, also lost a very large number.[2]

The difference of loss from this cause of battalions in the same brigade, where the circumstances were presumably similar, is still more remarkable. In Catlin Craufurd's brigade—36th Regiment, 71st H.L.I., and 92nd Highlanders—the loss per cent. was respectively 30·3, 18·1, and 14·2.

I have not included R. Craufurd's brigade in the comparison, because they separated from the army on December 31st and went to Vigo, and thus

[1] The Second Battalion 1st Guards picked up invalids from the base hospital at Corunna, so that they actually disembarked more men than they had on December 19th, though short of their strength in October. The Guards had landed in the end of October at Corunna, from whence they advanced with Baird to meet Moore's army.

[2] The proportion of *missing* would no doubt be affected by the number killed in action (the sick and wounded, except the worst cases, were embarked), but unfortunately many commanding officers lumped together the killed with the missing in their returns. Most battalions, however, who bore the brunt of the fighting sent in proper returns. Of those mentioned above it appears the loss was :—

	Officers and Men Killed.		Officers and Men Killed.
First Batt. 1st Guards,	5	First Batt. 52nd,	5
Second " "	8	" " 79th,	0*
First " 92nd,	4	" " 4th,	not given
" " 38th,	0*	" " 42nd,	37

* Fraser's division did not lose a man. The 50th and 81st lost heavily. The First Battalion 95th had 12 killed. Other battalions who sent in separate returns show figures such as 6 or 10 killed and wounded.—(From notes kindly sent by Mr Oman.)

were not in the worst part of the retreat or the battle. This brigade was composed of the First Battalion 43rd, Second Battalion 52nd, and Second Battalion 95th (Rifles). Their missing per cent. were respectively 9·5, 20·5, and 12·9.

II

PROHIBITION OF THE HIGHLAND DRESS

The Disarming and Diskilting Act of George II can be found complete in the "Historical Geography of the Clans," published by W. & A. K. Johnston, Edinburgh. It is a long Act, and contains a good deal besides a diskilting measure. I subjoin the clause referring to that subject.

Clause Restraining use of the Highland Dress :—"And be it further enacted by the authority aforesaid, that from and after the first day of August, One thousand seven hundred and forty-seven, no man or boy within that part of Great Britain called Scotland, other than such as shall be employed as officers and soldiers in His Majesty's forces, shall on any pretence whatsoever wear or put on the clothes commonly called Highland clothes—that is to say, the plaid, philabeg, or little kilt, trowse, shoulder belts, or any part whatsoever of what peculiarly belongs to the Highland garb ; and that no tartan or party-coloured plaid or stuff shall be used for greatcoats ; or for upper coats ; and if any person shall presume after the first day of August to wear or put on the aforesaid garments or any part of them, every such person so offending, being convicted thereof by the oath of one or more credible witness or witnesses before any Court of Justiciary, or any one or more Justices of the Peace for shire or stewartry, or Judge Ordinary of the place where such offence shall be committed, shall suffer imprisonment without bail during the space of six months and no longer ; and being convicted for a second offence before a Court of Justiciary or at the Circuits, shall be liable to be transported to any of His Majesty's plantations beyond the seas, there to remain for the space of seven years."

III

PRICE OF LABOUR, Etc., 1794-1798

(From the "Statistical Account of Scotland," Sir J. Sinclair.)

Appin, Argyle.—Day labourers to maintain themselves, 8d. to 1s. ; tailors and shoemakers, 6d. a day and victuals ; carpenters and other mechanics, 1s. to 1s. 2d. and victuals.

Kiltearn, Ross.—Ordinary ploughmen living in their own homes, £3 a year and six bolls meal ; exceptionally good ploughmen, £5 to £6 a year and 8 to 10 bolls ; day labourers, 6d., or if work very severe, something extra.

Urquhart, Inverness.—Wages greatly increased since commencement of war ; from 8d. and 9d. to 1s. a day in 1798 ; married servants, £3 a year, 6 bolls meal (9 stone to the boll), a cow's grass, fodder, and potato ground.

Black Isle, Ross.—Day labourers, 8d. summer, 7d. winter ; mutton, 2½d. to 4d. a lb. ; meal at various places, 14s. a boll.

HUNTLY, ABERDEEN.—Day labourers, 8d. to 1s. Masons, carpenters, slaters, 1s. 3d. a day. Best beef, 3½d. to 4d. a lb.

TOWN OF BANFF.—Beef and mutton in 1748, 1d. to 1½d. a lb., ; claret, 1s. a bottle. Beef and mutton in 1798, 5½d. to 6d. a lb. ; claret, 6s. a bottle ; meal in 1795, 21s. a boll (8 stones Dutch weight).

R.O., Bastia, February 5th, 1796.—Prices allowed to tailors in the 100th Regiment, as in the 42nd and other Highland Regiments.

For Officers.

	s.	d.
Making a jacket,	5	0
Turning ,,	3	0
Cocking and fitting the feathers in a bonnet,	1	0
Making a kilt,	0	6

For Soldiers and N.C. Officers.

	s.	d.
Sergeant-major or drum-major, making a coat,	3	0
Making a waistcoat,	3	0
Sergeants, for each,	2	0
Drummers, ,,	2	0
Rank and file, for each	1	6
Cocking a bonnet and feathering,	0	1
Making a kilt,	0	3
Making a pair of hose, including thread,	0	2

Thread, hooks and eyes, extra.

IV

RETURN of casualties of His Majesty's 92nd Regiment of Foot (or Gordon Highlanders), commanded by Colonel the Marquis of Huntly, from the 25th of June to 24th December 1799, both days inclusive.

Number of rank and file, 25th June 1799—767.

Recruits joined, 43 ; dead, 2 ; discharged, 5 ; deserted, 7 ; n.c officers and drummers appointed to the 11th Company, 5.

Killed, 57 ; died of their wounds, 30 ; drowned, 15 ; prisoners of war, 39.

Volunteers received and from what corps—9th Regiment, 4 ; 62nd Regiment, 15 ; 63rd Regiment, 2.—Total, 21.

Number of rank and file, 24th December 1799—710.

N.B.—Private Norman Buchanan, d.o.w. 3rd February 1800.

V

ORAN AIR BLAR NA H-OLAIND

AIR FONN—" *Alasdair à Gleanna-Garadh* "

Air mios deireannach an fhoghair,
An dara latha, 's math mo chuimne,
Ghluais na Breatunnaich bho'n fhaiche,
Dh'ionnsuidh tachairt ris na maimhdean ;
Thug *Abercrombaidh* taobh na mara
Dhiu le'n canain, 's mi ga 'n cluintinn ;
Bha fòirneadh aig *Mùr* [1] gu daingeann,
Cumail aingil ris na Fràngaich.

[1] General Sir John Moore.

Thriall *Abercrombaidh* 's *Mùr* na feile,
 Le 'n laoich éuchdach, thun a bhaiteìl ;
Tharruinn iad gu h-eolach, treubhach,
 Luchd na beurla ri uchd catha ;
N uair a dhlù na h-airm ri chéile,
 Dhubhadh na speuran le 'n deathaich ;
S bu lionmhor fear a bha 's an éisdeachd,
 Nach do ghluais leis fein an ath oidhch'.

Dh'fhag iad sinne mar a b'annsa,
 Fo cheannardachd Mhorair Hunndaidh,
An t-òg smiorail, fearail, naimhdeil,
 N an teannadh ain-neart ga 'r n-ionnsuidh ;
Le bhrataichean siod' a strannraich,
 Ri 'u cuid crann a damhs' le muiseag ;
S na fir a toghairt 's na Fràngaich,
 B' iad mo rùinse chlann nach diultadh.

Bha 'n leoghann colgarra gun ghealtachd,
 Le mhìle fear sgairteil là' ruinn ;
An Camshronach garg o'n Earrachd,
 Mar ursainn chatha 's na blàraibh ;
Dh'aontaich sinn mar aon sa bhaiteal,
 Le faobhar lann sgaiteach stailinn ;
Cha bu ghniomh le 'r laoich gun taise,
 Faoineis air an fhaich' le làmhaich.

Bhruchd na naimhdean le 'n trom làdach,
 Air muin chàich an àite teine ;
'N uair fhuair Sasunnaich droch chàradh,
 Phill iad o'n àraich n' ar coinneamh.
Ghlaodh Ralph uaibhreach ri chuid armunn
 Greasaibh na Gàëil n' an coinnidh,
'S tionndaidh iad an ruaig mar b' àbhaist,
 An dream ardanach, neo-fhoileil.

Grad air an aghairt 's an àraich,
 Ghluais na saighdearan nach pillte ;
Mar iolaire guineach, gun chaoimhneas,
 Nach b'fhurasda chlaoidh le mì-mhodh,
Thug iad sgrios na'n gathan boisgeach,
 Mar dhealanaich òidhche dhilinn ;
Ri sior iomain romp nan naimhdean,
 'S neul na fal' air roinn am pìcean.

'N uair a dh'ionndrainn a chonnspuinn
 Morair Gòrdon o uchd buailte ;
'S a chual iad gu'n robh e leòinte,
 Dh'ùraich iad le deoin an tuasaid ;

Mar mhaoim do thuil nam beann mòra,
  Brùchdadh bho na neoil mu'r guaillean,
Lean iad an ruaig le cruaidh spòltach,
  Gu fuilteach, mor bhuilleach, gruamach.

Bha Camshronaich an tùs a chatha,
  Air an losgadh mar an cianda ;
Leonadh an Ceann-feodhna sgairteil,
  Ri còmhraig bhaitealach a liath e ;
Gu sonraicht' coltach an dearcag,
  'S an fheoil nach taisicheadh fiamh i ;
Mu'n chrom a ghrian fo cleòc-taisgte,
  Phàidh sinn air an ais na fiachan.

Ged' bha na Rìoghalaich bho Albainn,
  Na fir ainmeil, mheamnach, phrìseil,
Fada bhuainn ri uair a gharbh chath,
  'S buaidh a b' ainm dhaibh ri uchd mhìltean ;
Ghreas iad air aghaidh gu colgail,
  'N uair a chual iad stoirm nam pìcean ;
Mo creach ! luchd nam breacan balla-bhreac,
  Bhi le lasair marbh na'n sìneadh.

Tha na Fràngaich math air teine,
  Gus an teannar goirid uapa ;
'S an mar sin a fhrois iad sinne,
  Ri deich mionaidean na h-uarach ;
Ach, 'n uair dh'fhaod ar laoich gun tioma,
  Dhol an àite buille bhualadh,
Bha roinn nan stailinne biorach,
  Sàthadh guineideach mu'n tuairmse.

Gu'm bi sin an tuairmse smiorail,
  Chinnteach, amaiseach, gun dearmad ;
Thug na leoghainn bhorba, nimheil,
  Bu cholgail sealladh fo'n armaibh ;
Ri sgiùrsadh naimhdean mar fhalaisg,
  A's driùchdan fallais air gach calg dhiu ;
'S bha Fràngaich a brùchdadh fala,
  'S an cùl ri talamh sa ghainmhich.

Mar neoil fhuilteach air an riasladh,
  Le gaoth a b'iargalta séideadh ;
Ruith nam baidibh ceigeach, lia'-ghlas,
  An deigh an cliathadh as a chéile :
Chìte na naimhde gun riaghailt,
  Teicheadh gu dian o uchd streupa ;
'S iad a leaghadh air am bialthaobh,
  Mar shneachd am fianais na gréine.

Ged' a phill sinn o ar dùthaich,
  Cha d' mhill sinn air cliù an cruadal
Bha sinn gach latha ga'n sgiùrsadh,
  Mar chaoirich aig cù ga'n ruagadh.
Dh'aindeoin an cuid slòigh gun chunntas,
  Tigh'n o'n Fhràing as ùr ga'r bualadh,
Bu leisg ar gaisgich gu tionndadh,
  'Nuair a chòrd an Diùc ri'n uaislean.

'N uair chuireadh am baiteal seachad,
  'S a dh-àireadh ar gaisgich threubhach,
Bha ioma Gàël 's an deachaidh
  Le miad am braise 's an streupa,
Fuil a ruith air lotaibh frasach,
  Bho luchd nam breacanan féilidh,
'S i sior thaomadh leis na glacan—
  'S truagh ! nach dh'fhaod ar gaisgich éirigh !

'S bochd gun sian orra bho luaighe,
  On a bha iad cruaidh 'na'n nàdur,
Fulangach gu dhol san tuasaid,
  Guineideach 'nuair ghluaist' an àrdan,
Cha robh math d'an nàmhaid gluasad,
  Dh'iarraidh buaidh orra' s na blàraibh,
Chaill iad air an tràigh seachd uairean,
  Tuilleadh 's na bha bhuain 'san araich.

'Nis o'n chuir iad sinn do Shasunn,
  Ghabhail ar cairtealan geamhraidh,
Far am faigh sinn leann am pailteas,
  Ged' tha Mac-na-praisich gann oirn
Olar leinn deoch-slainte' Mharcuis—
  Ar gualann thaice 's ar Ceannard ;
Tha sinn cho ullamh's a ãit leis,
  Dhion a bhrataichean bho ainneart.

The Poems are done into English metre by the Rev. Canon Hugh MacColl and Miss Helen Greenhill Gardyne, giving as nearly as possible the meaning of the originals, which are as remarkable for faithful detail as for vivid description ; but no translation can render the full force and beauty of the bard's expressive Gaelic.

BATTLE OF EGMONT-OP-ZEE

(ORAN NA BLAR NA H'OLAND)

In the last month of the harvest,
If my memory serve me truly,
On the second of October,
The British moved to meet the foemen.

Loud I heard the sounding cannon
Where Abercromby held the shore;
Stoutly pressed our brave Moore forward,
Firing hard upon the Frenchmen.

Noble Moore and Abercromby
Moved with brave men to the battle;
Skilfully they then led onwards
Those that speak the tongue of England.
With thick smoke the skies were blackened
When the armies neared each other.
Many then could hear the tumult
Who lay motionless by sunset!

He, the gallant Lord of Huntly—
Him we most desired to follow—
Is our leader, keen for battle.
From the standards dance defiant
All his fluttering silken banners;
And when storm of danger threatens,
And our men are keen to meet it—
Then my dear ones are the clansmen
Who in fight take no denial!

Fearless stood the stern-faced lion
With a thousand of his warriors;
Like a pillar in the contest
Was bold Cameron of Earracht! [1]
One in battle we united,
For not trifling was the struggle
With the keen edge of the steel blades,
Which our heroes now must handle.

Without firing rushed the foemen
On our troops with weight of purpose,
Badly fared the English speakers
When they turned back from the slaughter.
Proud Ralph shouted to his warriors—
"Hasten, sons o' the Gael, to meet them!"
High their spirit, loyal-hearted,
They once more shall turn the battle.

Swift the soldiers midst the slaughter
Onward pressed, nor danger heeded,
Like the eagles, fierce and vengeful,
By dismay not soon defeated.
As lightning on a night of tempest,
They dealt destruction's gleaming arrows;
With hue of bloodshed on their weapons
Firmly drove the foe before them.

[1] Colonel Cameron, 79th, who was wounded.

When Lord Gordon was found missing,
And the news came he had fallen
Wounded in the breast, the heroes
Furiously renewed the onset.
As a spate bursts from the mountains,
Rushing from the clouds about us—
So with frowning brows they followed,
Dealing heavy blows with bloodshed.

In the van the maddened Camerons
Raged to know their brave chief wounded ;
In the clash and stress of battles
Grey had grown the locks of manhood :
Yet his brown eye flashed undaunted
In the face no fear could whiten.
Ere beneath his folded mantle
Sank the sun, all debts were settled.

Famous were the men from Scotland,
Men of the Royals, brave and hardy ;
Far from us in midst of battle,
Though by thousands then confronted,
On they pressed with martial ardour—
Still their name is Victory !
Alas ! the men of chequered tartan
Were laid low by fire of cannon.

Skilful are the French in firing
Till you come to closer quarters ;
Then like hail their shot fell on us
For the sixth part of an hour.
When at length our gallant soldiers
Were allowed to charge the foemen,
Ready was the keen-edged steel point,
Surely aimed to thrust them backward.

Fierce were they as eager lions,
Proving that our men, when near them,
Must be strong, with mien defiant,
Bearing arms and striking boldly ;
Thus they lash the foe before them
Furiously, like heather burning.
Fast and thick upon the steel blades
Fell the dews of perspiration,
Blood of Frenchmen flows in torrents,
Who upon the sands have fallen.

Like blood-dyed clouds all torn and mangled
By a violent west wind blowing,
Rent apart and moving wildly
In shapeless grey and purple masses—

Thus the enemy, disordered,
Soon in flight were seeking safety :
Rapidly, as melts a snow wreath
In the sun's eye, they have vanished.

Fair fame of our native courage
Kept we bright in foreign countries ;
Day by day, as doth a sheep dog,
Like a flock we drove them forward,
In despite of countless levies
Coming fresh from France to strike us.
Loth to halt were then the heroes
When the Duke [1] made peace with Frenchmen.

When the fight at length was ended,
And the gallant men were numbered,
Many a Highlander had fallen
Through the greatness of his daring.
Blood was pouring from the shot wounds,
Flowing down upon the bodies
Of the men of belted tartans.
O ! that our heroes were permitted
To rise once more from field of valour.

Pity 'tis that men so fearless
From the shot were not protected !
Capable and proud by nature,
Vehement when roused in spirit ;
Hopeless was it for the Frenchmen
Such hardy foes in strife to conquer !
On the shore their dead lay scattered
Seven times more than ours in number.

Now, indeed, to winter quarters,
Back to England they have sent us ;
Where is ale in great abundance,
Though " Mac-na-praisich " [2] still is lacking.
Health we 'll drink unto the Marquis,
Our support and great commander,
Ready always when he calls us
To defend his flag from insult !

ORAN

AIR BLAR NA H-EIPHIT

C'arson nach tòisichinn sa chàmpa,
  Far na dh'fhàg mi clann mo ghaoil,
Thog sinn taighean Samhraidh ann,
  Le barrach mheang nan craobh,

[1] Duke of York.

[2] Mac-na-praisich, *lit.* son of fermentation (whisky).

Bu solas uaibhreach, ceannard,
A bhi gluasad ri uchd naimhdean ann,
'S a dh'aindeoin luaidhe Fhrangach,
B' aobhar dàmsha bhi ri 'r taobh.

Cha chualas ri linn seanachais,
Ann an cogadh arm na 'n strì,
Cuig mile-diag cho ainmeil ruibh,
A tharruinn airm fo 'n Rìgh ;
B' aobhar cliù an trèun-fhear Albannach,
A fhuair a chuis ud earbsa ris,
Nach cùbairean a thearbadh leis,
Thoirt gniomh nan àrm gu crìch.

Dh'iarr e moch dì-ciadain,
'S a' chiad diagachadh de 'n Mhàirt,
Gach *comisari* riarachadh,
Ar biadh a mach oirn trà ;
Rùm 'bhi air ar cliathaichean,
Gu h-ullamh mar a dh' iarramaid,
Nach faodadh iad air chiad-lungaidh,
Dol sios leis ann sa bhlàr.

'S ann air dir-daoin a dh'fhàg sinn,
Air sàr chablach fad air chùl,
Na 'm faigheadhmaid rian snàmha dhaibh,
Bu làidir iad na 'r cùis ;
Lean Mac-a-Ghobha [1] cairdeil ruinn,
'S gu 'm b' fhoghainteach a bhàtaichean,
A dh' aindeoin gleadhraich nàmhaid,
Chum e smàladh air an sùil.

Bha ar 'n àrd cheann-feadhna toirteil,
Ann san àm ga 'r propadh suas,
Bho dhream gu dream ga 'm brosnachadh,
Cha b' ann le moit na ghruaidh ;
Ghlacadh cuibhle 'n fhortain,
Ann san laimh nach tionndadh toisgeal i,
'S a dhùisgeadh sunnt gu cosnadh dhuinn,
Mar Fhionn a mosgladh shluaidh.

Thàirneadh na laoich shomalta
Na 'n comhlann throma, bhorb,
Bu tàrslach, làmhan, comasach,
An sradag fhonnidh falbh ;
A g' iarraidh àite an cromadh iad,
Na 'n tugadh nàmhaid coinneamh dhaibh,
Gu 'm fag-te 'n àrach tonn-fhuileach,
Le stàilinn thollach bholg.

[1] Sir Sidney Smith, R.N.

Bho nach tionndadh nàimh gu casgairt,
Bu dlù lasair air an deigh,
'N uair chunnacas gnùis nam Breatunnach
B'fhearr casan dhaibh na strèup ;
Thug iad an cùl gu tapaidh ruinn,
A shiubhal gu dlù astarach,
A sior dhion an cùl le marcaichean,
Chum lasachadh na 'm ceum.

Bha gillean lùghar, sgairteil ann,
Nach d' aom le gealtachd riamh,
Mar dh' fhaodadh iad ga 'n leantain,
Philleadh caogad each le 'n gniomh ;
Bu smaointean faoin d'a marcaichean,
Nach faighte daoine ghleachdadh iad,
'S na laoich nach faoite chaisleachadh,
Ga 'n caol ruith mach air sliabh.

Bu tric an còmhdach casgairt sinn,
Thug sud oirn stad na dhà,
Bhi gun eòlas ann san astar sin,
'N dùil mhòr ri gaisge chàich ;
Dh' fheuch *Ralph* gach doigh a chleachda leis
'S an dian-te sròil a thaisbeanadh,
'S a dh' aindeoin seòltachd dh' fhairtlich oirn,
An toirt gu casgairt làmh.

Bha sinn làidir, guineideach,
Dàna, urranta 'san strì,
Bha iadsan ràideil, cuireideach,
Làn thuineachadh 's an tìr ;
Ghabh iad àird na monaidhean,
Gu 'n dh' fhuair iad àite cothromach,
'S an dianadh làmhach dolaidh dhuinn,
Gu 'n toileachadh r'a linn.

Thairneadh gàradh droma leinn,
De dh' armuinn fhonnidh thréin,
Bho shàil' gu sàil' a coinneachadh
'N trà chromaidh air a ghréin ;
Bu daingean, làidir, comasach,
A phàirc ga m' fhàl na bonaidean,
Cha bu chadal séimh ga 'n comunn,
'S càch ma 'r coinneamh air a bheinn.

Stad sinn ré na h-oidhche sin,
Gu leir an cuim nan àrm,
Bha leannan fein, gu maighdeannail,
Fo sgéith gach saighdear, bàlbh ;

Na 'n tigeadh feum na faoineachd orr',
'S gu tugte aobhar bruidhne dhì,
Bu neamhail a spéic phuiseanta,
  Bho 'n bheul bu chinnteach sealg.

Dh' earbadh dìon an 'n anmanan,
  Ri Albannaich mo rùin
Fir nach tàirnnte cearbaich orra,
  'N àm tharruinn arm gu dlù ;
Rinn iad a chaithris armailteach,
Gu h-ullamh, ealamh, ealachuinneach,
'S na 'n deanadh nàmhaid tairgneachadh,
  Bha bàs allabharach na 'n gnùis.

Sinn ullamh air ar connspagan,
  Gu dol san tòir gu dion,
An treas madainn diag a shònraich iad,
  Le 'r ceannard mòr gu 'n fhiamh ;
An dà réiseamaid a b' òige againn,
Na Gréamaich agus Gòrdonaich,
A ruith gu dian an còmhdhail,
  Na bha dortadh leis an t-sliabh.

Cho ullamh ris an fhùdar,
  A bha dol na smùid ma 'r ceann,
Ghluais na gillean lù-chleasach,
  Air mhire null do 'n ghleann ;
Thug sinn le teine dùbailte,
Bristeadh as na trùpairean,
Bha Gréumaich nan éuchd fiùghantach
  'S cha d' éisd iad mùiseag lann.

Mar stoirm a b' iargalt connsachadh,
  A spionadh neòil a's chrann,
A riasladh fàirge mòire,
  Gu pianadh sheòl 's ga 'n call ;
Cruaidh dian bha buaidh nan Gòrdonach,
Bu lionmhor sguab a's dorlaichean,
A bhuain iad air a chòmhnard,
  Far an tug na slòigh dhaibh ceann.

Dhlùthaich ar n' arm urramach,
  Gu h-ullamh air ar cùl,
Lion iad an t-sreath fhulangach,
  Rinn guineideach gu smùis ;
Bu naimhdeil dian an gunnaireachd,
A dh'fhàg an sliabh 's nial fuileach air,
Bha cuirp na 'n riadhan uireasach,
  Fo 'n ian gun tuille lùis.

'N àm propadh ris an nàmhaid,
  Sinn g'an smàladh ann sa' cheò,
Las a bheinn mar àmhuinn ruinn,
  A bàrcadh na prais oirn ;
Shaoil sinn gur h-i *Vesàvius*,
A sgàin bho bonn le tàirneanaich,
Airm chaola b' fhaoineis làmh ridhe,
  'S craos na chaoir tigh'n' beò.

Bha craoslach nan geum neimheil,
  Gu brèun, aineolach, sa' cheò,
A bheist bu tréine langhanaich,
  Bu reusan sgreamh do dh' fheòil ;
Bu chaillteach dhuinn an dealanach,
'S a liughad saighdear bearraideach,
Bha 'n oidhche sin a mearachd oirn,
  Gu 'n anam air an tòir.

Dh' aindeoin a h-ard bhùrainich,
  Bha làidir, mùiseach, garbh,
Ga b' oil leis an cuid trùpairean,
  Am bruchdadh rinn an arm ;
Ge d' fhuair sinn beagan diùbhalach,
A laoghad cha do lùb sinn daibh,
Bu lionmhor marcach cùl-donn diù,
  Fo 'r casan brùite, màrbh.

Thug iad an cùl, 's cha mhasladh dhaibh,
  Chuir casgairt iad na'n teinn,
Sinn ga'n sgiursadh do 's na fasaichean,
  'S gach tùbh na las a bheinn ;
Thionndadh gach cùis taitneach dhuinn,
Bho bhon a cùil a căs-mhulaich,
Cha d' fhurich gnùis dhiu gleachda ruinn,
  Nach d' bhrùchd amach na still.

'S căs a throm an ruaig orra,
  Cho cruaidh 's chualas riamh,
Bha *Abercrombie* suas riutha,
  Le shluadh a dh' fhuasgail fial ;
Mar bhi'dh am baile bhuannaich iad,
Le canain air a chuartachadh,
Bha barachd dhiù 's na h-uaighichean,
  'S a dh' fhuaraich air an t-sliabh.

Thàirneadh gàradh làidir,
  'Dh' arm tabhachdach nach striochd,
Ma choinneamh *Alexandria*,
  Air airde *Aboukier* ;

'N uair rainig sinn an làrach sin,
'S a dhealaich mi ri m' chàirdean ann,
'S ann ghiùlain iad gu m' bhàta mi,
'S fuil bhlàth fo 'm an fhiar.

Tha 'n dà Bhaiteal àraidh
An deagh Ghàëlig ann am chuimhn',
Cha 'n e 'n treas fear bu tàire,
'S math a b' fhiach e bàrd ga sheinn ;
Tha mi sa' cheaird air mhàgaran,
Cha 'n fhilidh no fear dàna mi,
Na dh' innis mi cha nàr leam e,
Co chluinneas c' àit' an d' rinn.

SONG ON THE BATTLE OF EGYPT

In that far camp my song begins
In which I left my comrades dear,
Where with tender fronds of palm trees
We built the summer bowers.
Proud were our hearts to mark our chief
Meeting the foemen breast to breast ;
Such joy were mine—such as he feels
Who joins the dance—if by your side,
Despite French lead, I stood once more.

We have not heard in olden tales
Of clash of arms and battle-fray,
Of fifteen thousand men more famous
Who for the King have drawn their swords.
Far-renowned was Scotland's hero [1]
With such charge to be entrusted ;
Nor were cowards by him chosen
To end the mighty deed of arms.

It was early in the morning
Of the mid-day of the week,
When the teens of March were opening,
That the orders came for each
Of our commissaries to issue
Drink and rations to the army ;
So that strengthened for the battle
We might stand prepared betimes.

Upon Thursday we departed,
Our noble fleet left far behind.
Mighty was their strength to aid us,
But no channel there could float them.

[1] Sir Ralph Abercromby.

Staunch Sir Sidney Smith with valour,
As of old, our force supported ;
Shrill fury of the foe extinguished,
As a light is quenched in darkness.

Now was seen our great commander
Encouraging his men with skill ;
Each several company inciting
To valour, prompt in look and word,
Like Fingal mustering his host.
His strong hand seizing Fortune's wheel,
He swiftly gave the happy turn
That wakes the soul to victory.

The comely warriors were drawn up
In stern well-ordered companies ;
Strong are their hands, and skilled to wound
With the fine spark of tempered steel,
When seeking for a place to strike.
There, if the foe would stand to meet them,
Dark would be the ground with blood,
From havoc of the piercing sword-blades.

But since the foemen would not tarry
To dare the combat hand to hand,
Nor face the British line of battle—
With fire we pressed them, following closely,
Till they turned their backs upon us,
Showed their heels and fled with swiftness ;
Though trained horsemen in their rear
Protected still their flagging footsteps.

There were lads, swift-footed, hearty,
Yielding not whate'er the danger,
Pursuing still as they were able.
Fifty horse their skill turned backwards,
For in vain the riders fancied
That no man would dare to fight them,
Whilst the still unvanquished heroes
Chased them o'er the plain far-stretching.

Signs we left of deadly slaughter ;
Yet, because the ground we knew not,
More than once we needs must linger ;
Though Ralph, trusting to our comrades,
Tried all his well-practised methods
To reveal the foemen's banners.
Yet we failed, despite his wisdom,
To bring their forces to hand-fighting.

We were strong men, bold and hardy,
Sure of ourselves and fierce in strife ;
Wily the foe, and full of cunning,
With good knowledge of the country.
Therefore they sought among the hills
To find a fitting place of vantage
From which their guns could work us mischief
And success, though fleeting, capture.

A line of strong defence we formed
Of matchless, mighty warriors ;
Across the plain from base to base
They stretched, at setting of the sun.
Keen soldiers, steadfast, capable,
Fencing the stricken field with bonnets.
Peaceful sleep was not their portion,
For on the hill the foe still faced them.

Throughout the night, all stood to arms ;
Each soldier holding maiden-wise
His silent sweetheart 'neath his wing.
Silent until, when need or call
Of danger giveth cause for speech,
Her lively answer would be ready,
A stinging message from a mouth
That never missed her deadly aim !

On you, my Scottish comrades dear,
The safety of our lives depended.
Men who in the stress of arms
Would never bring misfortune on us ;
Good watch they kept throughout the night,
Aye ready, eager, standing steadfast.
Should challenge come from any foeman,
A sudden death would be his portion !

Prepared on all sides with defiance
To follow up the swift pursuit,
The thirteenth morning was the time
Fixed by our gallant great commander.
Then two, our youngest regiments,
Gordons and Græmes,[1] impetuously
Rush forth to meet the opposing hosts
That stream in numbers down the hill.

Quick was the powder blazing
About our heads in circling smoke—
So marched—so ran—our agile lads !
As when they hastened to the sport

[1] The 90th Regiment was commanded by Colonel Graham of Lynedoch.

In Highland glen—so now they break
With double fire the lines of horsemen.
Like heroes fought the Græmes, nor heeded
The murmur of the sounding sword-blades !

As when a storm, in strong contention,
With violence rends the clouds and trees,
Raising a sea most boisterous—
On fair sails sending peril grievous—
Hard was the Gordons' victory !
But when on that contested plain
They met the onset of the foe,
Abundantly they reaped the sheaves
And gleanings of Death's harvest-field !

Our gallant troops, intent on vengeance,
In readiness upon our rear,
March on to fill the dauntless line
Formed closely, firing straight and sure,
Well skilled in fierce Destruction's ways.
Thus on that blood-stained field they left
In piteous rows the wounded men,
Laid spent and weary 'neath the skies.

When thus we faced the enemy,
Thick mist had hidden all from view ;
Then blazed the hill as from a furnace
Pouring brass upon us there !
As though Vesuvius in thunder
Forth from her strong foundations burst !
Vain our small arms 'gainst that mouth
Whence fiery waves do issue still.

But the wide mouth muttered vengeance,
Speaking darkly through the mist ;
There that mighty engine roaring
Causeth horror to the flesh.
For by that lightning many a lad—
Light-hearted soldier from our midst—
Was missing from us on that night,
And not a soul to seek them out.

But despite the deep-toned wail
Of that strong voice, loud and deadly,
And in spite of all their troopers,
Fiercely rushed our army on them !
Though some loss we suffered, never
Did we yield one step before them !
Many of their brown-haired riders
Fell down dead beneath our feet.

No disgrace was theirs in fleeing—
By our fire their ranks were shattered—
As we drove them to their downfall
Where the hill was widely blazing.
All things now were in our favour,
Ours the hill from base to summit ;
Not a man remained to face us—
All fled from us like a torrent !

Never flying foe was pressed
By swift pursuit more fierce and hard !
Abercromby came up with them,
Brought his hosts to our relief.
Were it not for cannons frowning
To guard the town they now had reached,
Many more had found their graves,
Than on the field of battle stiffened.

In front of Alexandria,
Upon the heights of Aboukir,
A gallant and victorious army
Formed a strong, unyielding wall.
When we reached that field of battle,
I was parted from my comrades,
Who to the boat sore-wounded bore me,
Warm blood all the grass bedewing.

In good Gaelic my remembrance
Holds these special battles twain ;
And the third was no less famous,
Most worthy to be told in song.
Poet am I not, nor singer—
But no shame have I in telling,
So that all who list may hear me—
Of what chanced and where 'twas done !

VI

When the Experimental Rifle Corps was succeeded by the Rifle Corps and 95th Rifle Regiment, a Highland Company seems to have been continued, though without any distinctive dress. In an acquittance roll of Captain Alexander Stewart's company, Rifle Regiment, 1801, there are sixty-seven men, and their names show them to have been almost all Highlanders. A letter to him from his commanding officer in 1802 desires him to enlist as many as possible in his own country (Appin). Captain Stewart had been in the 92nd, and he seems to have been succeeded by another Gordon Highlander, for Colonel A. W. Cameron told me his father commanded the Highland Company of the 95th Rifles, and he is mentioned by Blakeney as commanding it at Corunna.

VII

EXTRACT FROM LIEUT.-GENERAL HILL'S DISPATCH AFTER THE ACTION OF ARROYO-DEL-MOLINOS, DATED MERIDA, 30TH OCTOBER 1811

"The 71st and 92nd Regiments charged into the town with cheers, and drove the enemy everywhere at the point of the bayonet, having a few men cut down by the enemy's cavalry. The enemy's infantry, which had got out of the town, had, by the time these regiments arrived at the extremity of it, formed into two squares, with the cavalry on their left; the whole were posted between the Merida and Medelin roads, fronting Alcuescar; the right square being formed within half musket shot of the town, the garden walls of which were promptly lined by the 71st Light Infantry, while the 92nd Regiment filed out and formed line on their right, perpendicular to the enemy's right flank, which was much annoyed by the well-directed fire of the 71st. In the meantime, one wing of the 50th Regiment occupied the town and secured the prisoners, and the other wing along with the three six-pounders skirted the outside of it, the artillery, as soon as within range, firing with great effect upon the squares.

"Whilst the enemy was thus occupied on his right, Major-General Howard's column continued moving round his left, and our cavalry advancing and crossing the head of the column, cut off the enemy's cavalry from his infantry, charging it repeatedly and putting it to the rout. The 13th Light Dragoons at the same time took possession of the enemy's artillery. One of the charges, made by two squadrons of the 2nd Hussars and one of the 9th Light Dragoons, was particularly gallant; the latter commanded by Captain Gore, and the whole under Major Busche of the hussars. I ought previously to have mentioned that the British cavalry, having, through the darkness of the night and the badness of the road, been somewhat delayed, the Spanish cavalry, under the Conde de Penne Villemur, was on this occasion the first to form upon the plain and engage the enemy, until the British were enabled to come up.

"The enemy was now in full retreat, but Major-General Howard's column having gained the point to which it was directed, and the left column gaining fast upon him, he had no resource but to surrender, or to disperse and ascend the mountain. He preferred the latter, and ascending near the eastern extremity of the crescent, and which might have been deemed inaccessible, was followed closely by the 28th and 34th Regiments, whilst the 39th Regiment and Colonel Ashworth's Brigade of Portuguese infantry followed round the foot of the mountain by the Truxillo road, to take him again in flank. At the same time, Brig.-General Morillo's infantry ascended at some distance to the left with the same view."

NOTE.—The left column was composed of the 50th, 71st, 92nd, and one company of the 60th.

EDINBURGH CASTLE, 15*th August* 1845.

SIR,—I have the honour to state that in the year 1829, when I preferred a claim on the part of the 92nd to bear certain honourable badges commemorative of its conduct in the field, I then confined myself to the great battles or general actions for which medals had been conferred on the officers in command, with the exception of "Almaraz," which having been granted to the other regiments engaged on that occasion, although not more distinguished than the 92nd, the withholding the grant from that corps would convey to future times that there had been some misconduct on its part. Acting on this rule, I did not then put forth any claim to the words "Arroyo-del-Molinos," lately granted to the 34th Regiment, whose services on that occasion I do not mean to question.

At the same time, and with all deference, I venture to assert that if there was one more pre-eminently distinguished than another in the attack and surprisal of the enemy at Arroyo-del-Molinos, on the 28th October 1811, that regiment was the one which I now have the honour to command, in proof of which I beg leave to refer to Lord Hill's dispatch, dated Merida, 30th October 1811, as well as to the casualties on that occasion, when, out of a total of seven killed, the 92nd had three, and out of a total of seven officers wounded, the 92nd numbered four of them. That these casualties were not increased (as generally happens) from any hesitation or delay in the advance and attack of the regiment will appear very evident from the fact that a wing of the 50th (which was held in reserve) was immediately required in the town and right suburbs to secure the prisoners overthrown by the 92nd in its impetuous charge through the main street ; the 71st at that time bearing a heavy fire upon the enemy's right from the garden walls, and the 92nd deploying in the open field in his front. While the 71st were being disengaged from the garden and suburbs on the left, the 92nd advanced and broke the enemy's right square, which then dispersed, as detailed in the dispatch, and was pursued by the 28th and 34th Regiments. When the enemy dispersed he threw away arms, drums, and everything likely to encumber him in ascending the Sierra. Many of these fell into the hands of the 92nd, in whose immediate front he broke ; but some also fell into the hands of the regiments ordered to pursue, while the 50th, 71st, and 92nd were recalled, preparatory to the movement to St Pedro the same day, and on the next to Merida.

Under these circumstances, I trust His Grace the Commander-in-Chief will recommend Her Majesty to grant the 92nd Regiment permission to bear the words "Arroyo-del-Molinos" on its colours and appointments, the same mark of distinction having already been conferred on the 34th Regiment engaged in the same service.—I have the honour to be, Sir, your most obedient humble servant, (Signed) J. MacDonald,

Colonel Commanding 92nd Highlanders.

The Adjutant-General of the Forces, etc.,
Horse Guards, London.

Lieut.-General Sir William MacBean, Colonel of the 92nd, addressed a letter to the same effect to the Adjutant-General, dated 18th August 1845, to which he received the following reply, dated Horse Guards, 26th September 1845 :—

Sir,—Having had the honour to submit to the Commander-in-Chief your letter of the 18th August last with its enclosure, I have it in command to express His Grace's regret that he cannot give effect to your wish that the 92nd Regiment be permitted to assume the words " Arroyo-del-Molinos " on its colours and appointments, as the officer commanding the regiment did not obtain a medal for that action.[1]

I am to add that this distinction was conferred by Her Majesty upon the 34th Regiment in lieu of that which it enjoyed before, namely, a red and white tuft, which is now worn by the infantry at large.—I have the honour to be, etc.,

(Signed) J. MacDonald, A.G.

Lieut.-General Sir William MacBean, K.C.B.,
United Service Club.

[1] The officer commanding the 34th does not appear to have obtained a medal for that action, and Colonel Cameron, the officer commanding the 92nd, did not receive a medal for " Almaraz " ; yet that name was allowed to his regiment in 1829.

VIII

LIST OF OFFICERS OF THE 92ND REGIMENT (BOTH BATTALIONS), JANUARY 1ST, 1813, SHOWING NATIONALITY AND CAREER AS FAR AS KNOWN

Rank.	Name.	Family or Residence.	County.	Remarks.
Colonel	Lieut.-General Honble. Sir John Hope, K.B.	Hopetoun	Linlithgow	Afterwards Earl of Hopetoun
Lieut.-Col.	John Cameron	Fassiefern	Argyll	Killed at Quatre-Bras, 1815
,,	John Lamont	of Lamont	Argyll	Retired major-general, 1819
Major	*Archibald MacDonell	Inch, Lochaber	Inverness	Died in Lochaber, 1813, when lieut.-colonel of Veterans
,,	James Mitchell	Auchindaul	Inverness	Retired lieut.-colonel, C.B. Died in Lochaber, 1847
,,	Archibald Campbell	Melfort (?)	Argyll	
,,	Donald MacDonald	Dalchosnie	Perth	Retired lieut.-colonel, C.B., 1818
Captain	John MacPherson	Knappach	Inverness	Killed at St Pierre as major, 1813
,,	James Seton	yr. of Pitmeddan	Aberdeen	Died as major at Cambo of wounds received at Garris, 1814
,,	Samuel Maxwell			
,,	Evan MacPherson	Ovie	Inverness	Colonel. Governor of Sheerness. Died 1823
,,	James Lee (afterwards Lee Harvey)	of Castle-Semple	Renfrew	Retired colonel, K.H.†
,,	George W. Holmes			
,,	*Ronald MacDonald	Inch, Lochaber	Inverness	Fort MacDonald, Ceylon, named after him

Note—Those marked * seem all to have been connected with the Keppoch family. Those places marked (?) are doubtful. † Knight of Hanover.

Rank.	Name.	Family or Residence.	County.	Remarks.
Captain	Dugald Campbell	Fort-William	Inverness	Rose from the ranks. Retired major
,,	William Phipps			
,,	Peter Wilkie	son of Quarter-master, 92nd		Retired major
,,	William Mackay			
,,	William Logie			
,,	William Charles Grant			Died at Brussels of wounds received at Quatre-Bras.
,,	Ulysses Burgh	Aghanville	Queen's County	Assist. mil. sec. to Wellington. Afterwards A.D.C. to the King. Surveyor-general of ordnance. Died, General Lord Downes, 1863
,,	Donald MacBarnet	Kingussie	Inverness	Rose from the ranks, retired captain. Died in Badenoch
,,	Roger Aytoun	of Inchdairnie	Fife	
,,	Alexander, Lord Berridale	son of the Earl of Caithness		
,,	Charles, Earl of March	son of the Duke of Richmond		
,,	John Hill			
Lieutenant	Samuel Bevan			Died of wounds received at Maya, 1813
,,	John Warren			
,,	Angus Fraser			
,,	William Baillie	Dunean (?)	Inverness	
,,	John Ross	son of a parish minister	Ross-shire	Killed at San Sebastian while doing duty with Portuguese regiment
,,	William Fyfe	Speyside	Moray	Retired captain. Died at Garmouth
,,	Donald MacPherson	Sherrobeg, Laggan	Inverness	Retired captain. Died about 1835
,,	John Cattanach	Kingussie	Inverness	Rose from the ranks, retired captain. Died in Canada
,,	*Ronald MacDonald	Gaskmore	Inverness	Colonel, C.B., K.H., adjutant-general Bombay Army. Died 1848
,,	John A. Durie			
,,	Claude Alexander		Renfrewshire	
,,	Hay Rose			
,,	James John Chisholm	Knockfin	Inverness	Killed at Quatre-Bras, 1815
,,	Robert Winchester	Aberdeen	Aberdeen	Retired lieut.-colonel. Died, Edinburgh, 1846
,,	Thomas Hobbs			
,,	Thomas MacIntosh			

Note—Those marked * seem all to have been connected with the Keppoch family. Those places marked (?) are doubtful.

RANK.	NAME.	FAMILY OR RESIDENCE.	COUNTY.	REMARKS.
Lieutenant	*Donald MacDonald	Garvabeg	Inverness	
,,	Andrew Will	Brechin	Forfar	Died, Jamaica, 1819
,,	Alexander Gordon	Loin-bhulg	Inverness	
,,	Duncan MacPherson			Killed at St Pierre, 1813
,,	William MacEachran			
,,	Hector Innes	Cullen	Banff	Died, Jamaica, 1819
,,	Andrew M'Kenzie Shaw	Pettulich	Inverness	Retired captain
,,	George Logan			
,,	Ewen Campbell	Sonachan	Argyll	
,,	*Richard MacDonell	of Keppoch	Inverness	Died, Jamaica, 1819
,,	Richard Josiah Peat			
,,	William Little			Killed at Quatre-Bras
,,	Patrick Leatham			
,,	George Mackie	Auldearn (?)	Nairn	Died, Jamaica, 1819
,,	Alexander MacPherson	Kingussie	Inverness	Rose from the ranks, retired captain. Died in Badenoch
,,	Ewen Cameron Ross	Kilmonivaig, son of parish minister	Inverness	Died at Fort-Augustus about 1866
Ensign	Luke Higgins			
,,	George Gordon	Loin-bhulg	Inverness	
,,	J—— Mackie	Auldearn	Nairn	
,,	John Grant	Corriemony	Inverness	Father of the author of "The Romance of War." Died captain at Edinburgh, 1861
,,	James Hope	Hopetoun	Linlithgow	
,,	Alexander MacDonald	Dalchosnie	Perth	Died lieutenant at Vittoria of wounds received at Maya, 1813
,,	Thomas Mitchell	Auchindaul	Inverness	Killed as lieutenant at St Pierre, 1813
,,	*Allan Macdonald	Cul-na-Coil	Inverness	Killed as lieutenant at St Pierre
,,	Ewen Cameron MacPherson	Cluny	Inverness	Died in the Hon. East India Company's service
,,	Ewen Cameron	Letterfinlay (?)	Inverness	
,,	James Pattullo			
,,	Ewen Kenedy	Moy, Lochaber	Inverness	Rose from the ranks. Killed at Maya, 1813
Paymaster	James Gordon	Croughly	Banff	Died at Nairn, 1867
,,	John Philip			
Adjutant	Claude Alexander			
,,	Ewen Campbell			
Q'ermaster	Duncan MacFarlane			Rose from the ranks
,,	George Wallace			Rose from the ranks

NOTE—Those marked * seem all to have been connected with the Keppoch family. Those places marked (?) are doubtful.

LIST OF STAFF-SERGEANTS OF THE 1ST BATTALION 92ND REGIMENT AT MAYA, JULY 1813.

Sergeant-major	Duncan MacPherson.
Quartermaster-sergeant	John MacCombie.
Paymaster-sergeant	John Grant.
Schoolmaster-sergeant	John MacEwen.
Armourer-sergeant	Dundas Stevenson (came from the 42nd).
Drum-major	Colin MacKenzie.

Ninety per cent. of the men were Scots according to the Return of 25th July 1813. Napier acknowledged, in the *United Service Journal* for October 1840, that it had been proved to him that seventy per cent. were Scotch Highlanders, and this is borne out by Colonel Cameron's letter of four years earlier, in which he says two-thirds of 2nd Battalion could speak little but Gaelic, and the other third nearly all Scotch ; therefore, it appears that twenty per cent. were Scottish Lowlanders, and ten per cent. English and Irish.

The Clans most numerously represented on the prize roll for Vittoria, made out just before Maya, were in order of numbers—Clan Chattan, Ross, Clan Donald, Mackenzie, Fraser, MacKay, Cameron, Munro. Campbell, Gow or Smith, Murray, MacLeod, MacLean, Grant numbered each from seventeen to fourteen.

IX

CUMHA DO CHOIRNEAL IAIN CAMSHRON

A thuit ann am blar Bhatarlaidh, agus chaidh a chorp a thoirt dhachaidh do Chill-a-Mhaillibh 'n Lochaber

AIR FONN—" Gur muladach mi lium f'hin,
'S gu'n duine mu 'n cuairt."

'S lionmhor caraid 's fear daimh,
Nach gearain a cheann' bhi tinn,
Chaidh a leagadh 's an Fhraing,
'S a chuir Bonaparte thall d'ar dith ;
Ged bha Wellington ann,
'Chuir scapadh 'na champ bho thir,
'S leir ri fhaicinn ar call,
Dh'fhag na Gaidheal cho gann ri 'r linn.

'N sgeul a thainig as ùr,
Dh'fhag na h-Abraich fo thùrsa bron,
'S iad gun mhire, gun mhùirnn,
Gun aidhear, gun sunnt ri ceol ;
Chaill iad caraid 'sa chuirt,
'S treun bharrantas cuil air sloigh ;
'S lionmhor Morfhear a's Diuchd,
A bha cràiteach mu 'n diubhal mhoir.

Chunnacas long anns a chaol,
  'S i 'seoladh le gaoth bho 'n iar,
Dol gu fasadh nan craobh,
  'S ann leamsa nach b-fhaoin an sgial ;
'S corp aluinn an laoich
  Air a clar, 's gun bu daor a thriall ;
'S dh'fhag sid bron agus gaoire,
  Aig mnathan da thaobh Loch-iall.

Latha blair Bhatarlaidh,
  Thuit 'n t-shaoidh sin fhuair cliu 's gach tir,
Mor Chamshronach ar,
  Choisinn urram an cuirt an Righ ;
Fhir nach tionndadh do chul,
  'N uair a rachadh a chuis gu stri,
'S tric a dhearbh thu do thùrn,
  'S a fhuair sinn ort cunntas fior.

D' fhuil 'ga dortadh air mhire,
  Dh'fhag a' caoineadh do chinneadh gu leir
A's do chreuchdaibh a sileadh,
  'S cha robh doigh air a tilleadh le leigh,
'S gun ad' choir ach do ghille,
  'S e bronach le teum às do dheigh ;
A's b'e leon an cul slinnein,
  Leis 'n do thuit thu 'san iomairt, 's le 'm beud.

Bu tu 'm buachaill air treud,
  Ga 'n gleidheadh bho bheud gun chall,
Gus an cuirte orra feum,
  'S iad uil air dheadh ghleus a' d' champ ;
Neartmhor, fulanach, treun,
  Fad 'sa sheasadh tu fein air an ceann ;
Bhiodh do naimhdean lan chreuchd,
  'S tu cur an ra-treut na deann.

Sar churaidh gun chealg,
  Cait an cualas fear t-aimn rid' linn ;
A dh'fhuirich no dh'fhalbh,
  Thug ort urram le dearbhadh fior ;
Ann an cumadh, 's an dealbh,
  No 'n cumasg nan arm dol sios,
Leomhan fuileachdach, garg,
  'S lan ghaisgeach gun chearb 'san stri.

Fear do choltais le cinnt,
  Cha 'n fhaicte 'n cuig mil' air sràid,
Gun chron cuim' ort ri inn's,
  Bho mhullach do chinn gu d' shail ;

'S do dh'aoine gasd' aig do shlinne,
  Leis an reachadh tu 'n tionnsglaidh blair,
Ann an cogadh no 'n sith,
  'S tu bhuidhneadh a chìs thar chach.

Bu ghlan ruthadh a d' ghruaidh,
  Air each aigionnach, luath, 'chinn aird,
'S tu air thoiseach do shluaigh,
  Nuair a tharruinneadh suas *brigade* ;
Claidheamh nochda gun truaill,
  Leis an coisneadh tu buaidh a d' laimh,
Lann thana, gheur, chruaidh,
  Scathadh chlaignean a's chluas gu lar.

'S mairg a spionadh dhiot calg,
  'N uar a lasadh do mheamna ad' shroin,
Lamh dheas air chul arm,
  Leis an reachadh tu an sealbh a ghleois ;
Ghleidheadh do'n righ,
  'S cha leigeadh tu dhiot a choir ;
'S goirt do chairdean' gad' dhith,
  'S nach d' fhan thu a dh' innse sgeoil.

Sid an sgeula 'bha goirt,
  Dh'fhag Sir Eoghan na thosd, gun sunnt,
'S beag an t-ioghnadh a sprochd,
  'S deoir bhi sileadh bho' roisg' gu dlu ;
A dheadh mhac oighre gun spot,
  A dh'fhoillsich le phosta cliu,
Dhol gu bas le trom lot,
  Air a charamh a nochd 's an uir.

An latha mor sin chaidh crioch
  Air a chogadh, 's gach rioghachd thall,
S iomadh luach bu mhor pris,
  A thuit leis an stri 'san Fhraing,
Phaigh cloinn Chamshroin a chis,
  'S cha d'thig iad air tir gun chall :
'S ged a thainig an t-shith,
  'S daor a h-eiric 'san diola bh'ann.

An la fhuair iad bho'n Fhraing
  Do chorp priseil a nall thar chuan,
'N ciste ghiubhais nam bord,
  Ged 'bha 'n fhailt ud cho bronach, fuar ;
Dh' arduich d' onoir cho mor,
  'S nach bu mhuillean do 'n or a luach
Do thoirt dachaidh le coir,
  'S do thasgaidh fo'n fhoid far 'm bu dual.

'N Cill-a-Mhaillibh nam feirt,
Chaidh an laoch bu mhor neart fo dhion,
'Na uir dhuchasaich cheart,
Ann an tur na 'n clach snaighte, grinn ;
Ge b' e ghabhas dur-bheachd,
Air sgriobhadh na'n leachdan slinn,
'S leir an sid gur ceann feachd,
Fhuair urram le ceartas righ.

Fad 'sa shiubhlas a ghrian,
Dol deiseal na nial gu h-ard,
Gus an leagh le teas dian,
Na beanntuinnean sios gu lar,
Cluinnear iomradh do ghniomh,
Gus an teirig gach sliabh 's gach traigh,
Seasaidh fianuis do bhuadh,
'S do chuimhneachan suas gu brath.

'S ann an Lunnuin nan cleochd,
Dhealbh iad ioghnadh ro mhor mu d' chas,
Ris 'n do chosdadh an t-or,
Obair innealta, sheolt, lamh ;
'S nam biodh iomairt na'n dornn,
A bheiredh tu beo bho'n bhas,
Cha leigeadh crùn Dheorsa
Thu laidhe fo'n fhod cho trath.

LAMENT for Colonel John Cameron, 92nd Regiment, who fell at Quatre-Bras, and whose body was brought home to Kilmallie, in Lochaber, by "Ailean Dall," or Blind Allan, the Glengarry bard. This almost literal translation from the Gaelic was written by Professor Blackie in 1882.

There is many a dear friend and relation
Who will not complain any more of headache,
Who was laid low in France
And driven to death by Bonaparte ;
And, though Wellington was there,
Who scattered the French host from the land,
And one may plainly see our loss,
A loss which caused our people to be few.

Sad news has recently come to us,
That made the men of Lochaber weep sorely ;
And now they are without smile and cheerfulness.
Fitful joy, without glee, without music,
They have lost a friend at court,
And a powerful warrant to back our people.
And many an earl and duke
Had anguish of heart at so great a loss.

A ship was seen in the Narrows
Sailing up with a west wind,
Sailing up to the leafy wood ;
And it was no light freight it brought to me.
It bore the fair body of the hero
On her deck, and his voyage was a sad one,
And it caused wail and lamentation
To the women on both sides of Loch Eil.

On the day of Waterloo
A hero fell whose fame has gone forth into every land ;
A great man of the Cameron Clan,
Who gained for himself honour at the court of the king.
The man who never turned his back
When the matter to the tug-of-war came,
And many a doughty deed he did,
Of which a true report came to us.

It was the pouring of thy blood in a stream
That made all thy clansmen so sorrowful ;
And the blood flowed from thy wounds,
Wounds which there was no surgeon to heal,
With no one near thee but thy squire,
And he sad for grief and loss of thee,
And that blow on the shoulder was not a light one
By which you fell in the struggle to thy harm.

Thou wast the shepherd of the flock
That led them through harm without loss,
Until there was need of them
They all remained in good trim in thy camp.
Stout were they, hardy and brave,
So long as thou wast at their head ;
Wound after wound they inflicted on the foe,
And swift was his retreat
From thy impetuous onslaught.

A hero he was without guile.
When was his like heard of in his time ?
He who waited and he who departed
Gave honour to him with true assurance.
In stature and in gait,
And in the tug of battle second to none ;
A lion mettlesome and fierce,
And a champion without blame in the fight.

A man like to thee, in sooth,
Was not to be found among five thousand.
Without blot or blemish from the crown of thy head
to thy heel.

With thy skilful men by thy side
Thou wouldst go into the crest of battle,
Whether in war or peace
The homage of others you would win.

Fresh was the red on thy cheek
When thou went mounted on thy horse,
Mettlesome, swift, and high necked.
And then at the head of thy clan,
When the brigade was drawn up,
Bright shone thy sword unsheathed
With which thy hand gained victory.
There was the blade, sharp and hard,
With which thou didst cleave the skull of the foe.

Woe be to the wight that lifted a spear against thee
When thou didst snort flame from thy nostrils,
With thy right hand wielding the weapon
With which thou wouldst enter into the strife ;
Thou wouldst protect the king,
Nor wouldst thou forget the right.
Sore do thy relations miss thee,
And sad are they that thou didst not live
To tell the story of thy deeds.

That was sad news indeed,
News that made Sir Ewen sit in silent sorrow ;
No wonder that he should grieve,
With the salt tears flowing from his eye.
His good son and heir without blemish,
Who, by the rank he attained to, showed how he merited fame,
Gone to death with a heavy wound,
Placed this night beneath the sod.

On that great day an end was put to the wars,
To the struggle with Continental despotism,
And many a hero of mighty worth
Fell in the strife with France.
The Clan Cameron paid its due,
And they didn't come home without loss,
And dear was the fine that they paid for peace.

To-day they have brought from France
Thy honoured remains across the sea,
In a coffin of fine wood.
Though that was a welcome sorrowful and cold,
This was an honour that raised you higher
Than millions of gold would have done,
When they brought you home with due homage
And placed you beneath the sod with your ancestors.

'Twas in Kilmallie, that noble shrine,
Where our doughty hero found his rest;
Here he sleeps in his own native soil
Beneath a monument of beautiful polished stone;
And if you mark attentively
The inscription on that flagstone,
It stands plainly written there, that the great captain
Found honour from the justice of the king.

As long as the sun shall travel
Through the sky from the horizon to the zenith,
Until by fervent heat of mid-day
The mountains are warmed to their base,
So long shall the fame of thy deeds last
Till the mountains and the shore shall be no more,
The witness of thy prowess shall stand,
And thy memory endure for ever.

It was in London, where the bells ring,
That they made a work to be admired in honour of thee,
A work in which much gold was spent,
The work of an ingenious and skilful hand;
And, in troth, if the craft of any human hand
Could have brought you alive from death,
King George, with the crown on his head,
Would not have allowed you to die so early.

X

FRAGMENT of the list of men who had joined with him, dated at Alexandria in 1801, and found in Colonel Cameron's pocket-book when he died, with notes opposite the names as to their career. Three of the number had got commissions and a fourth was offered one, but declined :—

Serg. A. Cameron, Achnacarry.
,, E. Cameron, Achnasaul.
,, D. Cameron, Barr.
,, P. Ferguson, Kinlocharkaig.
Corp. A. Cameron, Clunes.
,, A. Cameron, Glendessarie.
,, N. Cameron, Invermallie.
,, D. M'Eachern, Fort-William.
Pte. Charles Cameron, Connich.
,, John Cameron, Connich.
,, John Cameron, Murshirlach.
,, Ewen Cameron, Glensulich.
Pte. Alexr. Cameron, Moy.
,, Donald Campbell, Barra.
,, Ewen Kenedy, Moy.
,, Ewen Kenedy, Fassiefern.
,, Alexr. Kenedy, Clunes.
,, Ewen M'Millan, Kinlocharkaig.
,, Ewen M'Millan, Glendessarie.
,, Angus Henderson, Annat.
,, Allan M'Master, Glenmaillie.
,, John M'Phee, Glendessarie.
,, Duncan M'Kenzie, Ballachulish.
,, Donald Rankin, Fort-William.

The remainder of the list is lost.

XI

INSCRIPTION ON COLONEL CAMERON'S TOMB AT KILMALLIE, BY SIR WALTER SCOTT

Sacred to the memory
of
Colonel JOHN CAMERON,
Eldest son of Sir Ewen Cameron of Fassiefern, Baronet;
Whose mortal remains,
Transported from the field of glory where he died,
Rest here with those of his Forefathers.
During Twenty years of active Military Service,
With a spirit which knew no fear, and shunned no danger,
He accompanied or led,
In Marches, in Sieges, in Battles,
The gallant 92nd Regiment of Scottish Highlanders,
Always to honour, almost always to Victory;
And at length,
In the Forty-second year of his age,
Upon the memorable 16th day of June, A.D. 1815,
Was slain in the command of that corps,
While actively contributing to achieve the decisive
Victory of
Waterloo,
Which gave peace to Europe:
Thus closing his Military career
With the long and eventful struggle in which
His services had been so often distinguished.
He died lamented
By that unrivalled General
To whose long train of Success and Victory
He had so often contributed;
By his country,
From which he had repeatedly received marks
Of the highest consideration;
And
By his Sovereign,
Who graced his surviving family
With those marks of honour
Which could not follow to this place
Him whose merit
They were designed to commemorate.
Reader,
Call not his fate untimely,
Who thus, honoured and lamented,
Closed a Life of Fame, by a Death of Glory.

XII

Grant, in his "Romance of War," gives, incidentally, a fair estimate of Colonel Cameron's character; but in his "Cavaliers of Fortune," he attributes to him unrelenting and even spiteful acts, which his chivalrous disposition and kindness of heart, as shown in his correspondence, prove to have been foreign to his nature. For instance, Grant states that, having in early life had a quarrel with Lord Huntly he never forgot it, but carried it into his intercourse with the Duchess of Richmond, Huntly's sister, a statement which is sufficiently disproved by the letters of both, and by Cameron accepting her Grace's hospitality at Brussels. A story has also been circulated in Lochaber of a severe corporal punishment, ordered by him in a moment of passion on a soldier from that district; whereas (though the captain of a ship could, in former days, order a man to be flogged), no officer, of whatever rank, had such power in the army. The colonel could order a soldier, for certain crimes, to be tried by Court-martial; and the Court, if the prisoner was found guilty, could condemn him to corporal punishment according to the law as it then existed, but the colonel could not.

XIII

The War Office has not any record of the nationality of the three kilted regiments in the early part of 1815, *i.e.* before Waterloo, but the Return dated 22nd October 1815 gives:—

42ND REGIMENT—

	English.	Scots.	Irish.	Foreigners.	
Sergeants,	0	35	2	0	
Corporals,	0	26	2	0	
Drummers,	0	13	2	0	
Privates,	16	370	34	0	
	16	444	40	0—Total,	500
			English and Irish,		56
				Scots,	444=88·8 per cent.

79TH REGIMENT, 1ST BATTALION—

	English.	Scots.	Irish.	Foreigners.	
Sergeants,	7	46	0	0	
Corporals,	6	42	3	0	
Drummers,	3	5	2	1	
Privates,	67	552	54	1	
	83	645	59	2—Total,	789
		English, Irish, and Foreign,			144
				Scots,	645=81·7 per cent.

92ND REGIMENT—

	English.	Scots.	Irish.	Foreigners.
Sergeants,	0	42	0	0
Corporals,	0	34	2	0
Drummers,	1	12	3	0
Privates,	17	394	46	0
	18	482	51	0—Total, 551

English and Irish, 69

Scots, 482=87·5 per cent.

The following return of the 92nd Regiment, dated 25th May 1815, was obtained by the author from the Record Office :—

	Officers.	Sergeants.	Drummers.	Corporals.	Privates.
English,	2	0	1	0	20
Scots,	34	46	11	33	511
Irish,	4	0	3	3	53
Foreigners,	0	0	1	0	1
	40	46	16	36	585

XIV

ORIGIN OF THE GORDON HIGHLANDERS.—In 1913 the Banffshire Field Club published a highly interesting pamphlet by J. M. Bulloch, called "The Gordon Highlanders, their Origin." This gives many details about the actual raising of the battalion, quotes letters, explains difficulties, and relates progress.

It prints also the first muster roll complete, with names, trades, ages, places of origin, and usually date of discharge or death. Mr Bulloch is inclined to discredit the story of the Duchess's kisses, chiefly because he has not found mention of it in contemporary letters ; but his arguments seem open to question.

BOUNTIES (see pages 16 and 26).—See page 26, where (February 21, 1795) bounties are restricted to £15, 15s. 0d. Bulloch quotes several cases where £17, 17s. 0d. and £18, 18s. 0d., one or two where 19 or 20 guineas, and one where £24, 3s. 0d., were paid to Gordon recruits ; and he states that Regimental Headquarters at Gordon Castle paid in 1794, £2478, 12s. 5d. in bounties : and this figure does not include sums paid by officers commanding companies to men. Government only paid at the rate of £5, 5s. 0d. per man ; the balance came from the purchase price of commissions, which seems to have been £2000 for a Major, £1500 for a Captain, and £300 to £250 for Lieutenants and Ensigns.

INDEX

www.ingramcontent.com/pod-product-compliance
Ingram Content Group UK Ltd.
Pitfield, Milton Keynes, MK11 3LW, UK
UKHW021839270726
14058UKWH00002B/245